PoliTO Springer Series

Springer, in cooperation with Politecnico di Torino, publishes the PoliTO Springer Series. This co-branded series of publications includes works by authors and volume editors mainly affiliated with Politecnico di Torino and covers academic and professional topics in the following areas: Mathematics and Statistics, Chemistry and Physical Sciences, Computer Science, All fields of Engineering. Interdisciplinary contributions combining the above areas are also welcome. The series will consist of lecture notes, research monographs, and briefs. Lectures notes are meant to provide quick information on research advances and may be based e.g. on summer schools or intensive courses on topics of current research, while SpringerBriefs are intended as concise summaries of cutting-edge research and its practical applications. The PoliTO Springer Series will promote international authorship, and addresses a global readership of scholars, students, researchers, professionals and policymakers.

THE SERIES IS INDEXED IN SCOPUS

More information about this series at http://www.springer.com/series/13890

Francesco Baino · Massimo Tomalino ·
Dilshat Tulyaganov
Editors

Ceramics, Glass and Glass-Ceramics

From Early Manufacturing Steps Towards Modern Frontiers

POLITECNICO
DI TORINO

Springer

Editors
Francesco Baino
DISAT
Politecnico di Torino
Turin, Italy

Massimo Tomalino
DISAT
Politecnico di Torino
Turin, Italy

Dilshat Tulyaganov
Department of Natural-Mathematical Sciences
Turin Polytechnic University in Tashkent
Tashkent, Uzbekistan

ISSN 2509-6796 ISSN 2509-7024 (electronic)
PoliTO Springer Series
ISBN 978-3-030-85775-2 ISBN 978-3-030-85776-9 (eBook)
https://doi.org/10.1007/978-3-030-85776-9

This Springer imprint is published by the registered company Springer Nature Switzerland AG
The registered company address is: Gewerbestrasse 11, 6330 Cham, Switzerland

Preface

The multiple forms and applications of ceramics and glasses are becoming increasingly important in modern science, society and common daily life. On the other hand, we also have to consider that oxide-based inorganic materials have always played a key role since the early stages of civilization. In fact, man-made ceramics and glasses can be considered as old as human civilization and every prominent culture has generated intrinsic style and traditions about such materials. From an historical viewpoint, it is worth pointing out that glass and porcelain works have been produced worldwide, making the Eastern and Western civilizations joined through the Silk Road, which promoted the implementation of the economic, political, cultural and technical contacts between the two cultures.

Ceramics and glasses exhibit an exceptional versatility in terms of composition, structure and related properties, thus being suitable for taking part in countless applications from aesthetics to high-tech. This book provides a state-of-the-art overview and updated technical picture of the fundamental importance of ceramic, glass and glass–ceramic products. Strategic and impressive applications of new glasses and glass–ceramics in emerging areas such as human regenerative medicine and brand high-tech fields, including energy, photonics and waste management, can help answering the important question of why glass-based materials are and will remain among the most fascinating and requested products in man's life.

Moreover, another valuable aspect should be pointed out beyond the scientific and technical relevancy of this book. In fact, it is the result of the deep and solid collaboration as well as nice friendship developed among the three editors, despite their belonging to distant locations and different "worlds," as they serve in the TTPU—the Turin Polytechnic University in Tashkent, Uzbekistan (Dilshat Tulyaganov)—and in PoliTO—the Politecnico of Torino, Italy (Francesco Baino and Massimo Tomalino). This book is a practical demonstration of how much a collaboration can grow and how high the contribution level can reach in representing a fundamental pillar of a growing common process in the joint activity that these two international academic and scientific realities (TTPU and PoliTO) are carrying on.

Turin, Italy — Francesco Baino
Turin, Italy — Massimo Tomalino
Tashkent, Uzbekistan — Dilshat Tulyaganov

Contents

The Historical Development of Porcelain and Glass

Massimo Umberto Tomalino and Dilshat Tulyaganov

Abstract Ceramic materials are as old as human civilization with the most ancient industries going back thousands of years. The excellent pottery and later on porcelain developed in China was based on the large natural availability of raw materials including kaolin and petuntse deposits. Efforts to duplicate such most-desired features of Chinese porcelain as its whiteness and translucency were long unsuccessful, due to the complexities of the porcelain system. The history of glassmaking begun in rudimental furnaces around the third millennium BC in the regions of Mesopotamia and Egypt and it was based on the chemical combination of very common materials like sand, lime and soda. This book chapter has the main objective of offering an interpretative historical key and a technical overview of the fundamental importance of porcelain and glass in economy, science and technology across history.

Keywords Ceramics · Porcelain · Glass · Historical development · Silk road

1 PORCELAIN: General Historical Introduction

Along with lithic materials, pottery and ceramic fragments are the most widespread traces of human civilization. The worldwide production of ceramic products of art achieved the peak in Europe only after the discovery of the mysterious composition of Chinese porcelain. According to the archeology's findings, the earliest use of ceramic materials were very probably the natural rocks, flint and obsidian in particular. Both these materials show an irregular or *conchoidal* fracture like many modern ceramics, such as glass. This property enabled to obtain very sharp edges of the objects, necessary for the best performances as tools and weapons. During the latter period of Neolithic, pottery became important thanks to the high abundance

M. U. Tomalino
Dipartimento di Scienza Applicata e Tecnologia (DISAT), Politecnico di Torino, Corso Duca degli Abruzzi 24, 10129 Torino, Italy

D. Tulyaganov (✉)
Turin Polytechnic University in Tashkent, Tashkent, Uzbekistan 100095

F. Baino et al. (eds.), *Ceramics, Glass and Glass-Ceramics*, PoliTO Springer Series,
https://doi.org/10.1007/978-3-030-85776-9_1

of clay. In addition, once mixed with water it can be shaped and then hardened by heating according a simple process. At the beginning, the European craftsmen and researchers didn't dedicate so many efforts to achieve early inventions in pottery but on the contrary, they were mostly trying to copy Chinese and Near East ceramics.

However, it is important to underline that the real relevant European contribution was the process industrialization. Fired clay, one of the earliest synthetic substance ever obtained, as well as the first material that humans entirely created by employing heat, represent a milestone along the human civilization pathway and a pillar of Neolithic revolution.

2 Historical Trip of Porcelain: From China to Europe

Chinese culture played the same fundamental role in East Asia that, in late antiquity, Hellenistic culture did in the eastern Mediterranean and roman culture: in both cases, they represented a reference term for the elite traditions of other populations who could find their own expression. The ruling classes of Korea, Japan, and Vietnam even adopted the Chinese concept of a political and cultural Middle Kingdom, extending universal principles to people on the margins of high civilization.

The oldest known pottery in China is dated to around 12,000 BC and by 8000 BC fired pottery was widespread in western Asia by having been discovered independently in Japan. The high antiquity of the art of making porcelain and the perfection achieved in China ages before specimens arrived to Europe, are documented although the period of its first manufacture is involved in great obscurity. Certainly, the Chinese always inventive artisans with bronze and jade, became also excellent ceramists from the Neolithic onward, although nearly all of the eastern populated area of China have almost unequaled resources of suitable raw materials for pottery production.

The idea of employing loess (a clastic sediment, formed of clay, sand and limestone accumulated by wind-blown dust covering about 10% of the Earth's surface), determined the development of Chinese pottery in the Bronze Age (1700–500 BC). Consisting mainly of quartz, loess has a high melting point so that it is an excellent material for building high-temperature kilns. In addition, the low clay content causes loess not shrinking when dried and fired. Therefore, it is ideal for making ceramic piece molds suitable to cast bronze vessels. Employment of loess for ceramic molds demonstrates that pottery and metallurgy developed parallel in China.

The same people certainly labored as both bronze casters and potters since each craft depended on the other. Although proto-porcelain wares exist from the Shang dynasty (1600–1046 BC) and the gray porcelains surely based on kaolin were manufactured at the time of Zhou dynasty (1027–221 BC) by the time of the Eastern Han dynasty (206 BC-220 AD) glazed ceramic wares had developed into proper porcelain. In the Han period, potters started adding some lead to glazes as a flux in order to lower the melting temperature of the glaze. This can be considered a turning point in technology, for when iron-oxide pigments, so covered by the lead glaze, changed colors in becoming bright and clear on the smooth and reflective surface of the fired vessel. Lead glazes thus increased the need of painted decoration. Excellent glazes on Chinese porcelain help account for the extraordinary success and prestige of the commodity. They enabled the potter to imitate metalwork and gemstone surfaces, to create a variety of shades reminiscent of other substances, such as jade, bronze and tinted lacquer, and to paint vessels with polychrome decorations.

As soon as the porcelain manufacture began to meet with court prestige, it progressed rapidly and the name of many distinguished artists appear in the chronicle of the time. Consequently, porcelain seems to have varied in color according to the taste of the different historical periods.

A better production of porcelain in terms of quantity and quality was mainly due to the discovery and exploitation of the Kau-ling or Kao-ling ore (meaning *high ridge*), located in the province of Kiangsi (or Jangxi) in the South-Eastern China. The Kau-ling mine gave the name to the exploited kaolin rock, whose hard and no-plastic nature should be more similar to a feldspar-rock rather than the kaolin-clay properly named in Europe and recognized to be softer and with a less content in silica (Fig. 1).

Kaolin was employed as their primary clay being a mineral that satisfies all the requirements: it fires a striking white, shapes to fine tolerances, and resist to the hottest portion of the kiln without melting (Fig. 2). Throughout oasis communities, contacts and interchanges along the *Silk Road* played the role of conveying Indian and Persian pictorial methods to China. Thus rhythmic patterns, rotating arabesques, stylized flowers, geometric shapes, molded relief, interlaced designs and exuberant colors were adopted to decorate the porcelain objects.

Very early, porcelain was introduced into Persia and Egypt directly from China as an article of profitable commerce. A proof confirming the high antiquity of porcelain is the finding of Chinese bottles of porcelain in the Egyptian tombs of Thebes. It is supposed that this commerce could exist in the time of the Romans. For example,

Fig. 1 Chinese ancient manufacturing in a chromolithography of 1930 (MAGMAX museum)

Fig. 2 Kaolin in China in an eighteenth century paint (MAGMAX museum)

the *vasa murrhina,* mentioned by Pliny the Old in his *Historia Naturalis* of 77 were the fine colored porcelains of China, transported by caravans from Asia to Egypt. The Arabian literature reports that a present consisting of forty pieces of Chinese porcelain was sent to the Caliph of Syria by his lieutenant Saladin (afterwards the hero of the Crusades) upon his conquest of Egypt in 1171.

Porcelain manufactured during the Thang dynasty was exported to the Islamic regions, where it was highly prized. By the eleventh century, caused by the economic activity booming at both ends of Eurasia, Egypt became the great intermediary for exchange in the ancient world, like a sort of pivot of trade between the Mediterranean and East Asia. The creation of a global system after the beginning of this century was a consequence of the Chinese commercial expansion towards Indian Ocean and the spread of Islam into India and maritime Asia.

Eventually porcelain and the technique to produce it began to spread into other areas of East Asia. Transporting porcelain as ballast was so lucrative that Western merchants adopted the practice when they entered Asian trade networks. Transporting porcelain by the maritime route of Silk Road, meant that greater quantities of the ceramic reached Southwest Asia than ever before, by resulting in lowered costs to consumers.

After the Thang period, Chinese glazes had the great advantage of being lead-free, whereas low fired, lead-glazed pottery used everywhere else released minute amounts of the metal into cooked and stored food, exposing consumers to significant health risk, even death. By the Song period, porcelain became a central emblem of Chinese culture, an artifact that in all the other countries was tried to be emulated at any cost. The rich society members were used eagerly purchasing porcelain objects in surprising ways. In general, the most creative response to Chinese porcelain by other ceramic traditions arose in regions with a sophisticated culture (as characterized by density of urban life, organized religion, and some measure of literacy), while peoples in relatively less developed areas sometimes abandoned their own pottery styles altogether. As far as ceramics were concerned, Korea, Japan, and continental Southeast Asia shared something more fundamental than a collective cultural context. All the three Asian areas possessed the right raw materials, mainly clays, needed for making fine pottery. In taking into account as they were well situated to learn about Chinese kilns, glazes and craftsman-ship, they had the additional advantage of not having to develop a sophisticated ceramic technology entirely from scratch.

Although India was a lucrative market for porcelain, it never became a "China shop" abroad. Indian potters never adopted the innovations of Southwest Asian artisans, such as luster glazing and tin glazing, and they had no inducement to turn out imitations of porcelain. In striking contrast to everywhere else in the world, potters in India not only had no interest in stealing secrets from one another, but they had no secrets to steal.

In Persia, there is no authentic record respecting the manufacture of Persian porcelain. It has been reported that chinaware in Persia was equal to any Chinese, showing a similar grain and transparency. Painted tiles of earthenware, called *china,* appears not to have any difference between pottery and porcelain. On the other hand, the testimony cannot be deemed decisive as to the fact of porcelain having been made in Persia. Indeed that porcelain was imported by Chinese vessels into the Persian Gulf and the allusion to the use by the Persian Court of porcelain bowls and dishes, it might be inferred that porcelain was not of native manufacture, but imported from China.

In the middle ages, potters living in Central Asia crossed by the *Silk Road* were greatly attracted by beauty of Chinese porcelain. Imitations were able to recreate similar surface and white color of the Chinese porcelain. The technological process comprised the following main stages: (a) heating of limestone to produce quick lime, (b) crushing the quicklime, (c) adding water to quicklime, (d) adding crushed clay particles and quartz powder to a suspension of slaked lime. After thorough mixing of the ingredients, a plastic bed was cut to the pieces of required thickness and sizes. Pieces were dried and fired at a maximum temperature of 1200 °C. The fired pieces, mostly covered with transparent colored glazes, were fired at lower temperatures. This type of architectural ceramic widely used in facades and interiors of numerous historical monuments of Central Asian region still keeps its high quality and beauty. One reason explaining this high stability should be the body's high silica content putting the glaze into compression thus the glaze is less likely to fail and fulfills protection function perfectly throughout the centuries.

In spite of many different areas of the ancient world directly involved into a general deep practical knowledge of porcelain, the real first "scientific" mention of porcelain is due to the Arabian geographer and Egyptologist Abu Abd Allah Muhammad al-Idrisi al-Qurtubi al-Hasani al-Sabti, or simply Al Idrisi, who resided in Palermo at the court of Roger II, king of Sicily. In 1154, he published a geographical work by order of that prince, where he mentions the Chinese junks carrying iron, silk and porcelain among the typical products traded from far Orient.

Historically speaking, the most striking contrast between China and Europe is that the former was more than three thousand years ahead of the latter in the manufacture of high-fired wares. The exported Chinese porcelains were held in such a great esteem in Europe that in the English language, *china* became a commonly used synonym for the Italian term p*orcellana.* The first clear mention of porcelain in Europe dates back to 1271 in *Il Milione* written by the Venetian Marco Polo in the form of a diary of his travels containing a quite detailed description of porcelain by hypothesizing that shells were a main ingredient of the wonderful recipe. He was the first European traveler who visited China where, during his residence along twenty-six years, he was always well accepted by the Empire and could study the manners of the people, acquired four different languages in use in the country until he was appointed governor. This enterprising traveller returned by sea from China to Venice.

In trying to imitate porcelain, potters outside China followed the compromise of compensating for inferior resources with artistry and ingenious substitution. Lacking abundant timber for firing kilns, potters tried to use dried grass, weeds, straw, and animal dung. Lacking kaolin and china-stone for pottery, craftsmen since the ninth-century developed a pottery body made of crushed quartz, white clay, and ground glass (*frit*). The new ceramic recipe, now known as *frit ware*, turned out an unusually hard earthenware product that did not require firing at a very high temperature. Brittle and stiff, the material made possible very thin vessels but, like porcelain, it could be molded into fantastic shapes. As a logic consequence, at the end it became a standard material for fine pottery in many parts of the Islamic world.

Since the low-firing earthenware of Southwest Asian potters could not duplicate the lustrous white surfaces of porcelain, also a novel glaze technology was developed. The effect of Chinese white wares was imitated by adding tin oxide as an opacifier to a clear glaze so that a fine cloud of tin-oxide particles diffused through the lead-glaze coating by covering up the brownish tones of the earthenware with a soft and matte white. Practical and profitable, the technology of tin glazing became successful worldwide for centuries.

It is very important to underline that under the great Ming dynasty, the porcelain manufacturing greatly increased and the porcelain wares were being exported to Europe, where some of the most known Chinese porcelains and styles such as the coveted blue-and-white wares arrived. During this period, much of the porcelain trade was controlled and expanded to Asia, Africa and Europe via the *Silk Road.* In this way, blue-and-white porcelain achieved its full artistic development with the consequence of a consistent export trade in the ceramics field. For example, in 1447, ornamental pieces of Chinese porcelain manufactured in Jingdezhen reached the court of the French King Charles VII.

Apart for copying the Chinese porcelain in *faience* (tin glazed earthenware), the *soft paste Medici porcelain,* manufactured in the fifteenth century at Florence, can be considered the first real attempt in Europe to have reproduced it with some success.

For the first time, during the Renaissance period in Italy, high-quality pottery reached a good level of quality, which also began to define new standards for dining decorum. In the fifteenth century, *tin-glazed earthenware* became fashionable among the increasingly prosperous middling rank of people, who could not afford silver plate or pewter but who desired something better than terracotta and wood. A sort of "globalization" of material culture began in the era of Columbus and Da Gama, with chinaware motifs, colors, and shapes, although reflecting the acknowledged superiority of Chinese manufactured goods (Fig. 3). Therefore, Columbus sailed the Ocean Sea with a letter from the Spanish crown recommending him to the Grand Khan of China and the kings of India (Fig. 4).

Europeans gained direct access to Asian markets just after 1500. Along with Indian cottons and Asian spices, the Chinese merchandise mainly based on tea, silk fabric, lacquer furniture, hand-painted wall hangings was also extended to porcelains, now required by the most of the Europeans. Only since the seventeenth century, desperate to reduce the flow of silver to Asia to pay for the goods, European rulers promoted efforts to replicate chinaware. In the meantime and concretely since 1518, the Portuguese merchants, coming from their voyage round the Cape of Good Hope, began direct trade by sea with the Ming dynasty. At that time, the price of porcelain was so high that only the royal or very rich and noble families could afford it. However, despite the market for porcelain in Europe, massive quantities were not imported until the Dutch just entered the trade one hundred years after Vasco da Gama first journey.

Fig. 3 Porcelain at Jingdezhen

Fig. 4 Routes map of the Polo's travels

Just to give the idea of the porcelain real value, it is documented that in 1521, a Portuguese merchant in Antwerp gave to the German artist Albrecht Dürer three pieces of porcelain in payment for a portrait of himself. Aiming for a quick return on their investment, Portuguese monarchs kept prices high, thereby reducing the market for the pottery and providing no incentive for increasing supplies. In reality, the Portuguese Crown regarded the Asia trade as a royal monopoly, not as a merchant enterprise. In fact, Europe had to wait until after 1500 to have the same access to porcelain (and many other commodities) that the rest of the world already took for granted. The scarcity of porcelain in the European countries of this period was due to the peripheral position of that region in a world system dominated by Muslim commerce, Mongol power and Chinese economic might.

Early in the sixteenth century, and particularly after the establishment of the trading post of Macao in 1557, Portuguese traders returned home with the first few examples of Chinese porcelain (as opposed to earthenware) to find their way to the courts of Europe. In addition, the Portuguese merchants made the Europeans familiar with the samples of kaolin, which they discovered to be essential in the production of the Chinese porcelain wares. In spite of several and immediate attempts to create European ceramics of this kind, the Chinese techniques and compositions used to manufacture porcelain were not yet fully understood and therefore the Chinese secret proves hard to discover. Countless experiments to produce porcelain had unpredictable results and failed constantly. The translucent quality of porcelain suggested to run experiments for the most part involving mixtures of powdered glass and clay, with the result of obtaining a quite satisfactory imitation of true porcelain but slightly softer.

Certainly, the Chinese did the best for keeping the composition of porcelain a great secret and sought to lead the inquirers astray by all sorts of wonderful tales about the preparation of the materials. However, among some important scientists of the sixteenth century, the Italian mathematician and physician Jerome Cardan was able to make a reliable scientific hypothesis about the magic recipe of porcelain. Even if the hypothesis was accurate enough, the erudite speculation continued to link porcelain to agate, seashells, lobster shell, gypsum, mother-of-pearl, and precious minerals long.

Another interesting attempt was run in 1575 by Bernard Palissy, the royal French potter at the service of Catherine de' Medici, the Florentine mother of three French kings and wife of Henry II, king of France. Due to her fascination for porcelain, Palissy struggled for sixteen years to imitate Chinese recipe and his best achievement was his so-called *rustic ware*, typically used for high-decorated large oval platters featuring small animals in relief among vegetation. Imitations continued to be made in France until roughly 1800, and then revived considerably in the nineteenth century when Palissy's pottery became the inspiration for *Mintons Ltd's Victorian majolica*, which was exhibited at the London Great Exhibition of 1851. In conclusion, certainly he failed to discover the secret of Chinese porcelain but had the great acknowledgment of having invented a style of rustic pottery, called *Palissy ware*, for which he is still famous. Nevertheless, the research of the ideal porcelain continued at Medici's family also after Palissy and Catherine thanks to the efforts of Grand Duke Francesco I de Medici who founded a laboratory in Florence. The main objective consisted in combining various minerals into mixtures to be melted in order to reproduce the Chinese porcelain. In particular, the trials were various empirical attempts based on milling silicates, frit together with quartz-sand and white clay, to form a final frit porcelain. The so obtained *Medici porcelain* shows a microstructure consisting of rounded quartz grains, cristobalite and pyroxenes into a glassy matrix. The main issue was that such a composition was requiring a curing temperature higher than the maximum limit that could be obtained by means of the available kilns (1100 °C). Therefore, all the products were presenting clear signs of warping and the production stopped just after the patronage of Francesco de Medici's sudden death, very probably caused by an arsenic poisoning.

Coming back to the subject of porcelain trades, it is important ti remark that the Portuguese establish the first intense commerce not with China but Japan with the plan of getting gold and precious commodities like porcelain. By means of marriages within the richest and the most powerful families, a huge number of Japanese embraced Christianity. At the end, the ambition, intrigues and religious dissentions led the Emperor of Japan to the Portuguese expulsion in 1641. In reality, it is supposed that this catastrophe was caused mainly by the Portuguese missionaries' interference with the porcelain manufactories. By means of their converts, they caused the ware to be ornamented with subjects, copied from prints of Scripture histories and legends of saints instead of adhering to the ancient orthodox native patterns already existing since immemorial time. The innovation was highly resented by the Japanese Emperor who probably cared more about the subjects of his porcelain objects than the conversion of his people.

Just before the Portuguese-Japanese crisis, large-scale imports of porcelain to Europe happened since 1598 on Dutch initiative because the Portuguese had the bad luck to reach China at a time when the imperial government regarded foreign merchants and maritime trade with hostility. The monopoly of the Japanese porcelain trade enabled the Dutch to import exclusively immense quantities of this species of porcelain, which was to be found abundantly in every house in Holland and the exportation of it to different parts of Europe constituted a most profitable branch of commerce.

Although the porcelain of Japan bears a great resemblance to that of China, it can be distinguished by a practiced eye: it is of a more brilliant white and the clay is of a better quality while the designs and decorations are much simpler. The best Japanese production against the Chinese one is the fine vitreous porcelain, the paste of which is prepared with such extreme work as to say that human bones are an ingredient of its composition. Maybe it corresponds to the highest perfection in ceramic art.

In contrast to the Portuguese, Dutch merchants put their energy into dispatching commodities to their homeland. In consequence, the Dutch dominated world trade in the seventeenth century, and Amsterdam emerged as the world's first central, dominating port. The Dutch, upon the expulsion of the Portuguese, were the next nation to carry on an intercourse with India and Japan, as the Dutch did not interfere with the religion of the people or attempt to make proselytes. Consequently, they long kept the monopoly and imported large quantities of porcelain into the North of Europe. After having tried unsuccessfully several times to obtain access to the trade of China, in 1602, the Dutch East India Company was formed in Holland and made a settlement at Formosa. The primary market at first was in the United Provinces of Holland, the most dynamic economic region in Europe.

The industry did not really take off until two events took place around midcentury at opposite sides of the globe. The first arose from an extraordinary political decision in China to limit the foreign exportation. Using high-quality clay discovered at a short distance, the Dutch potters of Delft manufactured the vessels by covering the inside of them with a transparent lead glaze and the outside with decoration painted on tin glaze. With the drastic limitation of imported porcelain, Delft opened new potteries almost every year for decades. The second event was the establishment of a community of engravers, painters, and printers for labor and inspiration, which caused the potters to use decoration from chinaware. As an immediate consequence, Delft blue-and-white, often called *porcelyn* because of its glossy appearance, entered the international pottery market. In spite of the apparently limited Chinese economic policy towards abroad, it is quite clear that by importing huge amounts of the pottery, the East India companies thereby transformed the everyday life of a large part of the European population between 1600 and 1800. Chinaware, however, was not simply a neutral object to be used and admired, for it had a significant effect on European society as it was able to play a leading role in a sort of a consumer revolution. Porcelain and its imitations ranked foremost among the commodities newly available and from the seventeenth century, they proved central to altering viewpoints and practices in elite taste, daily life and social conventions.

The book *Tiangong Kaiwu* (*The Exploitation of the Works*) written by the Chinese scientist Song Yingxing, which can be considered the equivalent version of what the *Encyclopedie* of Diderot will represent in Europe more than one century later, contains the description of the first named "kaolin earth" (kaolinite).

Although the manufacture of porcelain reached its fame to a peak during the Qing dynasty (1644–1912), the conquest of China by the Tartars and the consequent change of dynasty became a favorable opportunity for opening new commercial relations. In particular, Europe offered a receptive and lucrative market for good tableware. The porcelain know how by the late seventeenth century was based on a formulation containing clay, although many people continued to believe that it must be of a recondite sort, closer to such marvels as exotic gems, horn and ostrich eggs than to common earth. Desired almost as much as the philosopher's stone—the fabulous magical substance able to transform base metals into gold—the *Arcanum* seemed every bit as elusive and enigmatic. In comparison with the objective of discovering the secrets of porcelain recipe, European suppositions seemed feeble and faltering, scarcely more than confessions of ignorance.

Eager to compete with Chinese potters, the Europeans from the late seventeenth century—soon after porcelains began to arrive in massive quantities—made a big effort to investigate its composition and tried any type of experimental formula in order to approximate it. However, the Chinese did not put the focus on the composition of porcelain itself simply because they regarded the material as a natural extension of their ancient craft of high-fired pottery. On the other hand, the Europeans investigated the nature of the ceramic because of the novelty, excellence, and expense of the exotic commodity. In addition, it is important to remark the growing contribution of chemistry and physics in exploring and explaining new phenomena. It was not by chance that the most of the more powerful reigns of Europe, and in particular Saxony, started rapidly following the developing model based on the economic and financial controls, the support of commerce and industry, the exploitation of natural resources and the launch of new manufactures.

In 1701, reports circulated around central Europe, telling that a journeyman pharmacist and self-proclaimed alchemist in Berlin, named Johann Friedrich Böttger, had succeeded in transmuting mercury and silver coins into gold. Immediately afterwards, the nineteen-year-old charlatan fled for nearby Saxony, where he fell into the hands of August the Strong, Elector of Saxony and King of Poland (1670–1733). Böttger's attempts of transmuting base metal into gold on behalf of the king August miserably failed and the unlucky alchemist was punished by forcing him to work on a formula for porcelain, the new *white gold*. Locked in a laboratory of the fortifications of Dresden, Böttger bitterly wrote over the door, "*a gold maker has been turned into a pot-maker*". He was obliged to work with other alchemists in the futile research for transmutation and was assigned to assist Count Ehrenfried Wather von Tschirnhaus, a German man of letters and scientist who was working for a couple of decades to discover the secret of the true formula of porcelain. The general idea of Tschirnhaus was mixing a refractory material with a small amount of an active flux so that by a variety of mixture and potential products, ranging from marbled bodies to artificial gems such as lapis lazuli and jasper, to red and brown stoneware and white porcelain

Fig. 5 Böttger's discoverage of porcelain's secret in a chromolithography of nineteenth century (MAGMAX museum)

was tested. After almost six years of vain experiments, in 1707, the collaboration between the two alchemists produced the first result of obtaining a red stoneware (Fig. 5).

Just the following years, it was possible to achieve a hard, white, translucent type of porcelain specimen with a combination of ingredients, including kaolin and alabaster coming from local ores. This became a closely guarded trade secret of the Saxon enterprise. With millions of porcelains arriving to Europe from China, Tschirnhaus declared that the purchase abroad of so great a quantity of goods represented a national loss to be adverted. It was very well known that Augustus II was the most illustrious victim of what critics of Asian luxuries termed *porcelain disease* (*la maladie de porcelaine* or *Porzellankrankheit*), a feverish desire to possess the ware. It is not a legend but an historical fact what happened to the king of Saxony and Poland when he gave an entire armed regiment of dragoons to the king of Prussia in exchange of 48 pieces of Chinese porcelain, right now preserved in the Johanneum museum of Dresden. On the other hand, Augustus's passion for porcelain perfectly matched the scientific interest of Tschirnhaus. A man of formidable intelligence and broad interests, Tschirnhaus seems to have known all the intellectual minds of Europe by having started his advanced education by studying natural science and medicine. The question of the secret formula for porcelain, known as the *Arcanum*, first gripped Tschirnhaus during his studies in Leiden, one of the Dutch towns most inundated with chinaware. Working for August II from the early 1690s, in parallel investigating the formula by experimenting with various minerals and clays to come up with the right formula, also he considered the process by trying with optic lenses to achieve the temperatures needed for creating porcelain. It has been proved that he also spied out the techniques used in the Delft potteries of Holland and those in the pottery manufactory of Saint-Cloud in France.

When Böttger was assigned to him, Tschirnhaus welcome the young alchemist and the team became the first research and development enterprise in history, driven by visions of enormous profit and haunted by fears of industrial espionage. Although clever and industrious, Böttger also was lucky when Tschirnaus suddenly died just in coincidence with the final breakthrough of creating a version of porcelain. This circumstance allowed the putative alchemist to claim all the credit for the achievement. The first wares produced in 1708 actually consisted of red stoneware, a material similar to that used for the much-admired Yixing teapots, a sample of which was belonging to August II's private collection. Within a short time, the kaolin ore discovered in Saxony improved the product with the perspective of increasing its production. Apparently, the historical hit occurred when the blacksmith Johann Schnorr proposed to Böttger to make face powder from a strange white earth found by chance under his horse's hoof and brought to the laboratory in a small bag. Even if the amount of the material was very limited, the acute alchemist got the opportunity of experimenting it into the porcelain's formula. Therefore, the use of the powder (a very pure quality of kaolin) in December 1708 allowed obtaining the first specimen of hard, white and vitrified European porcelain. The reported achievement of a hot white teapot removed from the kiln and immediately dropped into cold water without damage has been confirmed in the 1980s when the procedure was successfully repeated in an experiment at the Massachusetts Institute of Technology. Even if Tschirnhaus had supervised the experiment preparation, his death in October of that year gave the opportunity to Böttger to report to the king Augustus II in March 1709 that he could make porcelain. Then historical facts assign credit for the European discovery of porcelain to be traditionally ascribed to Böttger rather than Tschirnhaus. Scale-up production of porcelain begins in the Dresden laboratories in 1709 and Saxon Porcelain Manufactory was officially opened at Meissen by the king Augustus II (Fig. 6).

While most descriptions of the invention of European porcelain focus on its composition, the main secrets of Böttger's success were the concept of partially fluxing a white clay with lime, the processing of these suitable compositions with Tschirnhaus' lenses and the achieving the high kiln temperature necessary for a satisfactory firing. As a direct consequence, the microstructure of Böttger's porcelain, consisting of crystals of mullite (a rare aluminum silicate of post-clay genesis) dispersed in a lime silicate glass, gives to porcelain its traditional characteristics and in particular the typical extraordinary resistance to the thermal shock. After Böttger's death in 1719, the original Meissen formula was improved by adding some feldspar as a flux in order to achieve a somewhat lower firing temperature during the process.

A few years after Johann Friedrich Böttger had succeeded in creating the European version of porcelain, the king had himself portrayed on a vase viewing his Meissen manufactory in the guise of a Chinese emperor. Headstrong, compulsive, and fabulously rich, August II anticipated the modern recognition that science and industry

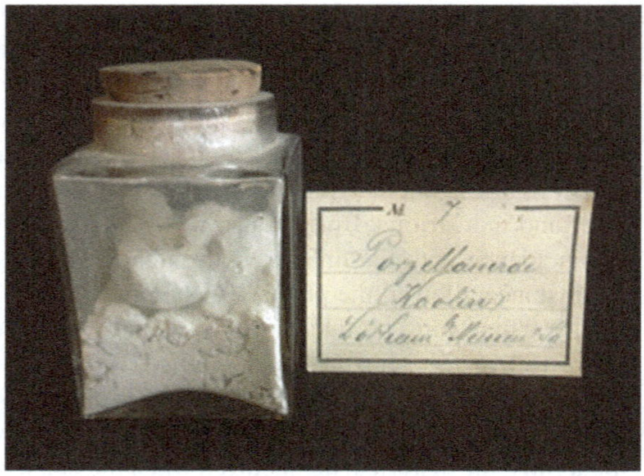

Fig. 6 Porcelain clay from the ore next to Meissen (MAGMAX museum)

necessarily are linked and, furthermore, that their coordination calls for strong state support.

The final set up of the Meissen factory was established after the development of a kiln able to reach firing temperatures of up to 1400 °C to guarantee the achievement of the right translucence of the final product. The first pieces of porcelain are put on sale at the Leipzig Easter Fair by 1710 were noted for their great resistance to thermal shock and for the red color known now as *Böttger's stoneware*. However, just three years later, Meissen is producing white delicate porcelain and colored glazes followed within the next few year. Böttger was passionately proud of his creations and therefore was able to inspire Augustus II with his vision of pieces designed by leading artists to outdo even the Chinese while Saxony achieved its greatest single distinction.

Even at the time when he became the Meissen factory's director, Böttger was confined in Dresden and although his house was luxury into the fortress, there were guards on the door until 1714 when his tyrannical employer finally released his alchemist. Although still in his early thirties, Böttger became extremely ill, probably from working with kilns and crucibles in an unventilated laboratory until he died in 1719 (Fig. 7).

The severe methods of guarding against spies and betrayal were not enough to preserve August II to fail keeping the secret of porcelain manufacture for himself. Within a few years, workers knowledgeable about the ceramic recipe and the creation of high-temperature kilns escaped from Meissen and the so confidential and precious information reached the ears of other princes. However, the difficult construction of effective kilns and the location of the kaolin's ores delayed Meissen's rivals. Only around 1760 some thirty porcelain manufactories appeared in Europe, about half of them in the states of Germany. Of course, the Chinese eventually suffered a great deal more than August the Strong from the spread of the secret. At the end, the lost of their profitable monopoly on porcelain was a direct consequence of the triumphal

Fig. 7 Stamps dedicated to Meissen porcelain (MAGMAX museum)

expansion of their ancient trade in pottery into the new, predatory marketplace of Europe (Fig. 8).

At the same time that August II made his breakthrough in producing porcelain, the wholesale destruction of silver in France meant there was more urgency than ever in the French research for the secrets of porcelain. In such a context, the French manufactory of Sèvres emerged as the dominant force in European porcelain field when Meissen was beset by its midcentury crisis.

It must be underlined that even with millions of pieces exported, porcelain was never the main Chinese export. In the early eighteenth century, it generally ran a distant third behind silk and tea. It wouldn't appear surprising that Jingdezhen also profited from making imitations of ancient ceramics and achieved domination of the global market in ceramics not only by virtue of the superiority of its product but also by the scale and organization of its production. In fact, it represented the climax of handicraft wholesale before the age of steam-driven machines of the industrial revolution. Many different kinds of laborers and craftsmen with their skill and dedication contributed to the incredible success of pottery production. Among them and besides kneaders of clay, it is important to list many more actors: clay miners, stone crushers, woodchoppers, basket weavers, rope makers, carpenters, ironmongers, barrel makers, brick molders, kiln builders and repairers, clay throwers and stampers, mold makers,

Fig. 8 Miner-captain statuette made in Meissen (MAGMAX museum)

wheel spinners, glaze mixers and appliers, kiln loaders, fire stokers, brokers, porters, and boatmen. On the other hand, it would be a mistake considering that the miracle of porcelain was due to enthusiasm and passion because it is necessary to remark that harsh treatment and miserly pay sometimes drove workers to sabotage, strikes, and rebellion. For example, in the early fifteenth century, four thousand crafts-men tried to flee Jingdezhen, but soldiers dragged them back to their work sheds. In addition, the history tells that following a devastating flood in 1540, hunger riots erupted in the city and brought work to a standstill. Revolts also took place in 1574, 1597, and 1602. Furthermore, in 1604, independent potters, forced by an imperial magistrate to produce more wares than usual, burned down kilns in protest. On the opposite, a sort of guardian spirit was adopted by many potters, who continued to trust in the miracles and magic that take place in the kiln by believing that porcelain is created out of the element earth, and combines in itself also the essential powers of the elements water and fire. According to such a philosophy so typical of the Ming period, the potter's clay molds can be made by man's hands but the magic changes that take place after they are fired in the kiln cannot be predicted. During the passage in the kiln, a transmutation was taking place when the magical nature of fire altered the shape of the molded clay for achieving an object with supernatural properties. Besides these believes, it is relevant to underline that many of the elaborate Chinese porcelain manufacturing secrets were revealed throughout Europe by Francois-Xavier d'Entrecolles at the beginning of the eighteenth century. He was the superior general in China of the French Jesuits, whose church was established just at Jingdezhen, the greatest porcelain center of the country. Independently of what he saw himself, he learnt many particulars from his Christian converts and by reading the Chinese works of the subject and he got the permission of observing directly the processes of the manufacture and easily obtaining abundant specimens of the two principal raw materials: kaolin and petuntse. D'Entrecolles sent a detailed report and even a sample of clay to France and both things arrived straight along to René-Antoine de Ferchault Réaumur, an innovator in the study of the nature of rocks and mainly in the research on heat and on a new system for measuring temperature just thought once he started exploring porcelain.

As many of his contemporaries, he found himself deeply involved in what was called the *furor for porcelain making*. Since 1715, Réaumur started searching for suitable clay materials in all the provinces of France. D'Entrecolles considered the specimens the basis of a series of experiments. When the proper combination of materials was discovered, they brought to the fine porcelain manufactured at Sevres. Réaumur successfully identified the principal components of the clay sent by d'Entrecolles and discovered that kaolin was infusible while petuntse fused readily so concluding that Chinese porcelain consisted of a mixture of an infusible earth with a fusible constituent. Further experiments showed that the transformation into porcelain occurring in the kiln could be explained by a crystals growth process in spite of the common assumption that crystals could develop only from an aqueous solution. However, Réaumur also believed that water was the basis of solid matter and so he concluded that some kind of *lapidifying juice* melted stony substance by generating crystalline extensions.

It is interesting to underline that the *Encyclopedia*, issued in 1751 by Denis Diderot, includes a chapter dedicated to porcelain, which can be considered the only significant Chinese product that the Europeans could aspire to imitate without saving big efforts and trials not only addressed to discover the right formula but also to investigate furnace transformation (Fig. 9). The pottery kilns also serving as handy laboratories for the study of volcanic action, allowed the Baron Josiah Wedgwood to invent in 1780 the pyrometer, simple equipment able to measure temperatures above 1000 °C.

The invention of soft porcelain in France and hard one in Germany, gave rise to the settlement and development of porcelain manufactures across Europe as a sign of prestige. Furthermore, the scientific and technological progress was coincident with the success of the chemical theories and analytical methods that culminated with the formal birth of the modern chemistry by Antoine Laurent Lavoisier at the end of the eighteenth century.

From the beginning of the nineteenth century, globalization of material culture proceeded rapidly in Europe, where the pottery manufacturers gained a relevant position among the most important innovators of new techniques and factory organization models. As a first global consequence, the industrial revolution caused a sudden collapse of Chinese porcelain in international markets by reaching the top in the late eighteenth century as a result of competition from European ceramics.

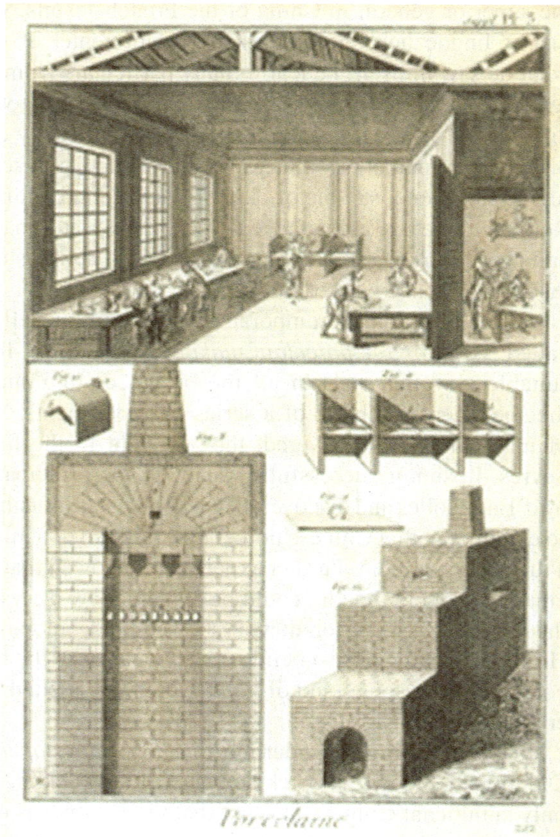

Fig. 9 Porcelain factory in Encyclopedie Diderot-D'Alambert (1751)

3 Historical Development of Porcelain Industry

The last decades of the eighteenth century marked the Chinese porcelain decline in Europe. The fall in popularity resulted mainly from changes in elite fashion, aesthetic sensibility and intellectual perspective. Although the pure white, the beautiful colors and the semitransparent brilliancy of the glaze of chinaware was still admired, the vessels manufactured in Europe were very well appreciated for combining those qualities with the aesthetic standards of classical antiquity. Therefore, the decline of Chinese porcelain was due to the success of the European potters and their ability of matching the European cultural traditions (Fig. 10). The more porcelain and *chinoiserie* were seen to clash with neoclassical norms, the more contempt increased against any Chinese aesthetic standards although during the first half of the eighteenth century, Chinese porcelain still enjoyed enormous prestige in Europe.

Indeed, potters in both Europe and China attained a level of artistic excellence and technical skill that has never been surpassed. While China's success was based on

Fig. 10 Historical manufacturing locations of porcelain in Europe: 1 (Sevres), 2 (Chelsea), 3 (Meissen), 4 (Baldissero), 5 (Saint Petersburg), 6 (Wien), 7 (Limoges), 8 (Worces ter), 9 (Capodimonte), 10 (Munich), 11 (Riga), 12 (Delft)

centuries of tradition, Europe's one was founded on a recent progress: in fact, only a couple of generations were necessary to the European potters to find sources of kaolin and other strategic minerals, to design and build high-firing kilns, to develop new ceramic formulas, to become able to copy the Chinese decorative techniques. In analogy with the exceptional achievements of Chinese craftsmen in the Song dynasty period, in Europe a large number of competing pottery centers stimulated experiment and innovation, thereby increasing the pool of industrial expertise as well as the variety and quality of production. Nevertheless, blue-and-white porcelain remained an exclusive Jingdezhen's creation and its triumph during the Ming period effectively crushed other ceramic centers.

The transition to a large-scale manufacturing industry properly said, occurred in Europe during the eighteenth century when the potteries from a craft institution evolved towards important manufacturing sites like Sevres in France, Meissen in Germany and Staffordshire in England. It can't be omitted the strong impact of greater scientific understanding, such as chemical analysis of raw materials, which changed the field of ceramics, in general, and porcelain, in particular. Consequently, the ceramic industry started taking off and playing an influential role in the industrial revolution and the development of factory systems in England and across Europe.

One of the first hit of a proper porcelain industrial production was the usage of powdered glass with the frequent addition of ash from charred bones, which began the specifically British tradition of bone china denominated soft-paste porcelain. It is less hard than true porcelain as it can be cut with a knife, and it is fired at a temperature of around 1200 °C instead 1400 °C. On the other hand, the producers of porcelain of this kind were well aware that the true porcelain of China was different and superior. In fact, the things rapidly changed as soon as it was discovered that the true porcelain contained two substances known from their Chinese names as kaolin and petuntse. When the secret was discovered, the two fundamental ingredients were at first imported from China, and then found and mined in Europe as well, under the names of *china clay* and *china stone* or *feldspar*, respectively.

Very soon ceramic became an important and growing export industry that stimulated entrepreneurs and engineers to develop modern production and marketing models. A leader in this revolution was Josiah Wedgewood who started producing by the end of the eighteenth century the two versions of an improved an unglazed black stoneware, *basalt* and the more famous *jasperware*. Consequently, Wedgwood won a monopoly on the production of pots and tiles for English dairies, providing wares with patterns simulating the grains of marble, jasper and porphyry (Fig. 11).

Other important innovations introduced by Wedgwood were the use of steam engine to ground flint, prepare enamel colors and mix clay and the adoption of the new transfer printing process, which involved inking a copper-plate engraving, printing it on paper, and then transferring the image to the ceramic piece. At the same time, Wedgwood promoted the kind of cost-effective transportation that linked Jingdezhen with the wider world. He sponsored turnpike construction, which enabled coal to be delivered to the Staffordshire region at reduced rates, and he figured as a leading champion of the Trent and Mersey Canal, cutting the first sod for it in 1766 and making sure that it ran past the front gate of Etruria Hall. Wedgwood supplemented Etruria's industrial innovations with new marketing methods that made its production highly responsive to consumer demand and stylistic changes. Under his own supervision, also techniques that became fundamental to modern commerce were introduced: market research, stock inventory, traveling salesmen, money-back guarantees, pattern books, sale catalogues, newspapers advertising and glamorous showrooms. Wedgwood's pottery style was purchased and copied everywhere from Portugal to Russia.

The general decline of Chinese porcelain in Europe and particularly the triumph of English pottery in international markets reflected a great inversion in the relationship between Europe and Asia. The former rejected an idealized image of the latter, and, at the same time, it began to exercise greater commercial and political dominion on the far side of the world. The high value of Chinese porcelain attributed by the European society, impelled European princes, potters, scientists, and alchemists to replicate it. However, the creation of European porcelain in the opening years of the eighteenth century marked the beginning of the end of the supremacy of Chinese porcelain and *chinoiserie* design (Fig. 12). After centuries of supremacy, Jingdezhen finally had encountered an adversary it could not defeat. Representing the climax of handicraft industry before the age of the industrial revolution, its labor-intensive methods and

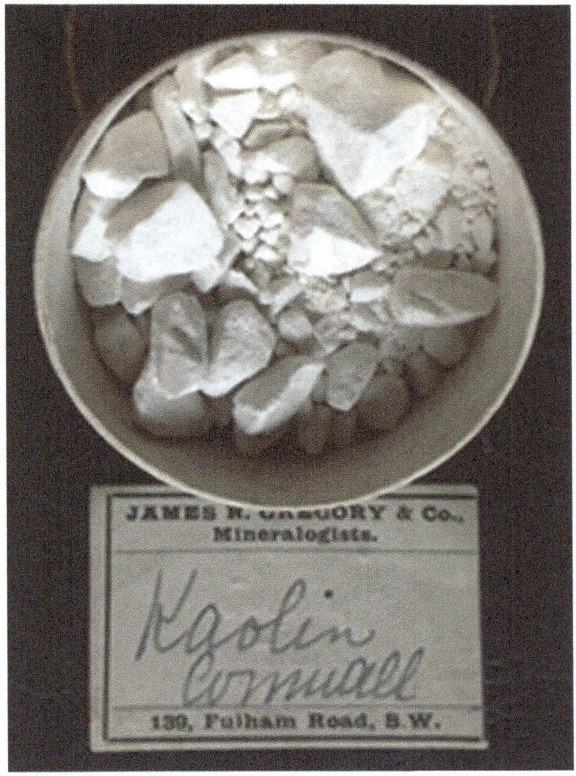

Fig. 11 Kaolin clay sample labelled from Cornwall (MAGMAX museum)

large-scale, decentralized structure served it excellently when its only challenges arose from relatively small kiln centers in Japan and continental Southeast Asia, potteries that did little more than copy the methods and products of the porcelain city.

In a long perspective, the influence of Chinese porcelain on Europe was an episode in the millennium-long history of the ceramic's worldwide history. Probably, the encounter with the West made the Chinese elite thinking that their culture was a model for the rest of the world, a way to impart virtue and civility to those whom they considered barbarians. Chinese porcelain had carried that culture to vast distances, reshaping ceramic traditions, circulating in societies in manifold ways and compelling the wonder of peoples everywhere. As Chinese porcelain lost its global markets, the *barbarians* of the West forcefully came to China, convinced of their superiority and bearing their cultural messages. Nevertheless, porcelain and its imitations provide the first and most widespread evidence for sustained cultural encounter on a world scale, perhaps even for intimations of truly global culture.

Not only German and English porcelain contributed to the decline of Chinese pottery but also the reality of the French manufacture of Sevres. In 1804, it reached its success peak under the supervision of Alexandre Brongniart, who decided to

Fig. 12 Visit of Napoleon I to the porcelain manufacture of Sèvres in a nineteenth century chromolithography (MAGMAX museum)

swap the production of soft-paste with the hard paste porcelain. After having studied chemistry with Lavoisier and become a mining engineer, Brongniart succeeded to Renè Just Haüy—the father of crystallography—as professor of mineralogy at the *Muséum National d'Histoire Naturelle* of Paris and then he dedicated himself to the science of ceramic chemistry and to the mounting of the famous *Musée Céramique.* Both activities were summarized in the *Traité des arts céramiques ou des Poteries* (Fig. 13).

The porcelain products manufactured at Sevres were soon considered as the best and the most artistic of the world by reaching a peak of fame lasting at least for the first half of the nineteenth century, also thanks to Napoleon, who became one of the most interested collectors of porcelain objects that he was used buying in large stocks.

In Italy, the first example of industrial porcelain was the Vinovo manufacturing located next to Turin, the capital city of the Savoyard kingdom. In particular, the quality of such a porcelain was enhanced by the discovery of an alternative to kaolin found by the chemist Giovanni Antonio Giobert, famous for having been among the first to spread Lavoisier's theory beyond the French Alps. In 1813, he suggested to use the found mineral, a variety of magnesite but in principle named Giobertite in his honor by Alexandre Brongniart, by achieving brilliant aesthetic results of the so obtained porcelain.

In Russia, the history of porcelain production is connected to the foundation of the *Imperial Porcelain Factory* in St. Petersburg, also known under its former name, Lomonosov Porcelain Factory. In 1744. It was the third factory in Europe to produce

Fig. 13 Raw materials for porcelain's preparation from Brongniart-Borson (Natural Science Museum of Turin)

hard porcelain since 1744, after the *Königlich-Polnische und Kurfürstlich-Sächsische Porzellan-Manufaktur* in Meissen in 1710 and the *Kaiserlich privilegierte Porcellain Fabrique* in Wien in 1718.

After many attempts, the secret of hard paste porcelain was discovered by Ivanovich Vinogradov, who studied metallurgy at Freiberg in Saxony and eventually invented the formula for the *Imperial Porcelain Factory* in 1746. The first Russian porcelain here produced was assigned solely for the Russian court while even the aristocracy was obliged to order services and works of decorative porcelain from Meissen. The manufacture was reorganized when Catherine II came to the reign (1762–1796), with the task *to please all the Russia with porcelain.* By the beginning of the twentieth century, the *Imperial Porcelain Factory* was one of the leading porcelain factories in Europe and in 1925, during the 200th jubilee of the *Russian Academy of Science*, the factory became well known as the *Leningrad Lomonosov Porcelain Factory.* While in 2005, the factory returns to its pre-Soviet name, the *Imperial Porcelain Manufactory* (Fig. 14).

In 1766, the English banker Francis Gardner founded the *Gardner Factory,* which was competing with the *Imperial Porcelain Factory* in terms of the quality of its production and artistic standards. The Gardner masters had a deep knowledge of the different techniques of porcelain decoration and besides underglaze and over-glaze design, they also employed transfer prints, colored lusters and relief compositions. Although the *Gardner Factory* continued to receive commissions from the imperial court in the nineteenth century, by the late 1880s, the manufactory was no longer profitable. During the Soviet era, the factory was renamed after the nearby town of Dmitrov as *Dmitrov Porcelain Factory.* Both mass-produced items and unique high-quality works were manufactured and still produced there. Nowadays the original name of *Gardner* is returned to the manufactory.

In 1832, the merchant Terenty Kuznetsov, was the first from Kuznetsov's *Porcelain Empire,* who started a business on porcelain manufacturing of his own building a

Fig. 14 Young minerals collector from *Imperial Porcelain Factory* of Saint Petersburg (MAGMAX museum)

small factory with six burning forges, located nearby of Moscow. High quality and relatively inexpensive china was produced for general consumers in Russia, Central Asia, Turkey and China. *Kuznetsovs* were granted the special right from the Russian emperor's court as the award for their contribution to use the state symbol, i.e. two-headed eagle, in their signet. Among other factories, *Dulevo* was one of the best in Europe, mostly due to the use of raw materials of the highest quality and the special treatment. In 1913, the plant in Dulevo had more than three thousand people and produced almost 20,000 items. After 1917, the most factories of the partnership were nationalized, like the *Riga Porcelain Factory.*

In North America, the origin of pottery production is associated to the availability of deposits of earthenware clay and of the wood needed in the kilns. For example, the abundance of these conditions were factors in the settling of *Jamestown* in Virginia, in 1607. Similar suppliers were available around 1620 in New England. The American potters used a lead glaze brushed onto the inside of the earthenware vessel to make the porous clay watertight. The most important pottery location in North America was the triangle formed by Bennington (VT), Trenton (NJ) and East Liverpool (OH). These locations had deposits of fine clay and river transportation, which provided easy access to markets. By 1840, there were more than 50 stoneware potteries in Ohio, earning Akron the tag *Stoneware City*.

It is important to point out that, as with many industries, the first stage of the industrial revolution did not result in a deterioration of working conditions. A partly rural craft-based skill, such as pottery making, became an injurious occupation only as industrialization progressed, bringing poor workers from the countryside into overcrowded town centers. Occupational diseases were prevalent in the potteries. The main problem was diagnosed at an early date lead poisoning. However, it has been necessary to wait for the year 1949, when the British regulations forbade the use of raw lead in glaze compositions. Before this date, there were about 400 cases of lead poisoning a year at the end of the nineteenth century. Although experiments with leadless glazes were recorded throughout the nineteenth century, lead was essential and the safe solutions adopted and approved early in the twentieth century was a lead glaze suspension out of fritted lead.

Pneumoconiosis was another serious health risk for potters: flint dust particles, when inhaled, caused gradual and often fatal damage to the lungs. It was a lingering disease, which took many decades to diagnose and control. Flint is still used as a component in the bodies of many traditional ceramic wares but the risk of pneumoconiosis has been drastically reduced through the installation of proper ventilation systems, the introduction of the cleanliness of workshops and the best practices of wearing protective clothing.

Finally, the commercial success achieved by the Europeans and American manufactures with their own pottery in ousting Chinese production from international markets foreshadowed western international dominance in the modern world. In the widest perspective, the decline of Chinese porcelain closely tracks the epochal decline of China in world affairs and the corresponding rise of the West to primacy worldwide.

Since the end of the world war, European nations have been forced to give up their colonial possessions, and they have lost the power to dictate military and political affairs across the board. There are also signs that China will contest the West for economic dominance in the modern-day, rapidly shifting world system. However, dependently on the fact that East or West (or neither) wins, porcelain will remain above the fight: if it no longer attracts the wonder and high regard bestowed on it for ages. While China has long since lost its monopoly, Jingdezhen, the town that once produced nearly all the world's porcelain, continues to turn out more than 300 million pieces annually still reproducing exactly some of the dazzling porcelains that captivated the world for so many centuries. The taste for collecting Oriental porcelain has been checked by the numerous imitations of the old ware coming from China. No-less than to the caprice of fashion, the reason why the amateurs historically prefer the old Dresden, Sevres, and Chelsea, is related to the limited quantity which contributes to make its value consequently daily increasing.

In the past, ceramic production was largely empirical. In order to maintain a certain uniformity, potters always obtained their raw materials from the same supplier by keeping secret any detail of their process. In reality, the main reason was that they were dealing with very complex systems that they did not understand. Today, after hundred years of ceramics research, processing and manufacturing are optimized, based on known scientific and engineering principles. Research in ceramic was spurred on by two main factors: (1) development of advanced characterization techniques such as X-ray diffraction and electron microscopy, which provides structural and chemical information; (2) developments in ceramic processing technology. In particular, the invention of X-ray diffraction around 1900 let crystallography finally reveal the internal structure of the clay minerals, responsible of making porcelain such a unique substance. Only around 1960 chemistry and physics would unite in the groundbreaking theory of plate tectonics, so making possible the explanation of the actual role of volcanoes in the continuous deformation and reconstruction of the planet's crust. The same theory also would clarify the link between the lava expelled by volcanic eruption and the clays fired to give porcelain.

In the early twentieth century, X-ray crystallography finally revealed that both kaolin and china stone contain alumina, a mineral oxide able to produce at high temperature microscopically fine crystals of mullite: that is the real main secret of the porcelain formula.

4 GLASS: General Historical Introduction

Glass can be considered one of the most important inventions of man and during the millennia of its history found and it is still finding enormous technological and artistic uses and applications.

Once in its molten state, glass can be shaped, modelled, rolled, pressed, cut and blown and after having been cooled, it can be polished, ground, punched, engraved, painted and perforated according to the final purpose of the different objects. Its properties are so peculiar that have emphasized its value in any time of the history: transparency, hardness, optical behavior, chemical and thermal resistance, cleanability, lightness, stability. The archeology has demonstrated that approximately the history of glassmaking begun in rudimental furnaces around the third millennium BC in the regions of Mesopotamia and Egypt and it was based on the chemical combination of very common materials like sand, lime and soda.

It is almost impossible to mention all the relevant modifications and innovative alternatives to this simple preliminary formulation skeleton but it is interesting to remark that the main recipes and the manufacturing processes of the ancient glassmakers have still survived. Of course, both of them have been continuously improved in order to get a huge gamma of varieties and applications so demonstrating the inexhaustible value of glass (Fig. 15).

The etymology of the word *glass* comes from the old English, whose modern root is the old Germanic word *glasam*, which is supposed to derive from the proto Indian-European term *ghel*, meaning "shining or glitter", while the words *mekku* and *ehlipakku* have an earliest origin in Mesopotamia before being introduced in Egypt as well. Pliny the Elder in his *Naturalis Historia* issued in 77, uses the Latin term *vitrum*. Much earlier to be named, glass and its uses predate recorded history.

Long before the ability to manufacture glass, early tribes discovered and shaped obsidian, which can be reasonably considered as a sort of *natural glass*, being the

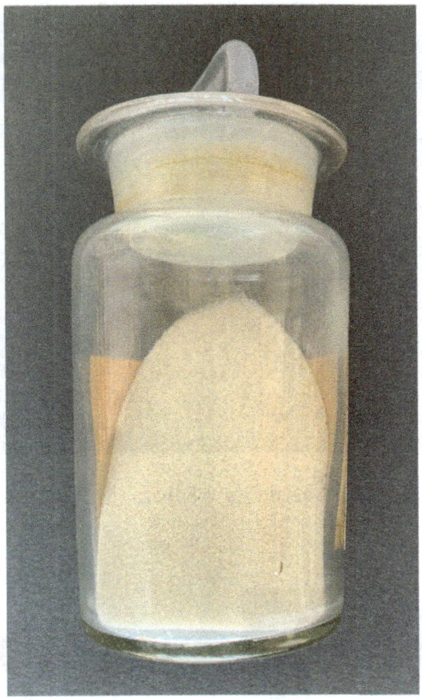

Fig. 15 Quartz sand for glassware (MAGMAX museum)

silica-based volcanic rock rapidly solidified from erupted lava. Because of its large availability, the prehistoric people found the useful properties of such a material once it is easily fractured in order to obtain sharp blades and arrowheads, fundamental hunting and working tools for the start of the human civilization. Another similar source of natural glass is the so-called *Libyan Desert Glass (LDG)* or *Great Sand Sea Glass,* which is a tektite originated by the impact of meteorites on the sand of Egyptian Sahara desert so causing it to melt suddenly and solidified again in a glassy material (Fig. 16). In comparison with the prehistorical era, the technology of synthetic glass is quite recent as it is supposed to date back to no later than five thousand years ago. Such an achievement of the human civilization has required a very slow and continuous development of both chemical formulation and technical process. By summarizing, three main types of glass represent the milestones of its history: soda glass, potash glass and lead glass. All the three kinds contain silica, i.e. quartz, as their major component (Fig. 17). In order to melt more easily, it is mixed with soda (sodium carbonate) and potash (potassium carbonate) so that the resulting material can be shaped rapidly after having reached a semi-liquid stage. The presence of some calcium oxide (often naturally contained in shells residuals) in the original blend is fundamental to prevent the action of water, which slowly would disrupt glass.

The typical composition of soda-glass is silica (73%), sodium oxide (17%, from soda), calcium oxide (5%, from lime or shells), magnesium or aluminum oxides

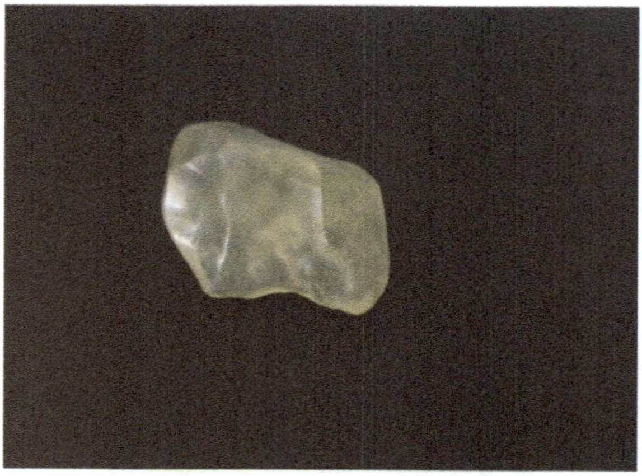

Fig. 16 Libyan Desert Glass from western Egypt Sahara (MAGMAX museum)

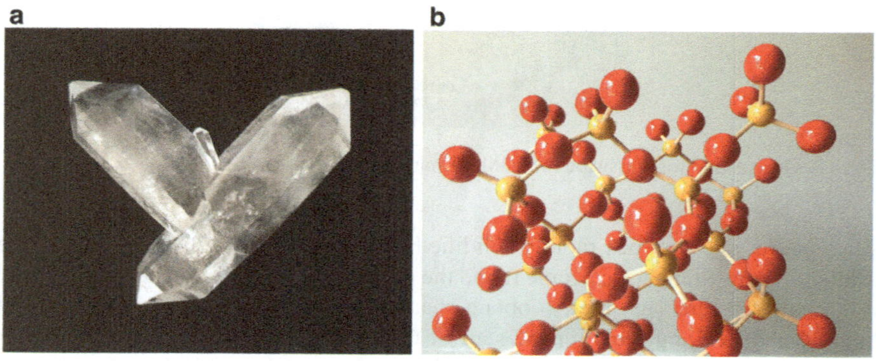

Fig. 17 **a** Quartz from Minas Gerais, Brazil, **b** Quartz crystal structure model (MAGMAX museum)

(5%). The soda deposits of Egypt granted the production of the eastern Mediterranean glass until the first millennium A.D. and becoming in short supply, potash replaced it because of its abundance into the wood ash obtained by burning forest and thus originating the so called *Forest glass*. Finally, the lead glass, based on a high percentage of lead oxide, was made in China around the tenth century and in Murano in the early seventeenth century. Its main characteristic is to show a very attractive appearance based on a high luster and it is easy to cut and polish on a wheel by allowing the easy imitation of gemstones once in color.

In formulation terms, the first relevant discovery happened around 2000 BC, when the Egyptian and Mesopotamian craftsmen realized that the melting point of silica (quartz pebbles and quartz sand) could be significantly lowered by adding a so called *flux* (from the Latin "fluxus" = flow). Such a system, probably applied in metallurgy, where a mixture of silicate and aluminate minerals was used to the smelting furnace to

Fig. 18 Natron from Wyoming, USA

render the slag liquid at the smelting temperature, was even fine-tuned by using soda (in particular under the form of sodium carbonate minerals) as the most suitable type of flux for glass (Fig. 18). On the other hand, the tight relationship between metallurgy and glassmaking should not be surprising.

The development of glass made a clear advance during the period included between the first and fourth century of the Roman Empire, known under the name of *First golden age of glass.* A second fundamental milestone occurred in the thirteenth century, when further advances in the production of glass were developed in Venice and Murano, so opening a *Second golden age of glass.*

5 Origin of Glass: Mesopotamia and Egypt

A rough estimation of the origin and the production of glass is respectively around 5000 and 2500 BC.

As an independent material, glass appeared in shaped objects like beads and vessels not before 3000 BC in considering the early glass objects unearthed by the archeologists in Syria and Mesopotamia and dated back to around 2500 BC.

Actually, the most famous and accredited record is due to the Latin naturalist Pliny the Elder. In his *Naturalis Historia*, he reports the tale of the merchants of a ship full of minerals, probably proceeding from the evaporitic deposits of Egypt, who moored in the north of the modern Israel. Thus, they used some lumps of such rocks just for preparing their lunch. Due to the action of fire, the combination of evaporate minerals (rich of soda) and sand (mainly silica), generated some transparent streams of the new born glass.

The current historical knowledge confirms that the real origin of glass is related to metallurgy or to an evolutionary sequence in the development of ceramic materials as the two hypothesis were deployed similar processes, which can be correctly considered precursors of glass making. Generally, it can be said that the original discovery of glass could have occurred by means of a few simple variations in the production of highly siliceous ceramics like faience. On the other hand, one more reason for supporting the thesis of a common background between glass technology and metallurgy is that both can be managed at higher temperature versus the cold-sintering process needed by faience production. A last but not less important consideration is that the origin of glass took the simultaneous contribution of both metallurgy and ceramic, which should not be seen antithetical.

Archeological evidences demonstrate that in the period 1450–1350 BC, Egypt became the primary glass producer by having installed a proper glass industry, which was extremely successful in the Mediterranean area for at least the next six hundred years.

In the third and second millennia BC, the technique of getting a colorless clear glass bulk was still unknown and the resulting material was like an opaque, semi-clear or even dulled paste. Thus, the early craftsmen were used painting glass in various colors by adding minerals like malachite, azurite, cuprite and chrysocolla.

Around 1500 BC the Mycenaean vitreous products and in particular jewelry became so famous to dominate the entire sphere of influence of Greece. Almost in the same period, glass appears in Italy and northern Europe presenting the characteristics of containing plant ash as flux.

Since the beginning of Iron Age (1400–900 BC) glass production was characterized by replacing the plant ash with natron, an evaporitic rock with a variable composition, consisting mainly of sodium and potassium carbonate, bicarbonate, sulfate and chloride. The further step in the glass industry development happened around 1200 BC when the Egyptian craftsmen invented the *cast glass* based on pressing glass into open molds, breakable away from the glass after it had cooled so that simple objects of the desired shape could be obtained. Such a *pressing in form* technique can be considered the oldest way of manufacturing glassware objects. During the first millennium BC, the glassmakers found the way of obtaining quite clear glass in spite of many efforts necessary to avoid a slight tone depending on the chemical contamination of the natural raw materials. Around 700 BC, for the first time the Egyptian glass industry suffered a decline period of almost a couple of centuries. The positive result of such a crisis was that the production was spread to new glass-making centers allocated in most of the countries along the eastern costs of the Mediterranean Sea. It is interesting to remark that the glass of the seventh century BC was characterized by the high content of antimony, obtained by its natural sulfide stibnite. The addition of such a substance to glass acted as an oxidizing agent to convert the strongly absorbing Fe (II) species to the fairly colorless Fe (III), thus removing the brownish darkening due to the iron impurities. From the sixth-fifth century BC to early Roman time, the variety of glass objects, from beads to bottles and vases that circulate in the Mediterranean area, Egypt, Near East and Europe became extremely wide and also the variety of colors and shapes increased significantly with respect to the Bronze Age.

Around 500 BC, the glass making techniques spread to East Asia and finally reached China.

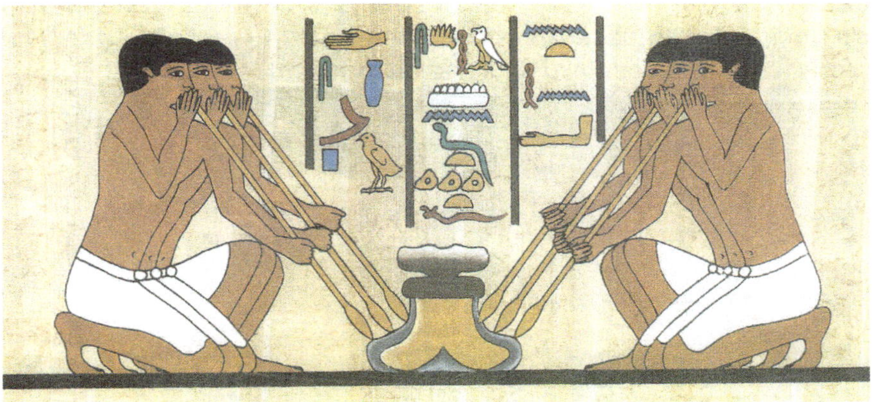

Fig. 19 Glass blowing pipe at ancient Egyptian times

By the end of the first century BC, much of Eurasia, starting from the Syrian coast, learnt how to obtain colored and plain glass. The revolutionary invention of *glass-blowing pipe*, based on the use of an iron tube about one-meter long to be dipped into the molten glass to pick up a lump, gave new possibilities of different uses rather than the only previous press and cast technology (Fig. 19). Furthermore, such a blowing technique simplified and accelerated the production process thus making the glassmaking art common to the entire ancient world.

In fact, during the first three centuries AD a huge amount of glass began to be blown not only within the Roman Empire but also far beyond its frontiers and a lot of innovative techniques of decoration and painting started being developed. Certainly some of the most important of them were the manufacturing of transparent glass and above all the imitation of the structure and the natural patterns of colored stones like agate and jasper.

6 Glass Industry: Roman Empire

After the conquest of Greece around 200 BC and the annex of Egypt by 30 BC, the Romans absorbed both Greek culture and the Egyptian technological knowledge of glassmaking (Fig. 20).

As an immediate consequence, the so-called *Roman glass* achieved its peak and consolidated its popularity during the period between the fourth century BC and the ninth century AD. The formation of the Empire and the pacification of the Mediterranean basin under the dominion of the emperor Augustus created a new world market, which contributed to the diffusion of the glass-blowing methods. The discovery that molten glass could be blown was closely related to the discovery that broken glass artefacts could be treated in a full *re-melting process*. According to

Fig. 20 Roman Empire at its peak (117 AD)

archaeological evidence, two main models for glass production in the Roman Empire can be stated. The early model of glass production was coming from the ceramics industry and it was based on the concept of using raw materials locally available so resulting in a large number of chemically distinct glasses being produced. A second model, based on the archaeological evidence of surviving furnace sites, suggests that the glass was made in a small number of primary production centers and the raw glass in bulk was then shipped to workshops all across the Empire for being shaped. Therefore, such a model would result in only a few chemically distinct groups of glass types. Certainly, both these two models for glass production and distribution in the Roman Empire were followed in the following Byzantine and Medieval periods. Furthermore, the question concerning the glass production in the Roman Empire is even complicated by the confusion between the precise definition of manufacture of raw glass and the secondary manufacture of finished goods. What clearly appears from the chemical analysis of the archaeological finds is that the glass manufactured during the Greek-Roman period belongs to the *natron glass* category based on the soda-lime-silica system: a silica matrix formed by quartz sand as a glass network, a stabilizer like lime and a flux like soda to lower the melting temperature of silica. It is obvious that at least the earlier glass factories were built next to rich sand sources since the transport of vast quantities of sand would have been costly also in the ancient times. The constant composition of the Roman glasses allows reinforcing the hypothesis that there was a limited number of productive sites from where the raw glass ingots were transported around the Empire for being re-melted, colored/de-colored and shaped into the different objects.

The industry of glass was strongly supported by the Romans, who engaged themselves not only to develop technical skills but also to promote glass as a strategic material for the internal and external civilization process. As a confirmation of this statement, it can be pointed out that the Roman glass technology remained in many ways unrivalled until the ninth century. Without forgetting that the introduction of the glass blowing technique made the glass production much easier and therefore cheap because of the large scale. During the Roman Empire, glass was such a versatile and beautiful substance that fine pieces became highly prized and symbols of wealth and social status symbol. Therefore, its diffusion and success made it a valid competitor to ceramic. It is not wrong to say that glass was used for a wider range of objects than at any other time in history, including the present. Among these uses, for its particular way of enhancing the attractiveness, it must be reported the huge fabrication of glasses and bottles for containing the most favorite Roman drink, wine, strictly to be preserved and served into transparent containers in order to maintain and thus appreciate its characteristic properties of flavor, color and brightness. On the contrary, outside the Roman Empire and western Eurasia, glass was mainly used in its colored and matt forms, just to imitate the natural precious stones. The long-term consequence of such a different concept of the glass' potentiality, like the perfecting of its clear version, was the development of a strategic material to be applied as a scientific investigating tool through the creation of mirrors, lenses and spectacles.

The application of glass to the construction of window panes deserves to be underlined as well (Fig. 21).

During the First Golden Age of Glass, it was considered an essential material and continued implementing its performances across the entire Roman Empire, either for practical skills and for their aesthetic achievements. However, since the fourth century, when the Roman Empire started declining and losing its central cohesion, the glassmaking sites were restricted more and more to their local peculiar formulations and manufacturing processes with a consequent limited diffusion of the general technical know-how.

Fig. 21 Roman glass windows

7 Art of Glass: Venice and Murano

After the sack of Rome in 310, the emperor Constantine moved the capital of the Roman Empire to Byzantium, which rapidly became the most important center of the world trade so that even *Roman glass* assumed the name of *Byzantine glass.* In the following couple of centuries, glassmaking technology suffered as well and a typical expression of such a crisis was the production of a dark green–brown forest glass, the so called German *Waldglas*. Its denomination derived from the use of the local silica and particularly of the flux agent, now becoming the tree ash in order to compensate the costly and complicate importation of soda from the traditional regions of Egypt. One of the main characteristics of *Waldglas* was to cool quickly by leaving a very little time to manipulate it, although it showed the advantage of being well suited to cutting and engraving (Fig. 22).

Despite the big problems caused by the incessant invasions in the eastern side of the Roman Empire, glassmaking continued surviving much better than in the west under the newly founded Byzantine Empire. In fact, the production of high status items, such as jewelry and tiny glass tesserae for mosaics, kept alive the great interest for glass. After the east was further divided in the Byzantine Empire and in the Islamic Empire in 630, glassmaking was not only ensured but also promoted and

Fig. 22 **a** Waldglas cup (fifteenth century), **b** Bayerisch waldglas vase (1930) (MAGMAX museum)

the quality of finished goods even enhanced as a consequence of the merging process between the traditional Roman and new Byzantine technical knowledge. In the period between eight to tenth century, the concrete results of this shared experience were the so-called *Islamic soda-lime glass*.

After the initial Crusades in the eleventh century, the center of glass manufacture gradually shifted from the Islamic glassmakers to the growing glass industry of Venice and Murano, since practically the tradition of glassmaking never died in Italy after the fall of Rome. Since the thirteenth century, the presence of the Venetian glass strongly influenced the rest of Europe. This initial industry of Venetian glass increased its importance thanks not only to the additional technology coming from the Byzantine glassmakers but also to the creation of the monopoly of Venice in 1277. In fact, it became almost total after the treaty signed between the Doge of Venice and the Prince of Anthioc in order to promote the importation of the raw glass and plant ashes from the Eastern regions, thus making Venice become the crucial center for glass in the known world.

Certainly, it was not by chance but due to several factors, eventually combined among themselves, that the Venetian glass industry exercised a global predominance, at least within the period between the fourteenth and the sixteenth century, to the point that it can be nominated t*he Second Golden Age of Glass*. Probably, the most important factors were the traditional experience of the original glassmakers, the skill and the raw materials from Syria, the growth of the local trade and commerce ability, the cultural context of the surrounding Italian Renaissance in all the arts, the creation of an intense network of schools and training centers dealing with the art of glassmaking, the quality of the raw materials available to the Venetian glassmakers. By the end of the thirteenth century, the whole glass industry was moved to the one-mile-distant from Venice island of Murano in order to overcome the dramatic free fires caused by the furnaces of Venice, blazing night and day to sustain the crazy rhythm of the glass demand (Fig. 23). In Murano, the furnaces and thousands of glassmakers could work twenty-four hours a day without any major risk and with the advantage of producing a huge amount of any sort of glass objects.

The Republic of Venice grew in wealth and importance with the clear awareness that the spread of the Venetian glass type was also becoming a spontaneous trigger for the external competitors. As a dangerous consequence, it became critical to maintain the secrets of the Venetian technologies and formulation. In order to protect the glassmaking monopoly, a lot of penalties and rewards were introduced. In a certain way, the allocation of the glass industry on the island of Murano not only removed the fire hazard, but also allowed the Government to better protect the hegemony of the Venetian glassmakers. On the other hand, very soon they became virtual prisoners of their own properties for being forbidden to leave their workshops in spite of the full recognition of their elevated social status.

1450 means the date of Murano's greatest technical innovation when, by refining the primary materials in an almost alchemical process, Angelo Barovier created *Latticinio*, or milk glass, imitating the pure opaque whiteness of porcelain, and *Calcedonio*, reproducing the effects of the mineralogical structures of chalcedony, agate and malachite (Fig. 24a).

Fig. 23 Glass-workers at Murano

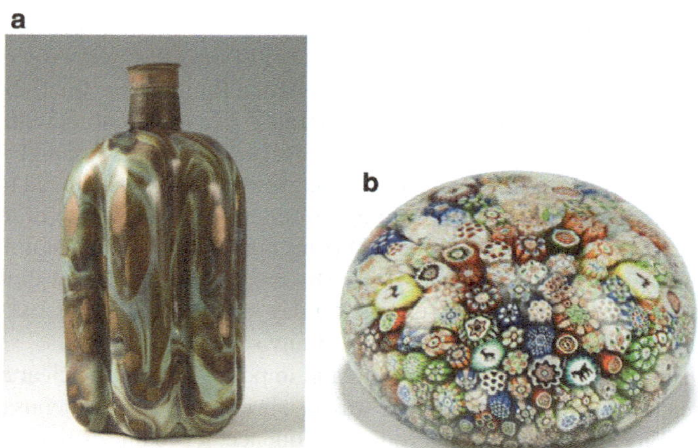

Fig. 24 **a** Calcedonio glass type, **b** Millefiori glass type

The Island of Murano developed two special techniques in glassmaking, both inspired to the cultural principles of the Italian Renaissance: M*illefiori* (a thousand flowers), obtained by embedding the glass matrix with thin canes of multicolored glass and C*ristallo*, a very elegant type exhibiting a deep purity, thinness, free of flaws and colorless (Fig. 24b).

One important reason for explaining the explosion of glass fashion in the Italian Renaissance might be the beautiful results obtained by using glass for imitating the precious substances produced by the mineral kingdom like Quartz in all its varieties and especially in the one known as *rock crystal*. In comparison to it, fine glass became a cheaper and more versatile substitute once the more and more skilled glassmakers learnt how to imitate and even to improve the beautiful transparency typical of the natural stone. In addition, the Venetian craftsmen also were excellent in reproducing many of the other beautiful rocks and minerals like agate, jade, jasper and lapis lazuli. The deeper attention paid to the mineral resources brought to do a more careful selection of them, besides silica. It is important to remark the biggest difference between the glass compositions of the previous Roman period and the improved Venetian material: it mainly consists in the alkali source, based on vegetal ashes instead of natron, traditionally used by the Romans. Furthermore, the Venetians used to add some lead, under the form of its minerals, in order to stabilize the formula. Not by chance, from the end of the sixteenth century, among the many innovations introduced in the glass industry, there was also the development of the lead glass known as *Crystal* or *Flint glass*.

8 Glass Along the Silk Road

From an historical point of view, in the late nineteenth century, the German geographer and geologist Baron Ferdinand von Richthofen adopted the term *Silk Road* as a general expression for the transportation routes linking China with the western regions (Fig. 25).

Later, it became a synonymous of the main artery connecting Asia and Europe, which promoted the implementation of the economic, political, cultural and technical contacts between the two cultures. Certainly, China was the center of the Silk Road in Asia but not the terminal because it extended to Korea, Japan and South East region of Asia and along the centuries it expanded in many more areas around this main basin. In fact, the Silk Roads are estimated to eventually extend for over four thousand miles from central China to Europe and started connecting on a regular basis East Asia, Central Asia and Mediterranean region nearly one thousand years ago. The extension network not only resulted in trading off goods but also corresponded to a relevant cultural exchange and historical experience as well.

The *Shan-hai-jing* (Book of Mountains and Seas) written by several authors along the third century BC can be considered the first historical record describing the early trades between the western and the eastern parts of the world. Although there are proves that the Chinese silk became the favored cloth of Hellenic top class people from the sixth to the third century BC, it can be assumed that the early cultural contacts between China and the West dated back to the eleventh century BC. It is credible to believe that the Silk Road peaked during the first millennium under the leadership of the Roman first and then the Byzantine Empires in the West and the Tang dynasty in China. The connection kept going fluently until the Crusades of the fifteenth century as well as the advances of the Mongols in Central Asia because both events dramatically affected the Silk Road. In fact, by the sixteenth century,

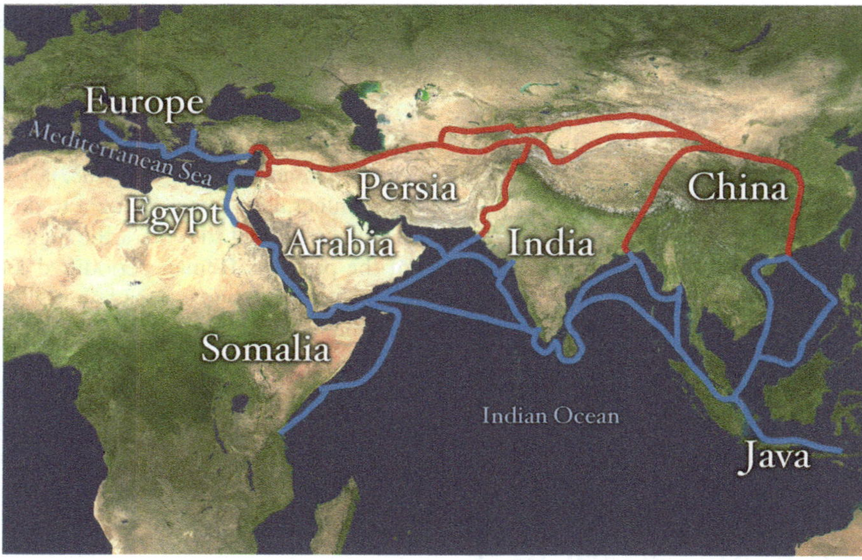

Fig. 25 Silk Road map

commerce with Europe had largely shifted to maritime trade routes, which resulted safer, cheaper and faster.

It would be possible to point out four main routes of the Silk Road briefly described as follows:

(1) *Steppe (northern) Silk Road*: Barbarian nomadic tribes wandered along this large grassland and thus playing an important role in promoting the contacts among ancient peoples. During the late Neolithic Age (fifth-fourth millennia BC), the Indo-European peoples entered Central Asia and moved westwards to southern Russia and even reached the central part of Europe by reaching the Indian subcontinent. A few ancient glass-made relicts were found along this route.

(2) *Oasis (north western) Silk Road*: it has been the main channel between China and the rest of the world. Most of the ancient glass artifacts along the Oasis Road after the Han dynasty are dated more than one thousand years later than the ancient glasses in Mesopotamia and Egypt. During the second millenium, contacts and exchanges between China and the West surely existed. However, the earliest documentation telling of caravans on this Oasis Silk Road appeared in the second century BC as soon as China opened officially its borders for trade. In the fourth century, the Sogdians settled in the fertile valleys surrounded by deserts, in today's Uzbekistan and Tajikistan, established their trading colonies becoming the greatest traders of inner Asia and constructed such famous towns as Samarkand and Bukhara in order to potentiate the eastern line of the Silk Road.

(3) *Buddhist (south western) Silk Road*: it was opened around 100 BC by the emperor Hanwu along the cultural intercourse line between China and India, already existing during the Neolithic age. This road was not particularly suitable for transporting such fragile products as glassware due to the hardship of climbing so many high mountains. The different types of unearthed glass samples reflect the multiple provenance of ancient glass artifacts and the relationship of glass culture and technology along the ancient Silk Road. The Buddhist Road was used earlier than the Oasis and Sea routes and it can be considered the main land communication between India and China. Part of the caravans went across Fergana Valley and Tashkent oasis to Samarkand, Bukhara, Khoresm and further towns on the Caspian Sea coast. Therefore, some cities of modern Uzbekistan became strategic and important milestones on the Silk Road and the country was largely enriched of historical artifacts since trade and crafts developed rapidly. This particularly favorable situation might have occurred because of the position in the oasis right in the center of the Silk Road main routes. The prosperity of towns like Samarkand and Bukhara in the current Uzbekistan and Balkhab in Afghanistan were described for the first time by the Chinese ambassador Zhan Qian around 113 BC.

(4) *Sea (southern) Silk Road*: it was a maritime transportation line from ancient China to western countries and a great sea-network connecting Europe, Asia and Africa. It started from the South China Sea, ran through the Indian Ocean, the Red Sea and the old canal of Egypt, into the Mediterranean Sea along the Nile River. A Greek writer of the late first century reported that glass was one of the main items transported along this cultural and commercial road. As the glass products were fragile and therefore difficult to transport along the land Silk Road, obviously the Sea Silk Road played an even more important role.

In general, the highest development of the Silk Road should be ascribed to the existence of four great empires approximately dated back from the first to the fifth century: the Han dynasty in the East, the Roman Empire in west Europe and west Asia, the Persian Empire in central Asia and the Kushan Empire in south Asia (located in the current areas of Afghanistan, Pakistan and northern India). All of them were powerful at that time and the contacts were strongly facilitated along the Silk Roads without blockades so that the exchanges between China and the rest of the world could proceed and grow quite smoothly. The ancient western glass-making technology, and particularly the glass-blowing technique, achieved its peak during the Roman Empire.

Nevertheless, it is interesting to remark the documents demonstrating the existence of a vast amount of trade between Central Asia and China even before the production of silk. Among the huge amount of goods exchanged along the Silk Roads, the exchange of ancient glass products was pursued by means of bartering. The typical locations, where the caravan traders used to meet, were characterized by the presence of water and animal grazing so that these places soon became important towns and nowadays they are included in amazing touristic tours like Samarkand, where a top quality glass was made and traded as luxury goods since the earliest times towards the Far East.

In the fifth century, after the collapse of the Roman Empire, the center of glass-making shifted back to the eastern Mediterranean, the current area of Syria, Egypt, Iran and Iraq where, on the other hand, glass had first been discovered and developed.

The influence of the Islamic culture on the history of glass is reflected by the distribution around the world of the Islamic glass, characterized by the introduction of new techniques and the innovation of old traditions.

In particular, by the ninth century a distinctively Islamic style had been established and became famous for its exquisite luxury ware. This glass was traded all over Eurasia and became as important in the vast Islamic region as it had been in the Roman Empire. Besides being found in areas of Scandinavia, Russia and East Africa, also in China there was evidence of its presence, thus confirming the very high degree of cultural and economic exchange between West and East through the Silk Road.

The most beautiful decorative glass manufactured in the thirteenth and fourteenth centuries was coming from the factories of Aleppo and Damascus, both located in Syria. On the other hand, very little is known about the glass for functional purposes although a quite wide production of scientific instruments like distilling alembics besides some mirrors and planar-convex lenses for magnification has been proved.

The Arabic contribution to the development of science, in general, and optics and chemistry in particular, resulted in the glass application to make instruments and equipment (Fig. 26). On the contrary, there was a very limited production of window glass probably because it was not so often used in the traditional houses of the Middle East where it was important that air could circulate in the hot season. The rapid and sudden collapse of the glass industries within the Islamic world can be only explained with the Mongol invasion of the thirteenth century.

Fig. 26 Glass chemical flasks in an Arabic manuscript

Fig. 27 Bohemian glass imitation, Turnov School around 1930 (MAGMAX museum) **a** precious stones, **b** semi-precious rocks

In the fifteenth century, the golden period of the Islamic glassmakers practically ended because of either the lack of court patronage for glassmaking and of the strong competition of the highest quality glass coming from Venice. Furthermore, also the spread of plague and the deportation of glassmakers should be considered.

On the Indian side, it is important to note that the early craftsmen made glass beads, ear ornaments, seals and discs of fairly good quality demonstrating a keen knowledge of the blowing glass technique. In spite of such a full knowledge of all the glassmaking techniques, a solid glass industry never developed in India. The possible explanations could be the shortage of strategic raw materials like natural alkali and the relegation of the glassmakers to the bottom of the caste-system. In addition, the outstanding classes of rich or religious people never considered glass as a high status material by limiting its main use only to the imitation of jade, precious stones and porcelain (Fig. 27).

The history of glass in China followed a different and more complex pathway. However, it is surprising that a civilization able to produce some of the most creative craftsmen in the history of world, excelling in pottery, metal-working, print-making and weaving, contributed almost nothing to glass development, especially in comparing with what occurred in the western part of the globe. Before glass-making spread in China by approximately the sixth century, the primitive people made faience and frit, both of them, like glass, used for imitating jade (Fig. 28). The earliest Chinese glass, that is a material assuming an essential glassy state, was an *alkali lime silicate glass* dating back the pre-Qin dynasty.

In order to increase the transparency and lower the melting temperature of glass, the Chinese glassmakers during the Han dynasty period tried to improve the flux agent by using minerals like galena, barite and saltpeter, respectively containing lead, barium and potassium as easily available and well-known materials in the traditional Chinese medicine. In particular, the knowledge that lead could decrease the melting temperature of a mixture and contemporary increase its flow, was coming

Fig. 28 Chinese glass beads

from the more ancient practice of the Chinese bronze metallurgy, which also brought the mold-casting methods into the glass making process.

During the Han period, the Chinese glassmakers started understanding more and more successfully the relationship between chemical composition and physical properties and in the seventh century under the flourishing Tang dynasty, the contact and exchange between China and the other countries became more frequent and extensive so that in particular the Islamic glasses started being imported to China. On the reverse, the typical amber, brown and green lead-glazes wares transported across the Silk Road served as inspiration to the Iranian glassworkers, who were able to reproduce it under the name of *Splash ware*. *Soda lime silicate glass* had been made and applied in inner China since the Tang dynasty when the glass blowing technique was introduced from the west. Although this type of glass became the most popular composition for worldwide glassmaking from the fifteenth to the nineteenth century, it was never so common in China, in comparison with the *potash lead silicate glass* or the *potash lime silicate glass*.

At the end, the status of the Chinese glassmakers was very similar to that of Indian ones and in terms of glass used for practical purposes, porcelain can be considered the main reason for the absence of glass in China. Pottery in general, but also porcelain—considering its high diffusion in China and Japan—was much cheaper than glass and could hold hot liquid very well. About the use of glass for making windows, it is obvious that the above mentioned warmer climate conditions would have encouraged the usage of oiled paper rather than glass. In considering that China started adopting a proper glass technology not before the 1670s, the traditional and common western usage of glass for ordinary tools was practically absent or deeply different in popularity. As a crucial consequence, such a scarce development of the glass technology hardly impacted the development of scientific tools necessarily made with glass. As an historical consequence, within the period 800–1650, when there was the rush of glass technology first in Islam and then in west Europe, glass technology hardly developed in China. The Chinese did not fully understand the manufacturing process of glass from the crude materials and their correct balance into the formulation. On the other end, the prevalent commercial strategy was that it was more convenient to import the necessary glass objects from Europe and mainly from Venice rather than leading an internal manufacturing enterprise. Such an initial lack of the European glass instruments of measurement and sight for the different scientific and artistic purposes had a dramatic impact in the development of the future Chinese progress. The comparison between the results achieved by the ancient Chinese optics, based on empirical observations, and those of the Arabic and western optics, based on experiments done by means of glass instruments, allow understanding why from the seventeenth century the foundation of the Chinese optics was altered.

In Japan, things were not running so differently. By the middle of the sixteenth century, at the time of the peak of the glass expansion in Europe, although the Japanese showed a great ability in manufacturing wonderful glass, its use was only restricted to ornamental purposes. After the Meji restoration of 1868, there was a flood of new technologies and also the use of glass for windows, lamps and many other purposes became well known and practiced. Probably, the main reasons for glass was absent along the most of the Japanese history, were climate and geology. The former made glass windows undesirable for the very hot and humid half of the year by making the tiny house interiors very oppressive and so uncomfortable. The latter made glass windows very dangerous because of the almost daily earth tremors and frequent quakes in the most of the Japanese territory. On the other hand, it must be noted that the Japanese, like Chinese, developed the excellence of porcelain and pottery, based on the large natural availability of kaolin and petuntse deposits close to each other, so that there was no convenient reason for pursuing glass. In addition, the higher fuel consumption in producing glass versus porcelain made the most of workshops and potters argue against a much more expensive technology.

In summary, while Rome and Medieval Europe opted for pottery and glass, India, China and Japan for porcelain and paper. Both the opposite questions about why the Japanese, Chinese and Indians did not develop clear glass technology or about why the Romans did not invent porcelain have a coincident answer based on random circumstances like simply the presence of certain raw materials on the territory and their ores availability and concentration. Certainly, the impact on such a different development between the two civilizations was immense. The general conclusion on the comparison between the two different civilizations within the period sixteenth-nineteenth centuries is the fact that the increasingly glass-dominated western part of the world encountered the eastern one, which had effectively given up the use of glass.

References

1. Agricola Bauer, G.: De Re Metallica, Book XII. Translated by H.C. Hoover and L.H. Hoover. Dover, New York (1950)
2. Aitchison, L.: A History of Metals, 1st edn. Interscience Publisher, New York (1960)
3. Asimov, I.: Breve storia della chimica, 1st edn. Zanichelli, Bologna (2008)
4. Baino, F., Quaglia, A.: Evidences of glass-ceramic white opaque tesserae from Roman age: a thermo-analytical approach. Mater. Lett. **74**, 194–196 (2012)

5. Battaglia, M.: L'industria del vetro in Italia, 1st edn. Roma (1935)
6. Beck, H.C.: Glass before 1500 BC. Ancient Egypt and the East, 1st edn. Corning, New York (1962)
7. Biringuccio, V.: De Pirotecnia, 1st edn. Siena (1550)
8. Boulnois, L.: La route de la Soie, 1st edn. Arland, Paris (1993)
9. Brard, C.: Mineralogie appliquée aux arts, 1st edn. Levrault, Strasbourg (1821)
10. Brill, R.H.: Ancient Glass. Sci. Am. **109**(5), 120–130 (1963)
11. Brill, R.H., Martin, J.H.: Scientific research in early Chinese glass. In: Proceedings of the Archaeometry of Glass Session of the 1984 International Symposium on Glass. The Corning Glass Museum, New York (1991)
12. Brongniart, A.: Traité des arts céramiques ou Des poteries considerées dans leur histoire, leur pratique et leur théorie, 3rd edn. Atlas, Paris (1877)
13. Brown, E.: Des chimistes de A à Z. Ellipses, Paris (2002)
14. Charleston, R.J.: Our forefathers in glass. Glass Tech. **21**, 27–36 (1980)
15. Cummings, K.: A History of Glass Forming. A & C Black, London (2002)
16. D'albis, D.: Sèvres 1756–1783. La conquête de la porcelaine dure. Dossier de l'Art, Paris (1999)
17. De la Beche, H., Reeks, T.: Catalogue of Specimens Illustrative of the Composition and Manufacture of British Pottery and Porcelain, from the Occupation of Britain by the Romans to the Present Time, 1st edn. Eyre & Spottiswoode, London (1855)
18. De Mely, F.: Les lapidaires de l'antiquité et du moyen âge: les lapidaires chinois, 1st edn. Ernest Leroux, Paris (1896)
19. Derry, T.K., Williams, T.I.: A Short History of Technology, from the Earliest Times to AD 1900. Oxford University Press, New York (1991)
20. Dimbleby, V., Turner, W.E.S.: The relationship between chemical composition and the resistance of glasses to the action of chemical reagents. J. Soc. Glass Technol. **10**, 304–358 (1926)
21. Dreyfus, J.: The invention of spectacles and the advent of printing. The Library (1988)
22. Fassin, G.: Something about the early history of the microscope. Sci. Mon. **38**, 452–459 (1934)
23. Gan, F.X., Cheng, H.S., Li, Q.H.: Origin of Chinese ancient glasses studies on the earliest Chinese ancient glasses. Sci. China **49**, 701–713 (2006)
24. Giobert, G.A.: Chimica del cittadino professor Giobert. Manuscript, Torino (1804)
25. Grashdankina, N.D., Rakhimov, H.K., Pletnev, I.E.: Architectural Ceramics of Uzbekistan. UNESCO, Tashkent (2006)
26. Harrè, R.: Great Scientific Experiments: Twenty Experiments that Changed Our View of the World. Phaidon, Oxford (1981)
27. Hellemans, A.: The Timetables of Science. Simon and Schuster, New York (1988)
28. Hobson, R.L., Hetherinton, A.L.: The Art of the Chinese Potter. An Illustrated Survey. Dover Publications, New York (1982)
29. Holmyard, E.J.: Alchemy. Dover, New York (1990)
30. Huertos, E.G.: Mineralogia aplicada. Editorial Sintesis, Madrid (2003)
31. Jacoby, D.: Raw materials for the glass industries of Venice and the Terraferma, about 1370-about 1460. J. Glass Studies **35**, 65–90 (1993)
32. Jones, M.: Applied Mineralogy. Graham, London (1987)
33. Kraemer, H. (Stella A.): L'uomo e i minerali. Vallardi, Milano (1930)
34. Kurkjian, C.R., Prindle, W.R.: Perspectives on the history of glass composition. J. Am. Ceram. Soc. **81**, 795–813 (1988)
35. Leach, B.: A Potter's Book. Faber & Faber, London (1976)
36. Lee, I.S.: The Silk Road and ancient Korean glass. Korea Culture **14**, 27–30 (1993)
37. Leicester, H.M.: The Historical Background of Chemistry. Wiley, Hoboken (1956)
38. Litzov, K.: History of Ceramic Technology. Ceramic Monographs, Handbook of Ceramics, Freiburg (1982)
39. Macfarlane, A., Martin, G.: Glass a World History. University of Chicago Press, Chicago (2002)
40. Malandra, G.: I vetrai di Altare. Cassa di Risparmio di Savona, Savona (1983)

41. Maritano, C., Gallo, L.M., Pistarino, A.L.: Matières à porcelaine nelle collezioni del Museo Mineralogico dell'Università di Torino. Boll. Mus. Reg. Sci. Nat, Torino (2016)
42. Mass, J.L., Hunt, J.A.: The early history of glassmaking in the Venetian lagoon: a micro-chemical investigation. Mater. Res. Soc. Symp. Proc. **712**, 303–313 (2002)
43. Marryat, J.: A History of Pottery and Porcelain, Medieval and Modern. John Murray, London (1868)
44. Massain, R.: Chimie et chimistes. Editions Magnard, Paris (1952)
45. Mazzola, F.: La porcellana. Faenza Editrice, Faenza (1989)
46. McCray, W.P., Kingery, W.D.: Introduction: toward a broader view of glass history and technology. Ceram. Civiliz. **8**, 1–13 (1988)
47. Mikami, T.: The Art of Japanese Ceramics. Weatherhill-Heibonsha, New York-Tokyo (1979)
48. Monro, W.L.: Window Glass in the Making. American Window Glass Company, Pittsburgh (1926)
49. Moretti, C.: Le materie prime dei vetrai Veneziani rilevate nei ricettari dal XIV alla prima metà del XX secolo; II parte: elenco materie prime, materie sussidiarie e semilavorati. Rivista della Staz. Sper. del Vetro **31**, 17–32 (2001)
50. Moretti, C.: Glass of the past. CHEMTECH (1985)
51. Neri, A., Merrett, C.: The world's most famous book on glassmaking, the art of glass. The Society of Glass Technology, Sheffield (2003)
52. Obstler, M.: Out of the Earth and into the Fire: A Course in Ceramic Materials for the Studio Potter. American Ceramic Society, Westerville (2000)
53. Philips, C.J.: Glass: The Miracle Maker. Pitman Publishin Corporation, New York (1941)
54. Pliny the Elder: Naturalis Historia. 36, pp. 65–66–67. Einaudi, Torino (1988)
55. Priestman, S.: The Silk Road or the Sea? Sasanian and Islamic exports to Japan. J. Islamic Archaeol. **3** (2016)
56. Rapp, G.: Archaeomineralogy, 1st edn. Springer, Berlin (2002)
57. Rasmussen, S.C.: Advances in the 13th century glass manufacturing and their effect on chemical progress. Bull. Hist. Chem. **33**, 28–34 (2008)
58. Reamur de Ferchault, R.A.: Delle differenti maniere onde si può fare la porcellana e quali siano le vere materie di quella della Cina. Memorie della Reale Accademia delle Scienze di Parigi recate in italiana favella. Pietro Bassaglia, Venezia (1756)
59. Reamur de Ferchault, R.A.: Sopra la porcellana o continuazione dei principj che debbono condurre nella composizione delle porcellane di differenti generi. Memorie della Reale Accademia delle Scienze di Parigi recate in italiana favella. Pietro Bassaglia, Venezia (1756)
60. Roger, F., Berad, A.: 5000 Years of Glass, 1st edn. Lippincott, New York (1948)
61. Rosen, E.: The Invention of Eyeglasses. J. Hist. Med. (1956)
62. Saldern, A.V., Oppenheim, A.L., Brill, R.H., Barag D.: Glass and Glass Making in Ancient Mesopotamia. Corning Museum of Glass Press (1988)
63. Sayre, E.V., Smith, R.W.: Compositional categories of ancient glass. Science **133**, 1824–1826 (1961)
64. Seligman, C.G., Ritchie, P.C., Beck, H.C.: Early Chinese glass from pre-Han to Tang times. Nature 138–721 (1936)
65. Seligman, C.G., Ritchie, P.C., Beck, H.C.: Far Eastern glass: some western origins. Bull. Museum Far Eastern Antiquities **10**, 1–50 (1938)
66. Shortland, A.J., Tite, M.S.: Raw materials of glass from Amara and implications for the origins of Egyptian glass. Archaeometry **42**, 141–151 (2000)
67. Silvestri, A., Molin, G., Salviulo, G., Schievenin, R.: Sand for Roman glass production: an experimental and philogical study on source of supply. Archaeometry **48**, 415–432 (2006)
68. Sonntag, H., Schuster, B.: Meissen & Meissen: Europe's Oldest Porcelain Manufactory. RV Verlag, Berlin (1991)
69. Tait, H.: Glass, 5000 Years. Abrams, New York (1991)
70. Taylor, F.S.: The Alchemists. Barnes & Noble, New York (1992)
71. The Chinese Ceramic Society: Proceedings of the 17th International Congress on Glass. Beijing, section: Archaeology of Glass (1995, 1996)

72. Tite, M.S.: Glass and related vitreous materials. Proc. Intern. School Phys. Enrico Fermi **154**, 369–376 (2004)
73. Tite, M.S., Ite, I.C., Freestone, M.: A technological study of Chinese Porcelain of the Yuan Dynasty. Archaeometry (1984)
74. Tomalino, M.U.: Una storia della Mineralogia (A History of Mineralogy). Geda-MAGMAX, Torino (2011)
75. Tomalino, M.U.: Chimica-mente (A History of Chemistry). CLUT, Torino (2013)
76. Tomalino, M.U., Tulyaganov, D.U.: Glass: From Sand to Glass Ceramics Through the Silk Road. CLUT, Torino (2019)
77. Tulyaganov, D.U., Tomalino, M.U.: Porcelain: From Tableware to Dental Restoration. CLUT, Torino (2017)
78. Turner, W.E.S.: The physical properties of glasses. The relationship to chemical composition and mode of preparation. J. Chem. Soc. **129**, 2091–2116 (1926)
79. Turner, W.E.S.: Studies in ancient glasses and glass making processes. J. Soc. Glass Tech. **40** (1956)
80. Verità, M., Zecchin, S.: Indagini analitiche di materiali vitrei del IV-XV secolo del territorio Veneziano. Rivista della Staz. Sper. del Vetro **38**, 20–32 (2008)
81. With, E.: L'écorce terrestre. Les minéraux, leur histoire et leur usages dans les arts et métiers, 1st edn. Plon, Paris (1874)

Mineralogy and Properties of Raw Materials and Crystalline Phases of Ceramics and Glass–Ceramics

Massimo Umberto Tomalino

Abstract Almost every human endeavor is influenced by minerals. Many natural resources used in the manufacture of goods on which civilization and our lives depend begin as minerals in the ground. Mineralogy and its companion discipline, crystallography, are fundamental sciences, which from their original observations of natural minerals and their practical applications of economic interest and aesthetic appeal slowly became core parts of the physical and chemical developments of science in the period following the industrial revolution. This book chapter has the main objective of emphasizing the importance of knowing and understanding the role of minerals—through their structure and properties—as fundamental raw materials for glass ceramics and their multiple applications.

Keywords Mineralogy · Minerals · Physical chemical properties · Glass–ceramics · Silicates

1 Mineralogy

Mineralogy concerns itself mainly with the study of naturally occurring inorganic solids called minerals; these may be of terrestrial or extraterrestrial origin.

Although it is difficult to formulate an exhaustive and succinct definition, a mineral can be defined as a naturally occurring solid with a highly ordered atomic arrangement and a definite (but not fixed) chemical composition. It is usually formed by inorganic processes.

The qualification *naturally occurring* distinguishes between substances formed by natural processes and those made in the laboratory. Industrial and research laboratories routinely produce *synthetic equivalents* of many naturally occurring materials. Since the beginning of the twentieth century, mineralogical studies have relied heavily on the results of synthetic systems in which the products are given the names of their naturally occurring counterparts.

M. U. Tomalino (✉)
Dipartimento di Scienza Applicata e Tecnologia (DISAT), Politecnico di Torino, Corso Duca degli Abruzzi 24, 10129 Torino, Italy

F. Baino et al. (eds.), *Ceramics, Glass and Glass-Ceramics*, PoliTO Springer Series,
https://doi.org/10.1007/978-3-030-85776-9_2

The qualification *solid* excludes gases and liquids. Thus, H_2O as ice in a glacier is a mineral, but water is not.

A *highly ordered atomic arrangement* indicates an internal structural framework of atoms (or ions) arranged in a regular geometric pattern. Because this is the criterion of a crystalline solid, minerals are crystalline. Solids that lack an ordered atomic arrangement are called *amorphous* (e.g., glass).

The statement that a mineral has a *definite chemical composition* implies that it can be expressed by a specific chemical formula.

According to the traditional and strict definition, which states that a mineral is formed by inorganic processes, it includes in the realm of mineralogy those of organically produced compounds that answer all the other requirements of a mineral.

Although the subject of mineralogy is most closely related to the discipline of inorganic chemistry, the mineral science is focused specifically on naturally occurring solid substances and encompasses the following sub-discipline [1]:

(a) *Descriptive mineralogy*. It involves measurement and recording of physical properties (parameters) that help identify and describe a specific mineral. Some relevant features of a mineral such as crystal form, hardness, color and specific gravity can be evaluated in hand specimen by using the five senses or some basic testing tools. Other, more objective criteria such as optical properties like refractive indexes and X-ray diffraction data require specialized techniques and equipment such as a petrographic microscope and X-ray diffraction system, respectively. [2]

(b) *Crystallography*. Prior to the discovery of X-ray radiation by Wilhelm Conrad Roentgen in 1895 and the subsequent X-ray diffraction experiment by Max von Laue (observing the regular pattern of X-ray diffraction effects when a crystalline substance is properly positioned in a X-ray beam), crystallography concerned itself mainly with the geometric form, external morphology and symmetry, and optical properties of crystals. A distribution of particles (atoms or ions) in three dimensions is said to form a *space lattice* if every point has identical surroundings and the space lattice continues to infinity to all directions. The intersections of the lines, called *lattice points*, represents locations in the space with the same kind of atoms or group of atoms with same composition, arrangement and orientation. The geometry of a space lattice is completely specified by the *lattice constants a, b, c* and the inter-axial angles α, β, and γ. The *unit cell* of a crystal is the smallest pattern of arrangement that can be contained in a parallelepiped, the edges of which form the *a*, *b* and *c* axes of the crystal. The unit cell edges lengths and axial angles are unique for each crystalline substance and they are called *lattice parameters*. It is convenient to classify lattices according to some basic symmetry groups corresponding to the 14 *Bravais lattices*, which are commonly used to describe lattice types (Fig. 1). Additionally the 14 Bravais lattices are grouped into the following seven systems with increasing symmetry: triclinic, monoclinic, orthorhombic, hexagonal, romboedric, tetragonal, cubic. Since 1912, its main emphasis has been and still is the investigation of the internal structure of crystalline materials, whether organic or inorganic in origin. Most commonly, an X-ray beam is used as the energy source for X-ray diffraction experiments. However, electron beams are also used for producing electron diffraction patterns. The ultimate aim of modern crystallographic techniques is the determination of the crystal structure. It provides information on the location of all the atoms, bond positions and bond types, internal symmetry (space group symmetry) and the chemical content of the unit cell. All such information is fundamental to concepts of crystal chemistry, which concerns itself with the interrelation of variability in chemistry and structure [3, 4].

(c) *Crystal chemistry.* The field of crystal chemistry relates the chemical composition, the internal structure and the physical properties of crystalline materials. A specific mineral is defined on the basis of its crystal structure, chemical composition and related physical properties. In many mineral groups, the overall pattern of the structure is relatively constant, whereas the chemical composition of such a group may be highly variable. This is commonly called a specific structure type, showing extensive chemical substitution (i.e. solid solution). The assessment of structure type, of the atomic bonding arrangement, of the variability in overall chemistry and of related changes in physical properties of a crystalline substance are the domain of crystal chemistry.

(d) *Classification.* By the mid-2020, there are approximately 5500 different mineral species certified by I.M.A., the International Mineralogical Association, which since 1958 promotes the science of mineralogy and standardizes the minerals nomenclature worldwide. To make sense of the divergent chemistries and structures represented by the minerals, it is customary to classify them according to a rational crystallochemical scheme. The classification is based on chemical composition (through the anionic group) and crystal structure of minerals.

According to the model of the German mineralogist Hugo Strunz (first announced in 1941) the following ten classes are formulated and listed in order of increasing complexity: (I) elements, (II) sulfides and sulfosalts, (III) haloids, (IV) oxides and hydroxides, (V) carbonates and nitrates, (VI) borates, (VII) sulphates, chromates, molybdates, tungstates, (VIII) phosphates, arsenates, vanadates, (IX) silicates and (X) organic compounds. [5, 6]

2 Physical Properties

The physical properties of a mineral are an expression of an interrelation of a mineral crystal structure and its chemical composition. Physical properties are generally more easily determined than the chemical composition or the structure of a mineral. A combination of visual observations followed by some simple physical tests may be enough to uniquely characterize a particular mineral [7].

1	: Triclinic	$\alpha \neq \beta \neq \gamma$, $a \neq b \neq c$
2, 3	: Monoclinic	$\alpha = \gamma = 90^{\circ} \neq \beta$, $a \neq b \neq c$
4, 5, 6, 7	: Orthorombic	$\alpha = \beta = \gamma = 90^{\circ}$, $a \neq b \neq c$
8	: Hexagonal	$\alpha = \beta = 90^{\circ}$ $\gamma = 120^{\circ}$, $a = b \neq c$
9	: Trigonal	$\alpha = \beta = \gamma \neq 90^{\circ}$, $a = b = c$
10, 11	: Tetragonal	$\alpha = \beta = \gamma = 90^{\circ}$, $a = b \neq c$
12, 13, 14	: Cubic	$\alpha = \beta = \gamma = 90^{\circ}$, $a = b =$

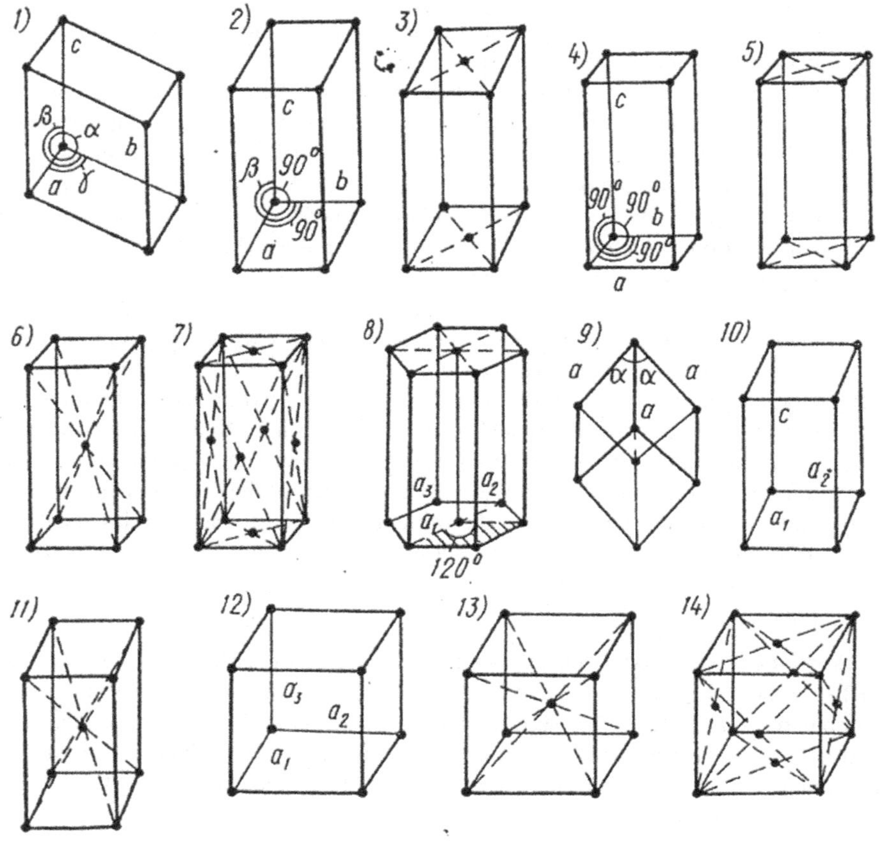

Fig. 1 The 14 Bravais lattices

(a) *Form and habit* of single or grouped crystals are responsible of the aesthetically pleasing aspects of mineral specimens in museum exhibits, in mineral shows or in private collections. Crystals are bounded by smooth plane surfaces and assume regular geometric forms. The regular geometric shape of a crystal, its *crystal form*, is not only pleasing to the eye but is also a diagnostic physical property. Mineral specimens that exhibit good crystal outline have formed under favorable geological conditions and are commonly the result of chemical deposition from a solution (or melt) in an open space, such as a cavity in a rock formation. The external form of crystals, or their morphology, is the outward expression of their mineral ordered atomic arrangement. In addition, the term *form* has a specific meaning in crystallography. In its most familiar sense, form is used to indicate general outward appearance. In crystallography, however, external shape is denoted by the word habit, whereas form is used in a special and restricted sense. A form consists of a group of crystal faces, all of which have the same relation to the elements of symmetry inherent in the crystal. Crystal habit, therefore, includes its general shape and irregularities of growth. A *unit cell* is defined as the smallest unit of structure that if stacked indefinitely in three dimensions would form the whole structure.

(b) *Luster* refers to the general appearance of a mineral surface in reflected light. The two distinctive types of luster are metallic and non-metallic, but there is not sharp division between them. Although the difference between these types of luster is not easy to describe, the eye discerns it easily, and after some experience, seldom makes a mistake. Metallic is the luster of a metallic surface, related to opaque materials to light, which cannot pass through even at very thin edges. Non-metallic luster is generally shown by light-colored minerals, which transmit light, if not through thick portions at least through their edges.

(c) *Color* of a mineral is easily observed and instantly evaluated. For some minerals it is characteristic and serves as a distinguishing property but it can be also a tricky property because color varies not only from one mineral to another but also within the same mineral or mineral group being also affected by the environment and different geological circumstances of formation and growth.

(d) *Cleavage*, parting and fracture are the response of crystalline material to an external force. When such a force is applied, a mineral is subjected to stress. If the internal structure of the crystalline structure is deformed, due to the stress, it is said to have undergone strain. The strength of a crystalline material is a function of its bonding mechanism and the presence or absence of structural defects. The type of bonding (ionic, covalent or metallic) is of major importance in a mineral's reaction to an applied force. If mineral contains structural defects along specific planes or directions, it will tend to deform along such features more easily than a mineral with a truly perfect or better-ordered structure. If the strain in the mineral is so great as to exceed its overall strength, it will break. Many minerals have planar directions in their structure that are systematically weaker than other directions. This is the result of planes in the crystal structure that are joined by fewer bonds per unit volume than are other planes in the structure, or are joined by weaker bonds. In particular, cleavage is the tendency of minerals to break parallel to atomic planes. Cleavage is a reflection of internal structure because within a structure the strength of chemical bonding is commonly different in different directions. This is especially shown by layer structures in which the bonding between layers is very strong but between layers is much weaker so that a perfect cleavage can be observed parallel to the layering.

Parting happens when minerals break along planes of structural weakness possibly resulting from pressure or twinning. Because it is parallel to rational crystallographic planes, it resembles cleavage. However, unlike cleavage, parting is not shown by all specimens but only by those that are twinned or have been subjected to the proper pressure.

Finally, fracture is the way minerals break when they do not yield along cleavage or parting surfaces without showing obvious crystallographic control. The common terms used to describe fracture surfaces, listed in order of increasing relief on the surface, are *conchoidal* (smoothly curved as on broken plate glass), *irregular* or *uneven* (a rough surface or one with random irregularities), *hacky* (a surface with sharp-edged irregularities), *splintery* (resembles the splintery end of a broken piece of wood).

(e) *Hardness* is the resistance that a smooth surface of a mineral offers to scratching. The degree of hardness is determined by observing the comparative ease or difficulty with which one mineral is scratched by another, or by a tool like a file or a knife. The evaluation of hardness is the assessment of the reaction of a crystal structure to stress without rupture. In relating the hardness of a crystal structure to its bonding, it must be noted that the structure's overall strength is a composite of all its bond types, whereas the hardness of that same structure is an expression of its weakest bonding. A series of ten common minerals was chosen by the Austrian mineralogist Friedrich Mohs (1824) as a scale. By comparison, the relative hardness of any mineral can be told. The following minerals, arranged in order of increasing hardness, comprise what is known as the *Mohs scale of hardness*:

1. Talc
2. Gypsum
3. Calcite
4. Fluorite
5. Apatite
6. Orthoclase
7. Quartz
8. Topaz
9. Corundum
10. Diamond

Hardness can also be measured by more quantitative techniques than a scratch test in leading to an absolute hardness scale. The relative position of the minerals is preserved but corundum, for example, is two times as hard as topaz and four times harder than quartz. Hardness is a vector property. Thus, crystals may show varying degrees of hardness depending on the directions in which they are scratched.

(f) *Tenacity* is the resistance or cohesiveness that a mineral offers to breaking, crushing, bending or tearing. The term *brittle* identifies a mineral that breaks and powders easily due to the presence of dominant ionic bonding. The term *malleable* is applied to minerals that can be hammered out into thin sheets;

sectile defines a mineral which can be cut into thin shavings with a knife; *ductile* is a term used for minerals that can be drawn into wire (these last three characteristics—malleability, sectility and ductility—are diagnostic of materials held together by metallic bonding); *flexible* refers to minerals which bend but do not resume their original shape when pressure is released; finally, *elastic* are the minerals that, after being bent, will resume their original position upon the release of the pressure.

(g) *Specific gravity* or relative density is a number that expresses the ratio between the weight of a substance and the weight of an equal volume of water at 4 °C. The specific gravity of a crystalline substance depends on the kind of atoms of which it is composed and the manner in which the atoms are packed together. I the particular case of the minerals identified as solid solutions series there is a continuous change in specific gravity with change in chemical composition.

3 Silicates

Synthetic equivalents of silicates are the key crystalline components of ceramics and then the most important glass-forming systems are based on silicate compositions [8]. Therefore, further sub-classifications are made and the most important for the scope of this book is just that one concerning the class of the silicates. In considering the structural (atomic) arrangements of the silicate tetrahedral, the following 6 sub-classes can be defined and listed in increasing complexity: (1) nesosilicates (single tetrahedral), (2) sorosilicates (double tetrahedral), (3) cyclosilicates (ring of six tetrahedrons), (4) inosilicates (chain of tetrahedrons), (5) phyllosilicates (sheet of tetrahedrons), (6) tectosilicates (framework of tetrahedrons). Dealing with the crystal structure of silicates, it is necessary to underline that the classification is organized in groups with many species and complex structures [9].

Neso-silicates*(single tetrahedral structure)*

Mullite [named after the discovery locality, Isle of Mull, Scotland] (1924) [10, 11].

Theoretical composition $Al_6Si_2O_{13}$ ($3Al_2O_3{\cdot}2SiO_2$)—Orthorombic.

Natural occurrence: Isle of Mull (Scotland), Lombardy (Italy), Eifel (Germany), Antrim (Ireland), Kimberley (South Africa), Amur (Russia).

Physical/chemical properties: colorless, white, yellow, pink, red, gray, transparent to translucent, vitreous luster, distinct cleavage. S.G. = 3.11–3.26 g/cm^3; H = 6–7; CTE (thermal expansion coefficient) = (4.5–5.6) × 10^{-6}/K. Melts congruently at 1910 °C.

Mullite forms as the final product of a series of reactions in clay minerals at temperatures higher than 1100 °C. Mullite phase is able to reinforce refractories, porcelain and glass–ceramic matrices by exhibiting enhanced mechanical properties. Thus, recently, two types of nanocrystalline glass-ceramics have been developed namely transparent glass-ceramics and tough, high-modulus glass-ceramics with precisely engineered surfaces. Transparent glass-ceramics are formed from certain aluminosilicate glasses based on mullite phase capable of efficient crystal nucleation and slow growth. The key crystalline phases includes mullite, which shows unique chromium-luminescence behavior.

Zircon [renamed from the Arabic (and, in turn, from the Persian "azargun") "zar", gold, plus "gun", colored, referring to one of the many colors that the mineral may display] (1783) [12].

$ZrSiO_4$—Tetragonal.

Natural occurrence: Saxony (Germany), Urals (Russia), Larwik (Norway), Sillinjarvi (Finland), Colorado, North Carolina, South Carolina New Jersey (USA), Ontario.

(Canada), Matura (Sri Lanka), Ampanobe (Madagascar)Mud Tank (Australia).

Physical/chemical properties: colorless, yellow, grey, reddish-brown, green, brown, black, transparent to opaque, vitreous to adamantine luster, indistinct cleavage, brittle, conchoidal fracture. S.G. = 4.60–4.70 g/cm^3; H = 7.5; CTE = (2.6–3.1) × 10^{-6}/K. Melts congruently at 2420 °C.

Zircon is traditionally been used as an opacifier in ceramics industry. Zircon is essential crystalline phase in a vitreous ceramic whiteware for technical application. Due to unique properties as high melting point, low CTE, very high strength it is ideal as high temperature construction material for furnaces and kilns as well. Interestingly, zircon contributes to achieve a desired spectrum of clinically relevant mechanical properties such as hardness, elastic modulus, fracture toughness and brittleness index, chemical solubility of glass ceramics. Such spectrum of properties can be utilised for developing a machinable ceramic for low stress bearing inlays, onlays and veneers.

Forsterite [named in honor of Jacob Forster (1739–1806), German mineral collector and mineral dealer] (1824) [13].

Mg_2SiO_4—Orthorombic.

Natural occurrence: Vesuvius (Italy), Ural mountains (Russia), More (Norway), Eifel (Germany), Arizona (USA), Myanmar (Burma), Basham (Pakistan), Mpapwa (Tanzania).

Physical/chemical properties: green, pale yellow or white, vitreous luster, imperfect cleavage, brittle, conchoidal fracture. S.G. = 3.27 g/cm^3; H = 7; CTE = (8.3–9.5) × 10^{-6}/K. Melts congruently at 1890 °C.

Forsterite ceramics might safely operate at temperatures up to 1000 °C, features excellent electrical insulating properties and good mechanical strength. It is used as radiation detection devices and in other vacuum and hermetic seal type applications.

A forsterite-based glass–ceramic was shown to be a candidate suitable for applications in low temperature co-fired ceramics (LTCC) field.

Fayalite [named by the German chemist Christian Gottlieb Gmelin (1749–1818) for the type locality, Faial Island (Fayal Island), Azores District (Azores Islands), Portugal] (1840) [14].

Fe_2SiO_4—Orthorombic.

Natural occurrence: Hartz (Germany), Colorado, Wyoming, Massachusetts, New York (USA), British Columbia (Canada), Raftsund (Norway).

Physical/chemical properties: greenish-yellow, yellow or brown, vitreous to resinous on fracture luster, imperfect cleavage, brittle, conchoidal fracture. S.G. = 4.39; H = 6.5–7; $CTE_{(20–800\ °C)} = 10.2 \times 10^{-6}$ /K. Melts congruently at 1205 °C.

Fayalite forms solid solutions series with forsterite: the former is iron- rich while the latter is magnesium-rich end-member of the olivine mineral group.

The formation of fayalite and another colorless FeO-containing alumino-silicates leads to improved whiteness of porcelain body. Furthermore, it favors the vitreous phase formation and widens the sintering temperature range to a certain extent while the vitreous phase, in turn, promotes a desirable mullite phase formation.

Monticellite [named after the Italian mineralogist Teodoro Monticelli (1759–1845), author of several works on Mount Vesuvius] (1831) [15].

$CaMgSiO_4$—Orthorombic.

Natural occurrence: Vesuvius (Italy), Antrim (Ireland), California, Arkansas, New York (USA), Quebec (Canada), Kivu (Zaire), Qala-Diza (Iraq), Kola (Russia).

Physical/chemical properties: colorless, grey or greenish, trasparent to translucent, vitreous to resinous luster, poor cleavage, brittle, uneven fracture. S.G. = 3.03–3.27; H = 5–5.5; CTE = 10.8×10^{-6}/K. Melts congruently at 1503 °C.

It has been demonstrated that monticellite ceramics possess high mechanical strength, an appropriate CTE, bioactivity and biocompatibility in vitro, and might be used as bioactive bone-tissue repair materials.

Willemite [named in honor of William I (Willem) (1772–1843), King of the Netherlands] (1825) [16, 17].

Zn_2SiO_4—Phenakite group—Trigonal.

Natural occurrence: Altenberg (Belgium), Tsumeb (Namibia), California, Arizona, New Jersey, New Mexico (USA), Quebec (Canada), Puttapa (South Australia).

Physical/chemical properties: colorless, white, pastel green, apple-green, light blue, azure-blue, yellow, burgundy-red, brown, mahogany-brown, gray, black, pink, trasparent to opaque, vitreous to resinous luster, indistinct cleavage, brittle, irregular to conchoidal fracture. S.G. = 3.89–4.19 g/cm^3; H = 5.5. Melts congruently at 1512 °C.

Willemite glass–ceramics is known as materials for electronic applications. It is also characterized by a low thermal expansion coefficient and is used for fabrication of glazes.

Soro-silicates*(double tetrahedral structure)*

Melilites

The melilites belong to the family of sorosilicates with general formula (Ca, Na)$_2$(Al, Mg, Fe^{2+})[(Al, Si)SiO_7] and exist in igneous, as well as metamorphic rocks. Most naturally occurring melilite minerals are solid solutions of akermanite [$Ca_2(MgSi_2O_7)$], gehlenite [$Ca_2(Al_2SiO_7)$], Fe-akermanite [Ca2($Fe^{2+}Si_2O_7$)] and Na-melilite [$NaCa(AlSi_2O_7)$]. These minerals crystallize from calcium-rich, alkaline magmas and form many artificial melts and blast furnace slags [18].

All the end members akermanite, gehlenite, Fe-akermanite and Na-melilite have been synthetically synthesized. The thermal expansion of melilites measured by powder XRD in the temperature range of 25–925 °C varied in the range of 10.0–11.0 × 10^{-6}/K [19].

Akermanite [after Richard Akerman (1837–1922), Swedish metallurgist] (1890) [20]. $Ca_2Mg(Si_2O_7)$—Melilite group—Tetragonal.

Natural occurrence: Campania and Lazio (Italy), California and New Mexico (USA) Quebec (Canada), Argyllshire (UK), Antrim (Ireland), Cape (South Africa).

Physical/chemical properties: colorless, yellowish gray, green, brown, transparent to translucent, vitreous to resinous lustre, distinct cleavage, uneven to conchoidal fracture, brittle. S.G. = 2.944 g/cm^3; H = 5–6; Melts congruently at 1454 °C.

The CTE values of the glass–ceramics in the system akermanite-gehlenite measured in temperature range of 200–700 °C vary in the intervals of (10.16–11.10) and (10.04–10.99) × 10^{-6}/K, for the GC samples sintered at 900 and at 1000 °C, respectively [21].

Akermanite-gehlenite sealants for SOFC (solid oxide fuel cells) demonstrate a density (2.91 g/cm^3) equal to that of bulk glass and mechanical strength (106 MPa) confirming the good densification of glass powder compacts [22].

Gehlenite [after Ferdinand Gehlen (1775–1815), German chemist] (1815) [20]. $Ca_2Mg(Si_2O_7)$—Melilite group—Tetragonal.

Natural occurrence: Trentino (Italy), Eifel (Germany), California, Colorado, New Mexico (USA), Argyllshire (UK), Antrim (Ireland), Durango (Mexico).

Physical/chemical properties: yellow–brown, colorless, greenish grey, transparent to translucent and opaque, vitreous to resinous lustre, distinct cleavage, uneven to conchoidal fracture, brittle. S.G. = 3.038; H = 5–6. Melts congrueantly at 1590 °C.

Gehlenite is an Al-rich end-member of the melilite group and forms complete solid solution series with akermanite with a general formula of $Ca_2Mg_xAl_{2(1-x)}Si_{(1-x)}O_7$. The isomorphous replacements in end members of melilite group may occur by the scheme $Mg^{2+} + Si^{4+} = 2Al^{3+}$ suggesting substitution of aluminum for Mg in the tetrahedral sites $MgO_4{}^{6-}$ and for Si in the di-ortogroups $Si_2O_7{}^{6-}$ [21].

Cyclo-silicates*(with tetrahedral ring structure)*

Cordierite [after Pierre Louis Antoine Cordier (1777–1861), French mining engineer and geologist] (1813) [23].

($Mg_2Al_3(AlSi_5O_{18})$) ($2MgO{\cdot}2Al_2O_3{\cdot}5SiO_2$)—Orthorhombic, pseudo-hexagonal.

Natural occurrence: Bavaria (Germany), Kragero (Norway), Antsirabe (Madagascar), Fungure (Zimbawe), Namib (Namibia), Leppavirta (Finland).

Physical/chemical properties: grey, blue, blue-violet, greenish, yellowish brown; colorless to very light blue in transmitted light, transparent to translucent, vitreous, imperfect cleavage, sub-conchoidal fracture, brittle. S.G. = 2.60–2.66 g/cm^3; H = 7–7.5. $CTE_{(25-300\ °C)} = 2.5 \times 10^{-6}$/K. Melts congruently at 1468 °C.

Cordierite phase shows low thermal expansion, low thermal conductivity and good electrical insulation compared to other ceramic materials. Therefore, cordierite based ceramics and glass–ceramics are used as kiln furniture (saggars, kits, shelves, shelf supports), gas burners, electro-ceramics (resistors, fusibles, flame guards), refractory fillers, honeycomb substrate for catalytic convertors, etc.

One of the earliest applications for a glass–ceramic was performed by Corning Glass Works, Corning, N.Y. through manufacturing radomes from a glass–ceramic material that is highly crystalline with cordierite constituting the predominant crystal phase.

Beryl [possibly from the Greek "beryllos" which referred to a number of blue-green stones in antiquity] (Antiquity).

$Be_3Al_2(Si_6O_{18})$—Beryl group—Hexagonal.

Natural occurrence: Bavaria (Germany), Kragero (Norway), Antsirabe (Madagascar), Fungre (Zimbawe), Namib (Namibia), Leppavirta (Finland).

Physical/chemical properties: colorless, green, blue, yellow, white, pink, etc., transparent to translucent, vitreous or resinous, imperfect cleavage, conchoidal to uneven fracture, brittle. S.G. = 2.63–2.97; H = 7.5–8. Melts incongruently to phenakite (Be_2SiO_4) and liquid of high viscosity at 1450 °C. [24]

Cordierite is isostructural with beryl where the rings $Si_6O_{18}{}^{12-}$ are symmetrically surrounded by Be^{2+} ions in fourfold co-ordination and $A1^{3+}$ in sixfold co-ordination. Be^{2+} ions have relatively high field strength. The term field strength was introduced by Dietze to define the coulombic attraction of cations at the centre of anion groups. Here Z/a^2 (Z = cation charge, a = cation–anion distance) values are used to express a relative measure of the ability of cations to compete for their coordination requirements. The field strengths of Be^{2+} and Al^{3+} ions are very similar and hence their interaction energies with O^{2-} and SiO^{4-} anions in the melt at high temperatures are likely to be similar.

Colorless Beryl minerals were used in the past to prepare optical glasses.

Ino-silicates*(with single tetrahedral chain structure)*

Magnesium metasilicate $MgSiO_3$

Magnesium metasilicate belongs to the group of clino-pyroxens and exists in three polymorphic modifications. Enstatite is a low temperature modification, which, during a thermal circle, transforms into proto-enstatite (during heating) and clino-enstatite (during cooling of proto-enstatite). CTE $_{(300\text{–}700\,°C)}$ of enstatite, protoenstatite, and clinoenstatite were respectively 12.0×10^{-6}, 9.8×10^{-6} and 13.5×10^{-6}/K [25].

Melts incongruently at 1557 °C to forsterite and liquid [26].

Pyroxenes are major constituents of the earth's crust and of the upper mantle to depths of 400 km. They occur in a wide variety of geological settings, both igneous and metamorphic. The chain silicate structure of pyroxenes enables incorporation of various cations in their structure. Pyroxenes have the general formula of $M1M2(Si,Al)_2O_6$, where M1 is usually Ca^{2+}, Na^+, Fe^{2+} and Mg^{2+} and more rarely Zn^{2+}, Mn^{2+} and Li^+, while M2 refers to smaller ions, such as Cr^{3+}, Al^{3+}, Fe^{3+}, Mg^{2+}, Mn^{2+}, Sc^{3+}, Ti^{4+}, V^{5+} and even Fe^{2+}.

Enstatite [from the Greek enstates meaning "opponent" for refractory nature under blowpipe flame] (1855) [27].

$Mg_2Si_2O_6$—Pyroxene group - Orthorhombic.

Natural occurrence: Zdar (Tcheck Rep.), Bamle (Norway), Cornwall (UK), Bavaria, Eiffel (Germany), Styria (Austria), Mbeya (Tanzania), Pennsylvania, Texas (USA).

Physical/chemical properties: white, yellowish green, brown, greenish white or grey, olive-green, translucent or opaque, vitreous or pearly, distinct cleavage, irregular fracture, brittle. S.G. = 3.2–3.9 g/cm^3; H = 5–6. Enstatite GCs was reported to be produced via surface nucleation mechanism and sol–gel route. When both ZrO_2 and TiO_2 were used as nucleating agents, the sintered GCs featured flexural strength of 250 MPa, fracture toughness of 3.3 MPa m$^{0.5}$ and fracture surface energy of 51 J/m^2.

These GCs have been used in electronic circuits.

Diopside [named in 1806 by the French mineralogist Renè Just Haüy (1743–1822) from the Greek διζ for "double" and οψτζ for "appearance"] (1800) [28, 29].

$CaMgSi_2O_6$—pyroxene group—Orthorhombic.

Natural occurrence: Tirol (Austria), Val d'Aosta (Italy), Otokumpu (Finland), Urals-Yakutia, Lake Baikal (Russia), Ontario, Quebec (Canada), Tennessee (USA).

Physical/chemical properties: light to dark green, blue, brown, colorless, snow white, grey, pale violet, transparent to opaque, vitreous or dull, distinct cleavage, uneven to conchoidal fracture, brittle. S.G. = 3.22–3.38; H = 5.5–6.5. CTE = (8.0–9.5) × 10^{-6}/K. Melts congruently at 1391 °C [30].

Ceramics and GCs based on pyroxenes attract increasing interest in several advanced fields such as electronics, biomedicine, immobilization of radioactive wastes and sealants for solid oxide fuel cells (SOFC).

Wollastonite [named by J. Léman in honor of William Hyde Wollaston (1766–1828), English chemist and mineralogist] (1818) [31].

$CaSiO_3$—Pyroxene group (Wollastonite sub-group)—Monoclinic or Triclinic..

Natural occurance: Banat (Romania), Sardinia, Campania (Italy), Antrim (Ireland), Kongsberg (Norway), Hartz (Germany), New York, California, Nevada (USA), Ontario, Quebec (Canada), Chiapas (Mexico), Hiroshima, Ibaragi (Japan).

Physical/chemical properties: white, gray-white, light green, pinkish, brown, red, yellow, transparent to translucent, vitreous or pearly, perfect cleavage, uneven fracture, brittle. S.G. = 2.86–3.09; H = 4.5–5. Melts congruently at 1544 °C.

Wollastonite is a polymorphic substance; low temperature triclinic form (β-$CaSiO_3$) converts to the high temperature form pseudo-wollastonite (α-$CaSiO_3$) at 1125°C. CTE of α-$CaSiO_3$ is 11.8 × 10^{-6}/K and β- $CaSiO_3$ is 6.5 × 10^{-6}/K [32].

Both the naturally occuring and artificial wollastonite reduce drying and firing shrinkage, promote lower moisture and thermal expansion in the fired product and increases firing strength. Wollastonite is used as a source of CaO and SiO_2 in glazes. Wollastonite ceramics have been proven to induce the formation of a hydroxyapatite (HA) layer on their surface after soaking in SBF and they are biodegradable and biocompatible both in vitro and in vivo.

GC precipitating crystalline apatite and ß-wollastonite (CaO·SiO_2), in a MgO–CaO–SiO_2–P_2O_5 glass, called Cerabone® A-W also forms a bone like apatite layer on its surface in the living body and bond to the bone through the apatite layer.

Spodumene [from Greek "spodoumenos", meaning "reduced to ashes", alluding to the grayish-white mass that is formed when the mineral is ignited] (1800) [33]. $LiAlSi_2O_6$—Pyroxene group—Monoclinic.

Natural occurrence: Vasterbotten (Sweden), Kuortane, Tammela (Finland), South Dakota, New Mexico, California, North Carolina (USA), Manitoba (Canada), Laghman (Afghanistan), Minas Gerais (Brazil), Anjanabonoina (Madagascar), Bikita (Zimbawe).

Physical/chemical properties: colorless, yellow, light green, emerald-green, pink to violet, purple, white, gray, transparent to translucent, vitreous or pearly on cleavage, good cleavage, uneven to subconchoidal fracture, brittle tenacity. S.G. = 3.03–3.23 g/cm3; H = 6.5–7. CTE of β-spodumene measured in the interval 20–1000 °C = 0.9×10^{-6}/K. Melts congruently at 1380 °C.

Naturally occurring low-temperature form α-spodumene belongs to the monoclinic system whereas the high-temperature β-spodumene crystallizes in the tetragonal system. α-spodumene converts to β-spodumene at temperatures above 900 °C. Spodumene is often referred to as lithium feldspar and used in ceramics as a source of lithia. It is a valuable component in glass and ceramic glazes. β-spodmene based GC's are particularly useful for cookware and counter-top cooking surfaces because of their low thermal expansion. Commercial compositions utilize TiO_2 as a nucleating agent, and have partial substitution of MgO and ZnO for Li_2O to improve the working properties of the parent glass while also lowering the materials cost.

Phyllo silicates*(sheet of tetrahedrons)*

This sub-class includes the following groups of minerals more specifically dealing with GC's application: Kaolinite group (kaolinite, dickite, nacrite), Smectite (montmorillonite), Talc group (talc, pyrophyllite) and Mica group (muscovite, biotite, phlogopite, margarite, lepidolite). In addition to all the minerals belonging to these groups or to some of them, also other minerals such as feldspar, quartz and carbonates can form the so called *clay mineral group.* In fact, clay is a rock term, which is made up by a number of different minerals in varying proportions. The term *clay* refers to a natural occurring material composed primarily of fine-grained hydrated aluminum silicates, which in general shows a certain degree of plasticity at appropriate water contents and will harden when dried or fired. In the clay minerals three of the four O-atoms in each SiO_4 tetrahedron are shared with other tetrahedrons giving a continuous sheet (Fig. 2) that can be extended indefinitely in the ***a*** and ***b*** directions, i.e. in the plane of the paper. In particular, the phyllosilicate class includes a very important group of minerals dealing with the different micas composed of a sequence of silica tetrahedron and aluminum octahedron with interlayer cations and little or no water molecules. This structure is responsible of a specific cleavage of the crystals into thin flakes, which are usually transparent without demonstrating plasticity or swelling. Besides these properties, the value of mica is derived by other physical properties like non-conductivity of heat and electricity, dielectric strength, resistively and stability.

Clay minerals readily take up water on wetting while the absorbed water is driven off by heating to 100–200 °C. Water sorption is an important characteristic of natural clay particles. Clay particles can absorb or lose water in response to changes in

Fig. 2 $[Si_2O_5]_n^{2n-}$ tetrahedral silica layer

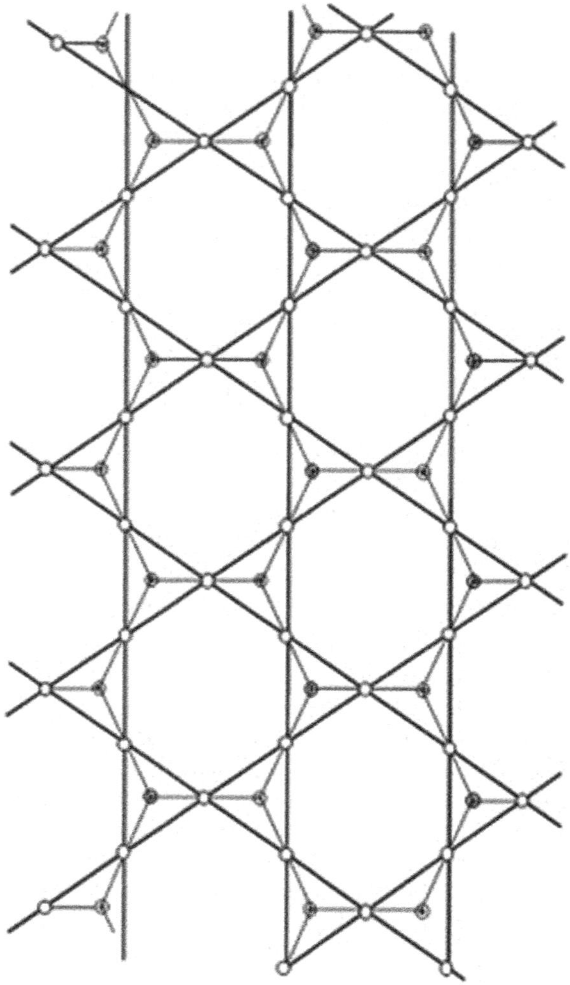

humidity content in the ambient environment; when water is absorbed, it fills the spaces between the stacked silicate layers. Clays become plastic, after standing for a short time, if they are mixed with liquids composed of polar molecules like water. On heating up to 150 °C, clay minerals lose the absorbed water while in the temperature range 400–900 °C the water bound up in the lattice as OH-groups is evolved and the crystal structure is partly destroyed or altered. In most cases, after the disintegration of the crystal structure into an amorphous stage, a temperature interval occurs where crystalline phases (mullite, cristobalite, etc.) are formed. With increasing temperature, a glass phase and a melt result. The main value of clay minerals for ceramics lies in the fact that when wet it can be easily molded in any desired shape, and when it is heated, part of the combine water is driven off, producing a hard durable substance.

Kaolinite [named in 1637 by Song Yingxing for the ancient Chinese type locality "Kaoling (Gaoling)", meaning "high ridge"] (1867) [34].

$Al_2(Si_2O_5)(OH)_4$—Triclinic—Kaolinite group.

Natural occurrence: Kiangsi (China), Cornwall, Devon (UK), Haute Vienne (France), Saxony (Germany), Donets (Ukraine), Georgia, South Carolina, Arkansas (USA), Quebec (Canada), Rio Arriba (Mexico).

Physical/chemical properties: white to cream and pale-yellow, also often stained various hues by impurities, transparent to translucent as single crystals, waxy pearly, dull, earthy lustre, perfect cleavage, plastic but inelastic. Kaolinite is the most refractory of the clays and has excellent firing properties. The basic structure shows that bonds between kaolinite layers are weak intermolecular Van der Waals' forces along with hydrogen bonds, while the bonds between octahedral and tetrahedral layers are formed through much stronger intramolecular bonds (Fig. 3). Therefore, kaolinite can develop its efficient double role of absorbing water and swelling during the ceramic cross-linking process and afterwards it shows the best high-temperature stability,

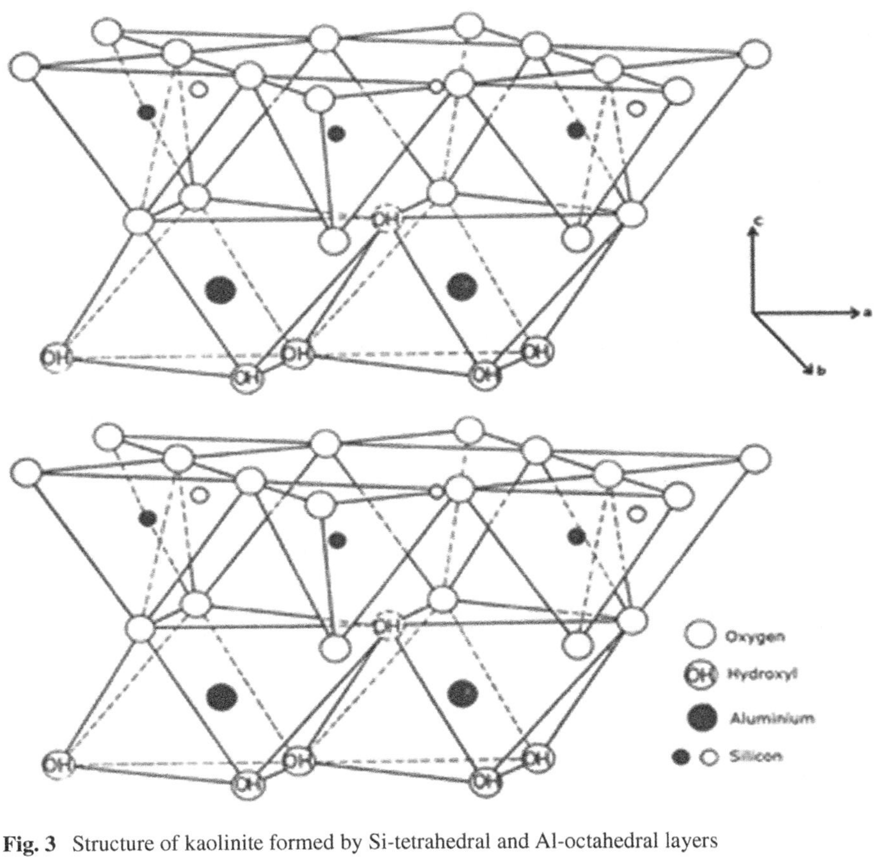

Fig. 3 Structure of kaolinite formed by Si-tetrahedral and Al-octahedral layers

being heated rapidly and shrinking very little. It is the ideal clay of porcelain fine china and the only clay that works well alone in pottery manufacture. The formation of small hexagonal flake-like equidimensional crystals after the curing temperature of ceramic materials give them a good plasticity. S.G. = 2.63 g/cm^3; H = 2–2.5.

Dickite [named after Allan Brugh Dick (1833–1926), Scottish metallurgical chemist, who first described the species. (1930) [35].

$Al_2(Si_2O_5)(OH)_4$—Monoclinic—Kaolinite group.

Natural occurrence**:** Wales (UK), Mad (Hungary), Rodna (Romania), Herault (France), Transvaal (South Africa), Colorado, Pennsylvania (USA), Chihuahua (Mexico).

Physical/chemical properties: white, or tinted by impurities, transparent, satiny lustre, perfect cleavage, flexible but inelastic. Dickite is a rare clay mineral having equivalent chemical composition to kaolinite but different structure. Therefore, the particular stacking of the atoms influence the properties in many ways including the color, hardness, cleavage and density. S.G. = 2.60 g/cm^3; H = 2–2.5.

Montmorillonite [for the type locality at Montmorillon, Vienne, Poitou–Charentes, France. (1847) [36].

$(Na,Ca)_{0.33}(Al,Mg)_2(Si_4O_{10})(OH)_2 \cdot nH_2O$—Monoclinic—Smectite group.

Natural occurrence: worldwide distributed. Vienne (France), New Mexico, Arizona, South Dakota, Wyoming, Virginia (USA).

Physical/chemical properties: white, buff, yellow, green, rarely pale pink to red, translucent, dull, earthy lustre, perfect cleavage, irregular and uneven fracture. S.G. = 2–3 g/cm^3; H = 1–2.

Concerning the montmorillonite's behavior versus water, it typically exhibits a gradual dehydration and phase change to a stronger nonexpendable clay. It is important that the basal spacing of montmorillonite due to the week van der Waals' forces hold montmorillonite sheets enable certain liquid molecules like water to penetrate between the sheets causing characteristic swelling. An extended surface area in montmorillonites determines their application in ceramic formulations, as absorbents and catalysts.

Talk [named by Georgius Agricola (Georg Bauer, 1494–1555) from Arabic "talq", pure, probably alluding to the color of its powder] (1546).

$Mg_3Si_4O_{10}(OH)_2$—Triclinic—Pyrophyllite-Talc group.

Natural occurrence: worldwide distributed. Tirol (Austria), Ticino (Switzerland), Trentino Alto-Adige, Piedmont (Italy), Ariege (France), Bavaria (Germany), Snarum (Norway), Siberia (Russia), Georgia, Pennsylvania, California (USA).

Physical/chemical properties: colorless, white, pale green, bright emerald-green to dark green, brown, gray, transparent and translucent, sub-vitreous, resinous, waxy, greasy, pearly lustre, perfect cleavage, fibrous fracture. S.G. = 2.58–2.83 g/cm^3; H = 1.

Talc's structure is based on combination of brucite $Mg(OH)_2$ layer and two silica layers and is formed by the replacement of aluminum in the octahedral layer by magnesium. Talc is widely used in the ceramics industry in both bodies and glazes.

In low-fire art-ware bodies, it imparts whiteness and increases thermal expansion to resist crazing. In stoneware, small percentages of talc is used to flux the body and therefore improve strength and vitrification. It is a source of MgO flux in high-temperature glazes.

Pyrophyllite [from the Greek for "fire" (πῦρ, pyr) and "leaf" (φύλλον, phyllos) for the way it exfoliates when heated] (1829) [37].

$Al_2Si_4O_{10}(OH)_2$—Triclinic—Pyrophyllite-Talc group.

Natural occurrence: Ural (Russia), Valais (Switzerland), Georgia, Arizona, California, South Carolina (USA), Nagano (Japan).

Physical/chemical properties: white, gray, pale blue, pale green, pale yellow, brownish green, translucent to opaque, pearly to dull luster, perfect cleavage, flexible but inelastic. Unlike kaolinite, pyrophyllite contains low alumina and high silica, which on heating mainly produces mullite and amorphous silica. Mullite crystallization from pyrophyllite is rather easy than that from kaolinite. The silica in turn yields large amount of viscous liquid at high temperature. Some studies suggest that pyrophyllite may be utilized in such compositions where mullite is a desirable phase by partial replacement of china clay, which is a viable alternative particularly in the background of depleting reserves of kaolinite and its continuous cost escalation. Additionally amorphous silica produced in the reaction system may acts as an in situ produced filler material that reduces the use of quartz in such system.

S.G. = 2.65–2.90 g/cm^3; H = 1–2.

Muscovite ["Muscovy glass" from Muscovy Province in Russia] (1794) [38]. $KAl_2(AlSi_3O_{10})(OH)_2$—Monoclinic—Mica group.

Natural occurrence: Ural (Russia), Kammerfors (Norway), Maine, North Carolina, Pennsylvania, Virginia, New Mexico, South Dakota (USA), Ontario (Canada), Minas Gerais (Brazil), Pradesh (India).

Physical/chemical properties: white to colorless, silvery-white, and tinged various colors by impurities, transparent to translucent, vitreous to pearly to silky luster, perfect in sheets cleavage, laminae flexible and elastic. S.G. = 2.77–2.88 g/cm^3; H = 2.5. Mica GC's are machinable and demonstrate high fracture toughness and flexural strength.

Phlogopite [from the Greek φλογωποζ "phlogopos" for "resembling fire", in allusion to the red tint of the original specimens] (1841) [39].

$KMg_3(AlSi_3O_{10})(F,OH)_2$—Monoclinic—Mica group.

Natural occurrence: Siberia (Russia), Kammerfors (Norway), New York, New Jersey (USA), Ontario, Quebec (Canada), Ticino (Switzerland), Trentino, Lombardia (Italy), Nancy Sound (New Zealand), Anpandrandava (Madagascar).

Physical/chemical properties: colorless, brown, gray, green, yellow, or reddish brown, transparent to translucent, pearly to submetallic on cleavage luster, perfect in sheets cleavage, thin laminae flexible and elastic. S.G. = 2.78–2.85 g/cm^3; H = 2–3. Phlogopite in association with apatite constitute a bioactive GC for bone substitution. This glass–ceramic has the advantage to be workable by the surgeon, if necessary, during operation. Different types of the material can be produced in dependence of the composition, nucleation, and crystallization of the basic glass.

Illite [from one of the co-type localities, in Illinois (USA)] (1937).

$K_{0.65}Al_{2.0}[Al_{0.65}Si_{3.35}O_{10}](OH)_2$—Monoclinic—Mica group.

Natural occurrence: Illinois (USA), Cornwall (England), Bohemia (Czech Rep.) Thuringia (Germany), Sarospatak (Hungary), Hodrusa (Slovakia).

Physical/chemical properties: gray-white to silvery-white, greenish-gray, sometimes stained other hues, translucent, waxy, greasy, dull, earthy luster, perfect cleavage, elastic and micaceous fracture. S.G. = 2.79–2.80 g/cm^3; H = 1–2.

Illite is usually an alteration product of muscovite and can be regarded as the mechanism where muscovite may be eventually altered to montmorillonite. During the alteration from muscovite to illite, the structure of illite may become "turbostratified" and the resulting layers of K-filled and K-deficient are randomly mixed, in addition to varying Al-Si substitutions in tetrahedral layers for charge balance.

Lepidolite [from the Greek words lepidos for "scale" and lithos for "stone"] (1792). [40].

$KLi_2Al(Si_4O_{10})(F,OH)_2$—Monoclinic—Mica group.

Natural occurrence: Urals (Russia), Maine, Colorado, South Dakota, California (USA), Manitoba (Canada), Rozna (Czech Rep.), Minas Gerais (Brazil), Bihar (India), Bikita (Zinbawe), Bity (Madagascar).

Physical/chemical properties: pink, light purple, light rose red, other colors possible but are rare, transparent to translucent, pearly to vitreous luster, perfect cleavage, elastic and micaceous fracture. S.G. = 2.80–2.90 g/cm^3; H = 2.5–3.5.

Lepidolite gives porcelains having a fine appearance and extreme whiteness. The most important property noted is an improvement of the bodies containing lepidolite in resistance to thermal shocks. Owing to its higher cost, it is unlikely that lepidolite will be used in porcelains except in small proportions with feldspar to produce bodies having special properties. As hitherto the most important applications of lepidolite will doubtless be in glasses and enamels where full benefit can be secured from its fluorine content and low fusion temperature.

Tecto-silicates*(framework of tetrahedrons)*

This subclass includes some of the most important rock-forming minerals and the most abundant ones of the earth's crust (approximately 64%), the feldspars, and widely distributed and ubiquitous quartz. This one is the simplest species, in composition an oxide (SiO_2) but included here because its structure consists of a framework of $(SiO_4)^{4-}$ tetrahedra with each oxygen linked to a silicon atom in a neighboring tetrahedron. This results is a stable, strongly bonded structure in which the ratio of Si: O is 1: 2. (Fig. 4). Except for quartz and its polymorphs all the minerals in this subclass are aluminosilicates, since in order to produce a net negative charge on the framework some of the Si^{4+} are replaced by Al^{3+}. The cations which balance the negative charge on these formula units are always large ions with coordination number eight or greater (Na^+, K^+, Ca^{2+}, Ba^{2+}). Both because of their ready availability

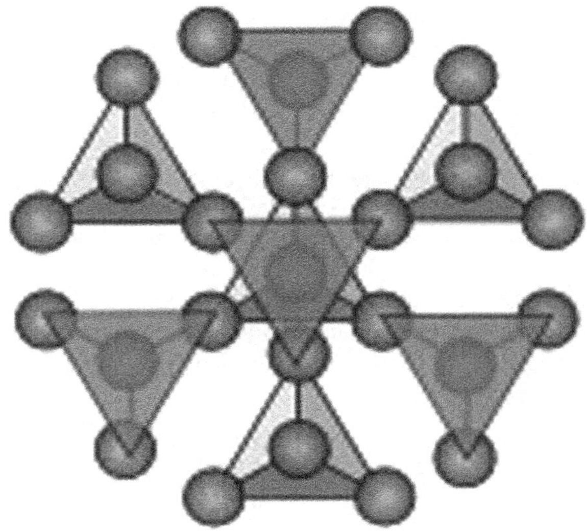

Fig. 4 Structure of tectosilicates

and useful properties, quartz and feldspars are widely used in manufacture of glass and ceramics.

Silica

The silica minerals include quartz, tridymite, cristobalite, coesite, stishovite, and several others. All silica minerals are tectosilicates having structures consisting entirely of SiO_4 tetrahedra that share all of their oxygen atoms with adjacent tetrahedra.

Quartz [known since pre-historic times and recorded by Theophrastus as κρύσταλλος or kristallos. The root words κρύοσ (ice cold) and στέλλειυ (to solidify) suggest the ancient belief that kristallos was permanently solidified ice. The earliest printed use of "querz" was anonymously published in 1505, but attributed to the German physician von Kalbe in 1527 while Agricola in 1530 used the spelling "quarzum" as well as "querze"] (315 BCE) [41].

SiO_2—Trigonal

Natural occurrence: Alps (Switzerland and Austria), Tuscany, Piedmont (Italy), Isere (France), Urals (Russia), Myanmar (Burma), Colorado, California, New Mexico, Montana, North Carolina, Arkansas (USA), Veracruz, Guerrero (Mexico), Yamanashi (Japan), Tamboholehehibe (Madagascar), Minas Gerais, Rio Grande do Sul, Bahia (Brazil), Artigas (Uruguay), Lake Superior, Ontario (Canada).

Physical/chemical properties: colorless, purple, rose, red, black, yellow, brown, green, blue, orange, etc., transparent to nearly opaque, vitreous and waxy to dull luster when massive, conchoidal fracture, brittle tenacity, though when massive, pyezoelectric and pyroelectric. Quartz is the most common of the silica polymorphs. Actually, there are two polymorphs that have the name quartz: α-quartz (low quartz) and β-quartz (high quartz). The introduction of nano-grained quartz powder with a much smaller specific surface area into the traditional CMAS-Na_2O–K_2O system showed that both glazes are characterized by a large part of the glassy phase, above 90% by volume. However, it is observed a higher arrangement of the continuous glassy phase structure in the glaze with the addition of nano-quartz. This glaze also shows significantly higher values for all measured mechanical properties. It seems that if cheaper methods for the production of nano-quartz are developed, it will be a new interesting direction of research aimed at improving the parameters of glazes and glass–ceramic materials of CMAS type. In terms of health risk, it is usually quite harmless unless broken or powdered. Broken crystals and masses may have razor-sharp edges that can easily cut skin and flesh. Handle with care. Do not grind dry since long-term exposure to finely ground powder may lead to silicosis. S.G. = 2.65; H = 7.

Tridymite [from the Greek "Tridymos", triplet, alluding to its common twinning as trillings] (1868) [42].

SiO_2 polymorph—Orthorombic.

Natural Occurrence: Veneto, Sardinia (Italy), Puy-de-Dome (France), Eifel Germany), Antrim (Ireland), Colorado, California, Oregon (USA), Hidalgo (Mexico), Kumamoto, Yugawara, Kanagawa (Japan).

Physical/chemical properties: colorless, white, yellowish white, transparent, vitreous luster, conchoidal fracture, brittle tenacity. Glass–ceramic uniquely suited for use as a disc substrate for utilization in a magnetic memory storage device exhibits a crystal phase assemblage comprised predominately of a mixture of lithium disilicate and tridymite. Both mineralogical phases are uniformly interspersed with a residual glass phase and form an interlocked microstructure with the glass. S.G. = 2.25–2.28; H = 7.

Cristobalite [named after the type locality of Cerro San Cristobal, Mexico] (1887) [43].

SiO_2 polymorph—Tetragonal pseudo-cubic.

Natural occurrence: Eifel Germany), Colorado, California, Oregon (USA), Guanajuato, Hidalgo (Mexico), Nagano, Shizuoka (Japan), Maharashtra (India).

Physical/chemical properties: colorless, white, yellowish white, transparent, vitreous luster, brittle tenacity. Porous fused silica ceramics which had been partially devitrified to cristobalite are stronger at 350 °C in the beta-cristobalite stability range, but weaker at 25 °C after the transformation to alpha-cristobalite. Micro-cracking as-sociated with the transformation to alpha-cristobalite contributes to the strength changes but cannot fully explain the impact of the beta–alpha transformation on strength. S.G. = 2.32–2.36; H = 6–7.

Feldspars** and **feldsphatoid

Orthoclase, albite and anorthite belong to this group of rock-forming tectosilicate minerals in which silicon ions are linked by shared oxygen ions to form a three-dimensional network. Feldspars are widely used in ceramic, glass and glass–ceramic production. Nepheline is a rock forming mineral in the feldspathoid group that occurs in intrusive and volcanic rocks with low silica, and in their associated pegmatites. It is used in glass and ceramic manufacturing and other industries.

Orthoclase [named "orthose" by the French mineralogist Renè Just Haüy (1743–1822) from the Greek orthos—"right" in allusion to the mineral's right angle of good cleavage] (1823) [44].

$K(AlSi_3O_8)$—Feldspars group—Monoclinic

Natural occurrence: Gotthard (Switzerland), Tirol (Austria), Piedmont (Italy), Bavaria (Germany), Cornwall (England), Urals (Russia), Colorado, Pennsylvania, California, (USA), Guanajuato (Mexico), Shiga (Japan), Betroka (Madagascar).

Physical/chemical properties: colorless to white, greenish white, grayish yellow, pale pink, transparent to translucent, vitreous and pearly on cleavages luster, conchoidal to uneven fracture.

Crystallisation of glasses based on the diopside anorthite eutectic system containing increasing amounts (0–30 wt. %) of orthoclase was studied. It has been found that increasing the content of the orthoclase phase in the glasses results in an increased bulk crystallinity when treated at 1100 °C for 3 hs. This effect is accompanied by a change in the texture of the glass ceramic specimens from coarse to fine grained. Microcline phase is formed with diopside and anorthite when the percentage of the nominal orthoclase reaches 20 or 30 mol %. S.G. = 2.55–2.63; H = 6–6.5.

Albite [from Latin "albus", white, alluding to its usual color] (1815) [45].

$Na(AlSi_3O_8)$—Feldspars group—Triclinic.

Natural Occurrence: Gotthard, Ticino (Switzerland), Tirol (Austria), Trentino, Piedmont (Italy), Savoie (France), Urals (Russia), Virginia, Connecticut, Alaska, California, (USA), Ontario (Canada), Minas Gerais (Brazil).

Physical/chemical properties: white to grey, bluish, greenish, reddish, transparent to translucent luster, brittle tenacity. A GC material based on albite has been synthesized by conventional ceramic process. High crystallinity (> 94% Vol.) is obtained by fast sintering, which allows energy saving processing. Albite is the main crystalline phase and tetragonal SiO_2 is a secondary phase. Dielectric characterization shows a non-Debye type dielectric behavior with low dielectric constant, 4.6 at 1 MHz, low dielectric losses, that it is the largest value reported in ceramic insulators. Those dielectric properties are attained by the low glassy phase content in the samples and their unique micro-nanostructure. All these properties make this material a very promising candidate in the market of ceramic electrical insulator, highlighting for high-voltage applications. S.G. = 2.60–2.65; H = 6–6.5.

Anorthite [from the Greek αν and ορθός, "not right angle" or "oblique", in allusion to the oblique triclinic form of the crystals] (1823) [46].

$Ca(Al_2Si_2O_8)$—Feldspars group - Triclinic.

Natural occurrence: Campania, Trentino, Piedmont (Italy), Urals (Russia), Sodermanland (Sweden), Lojo (Finland), Nevada, Alaska, California, (USA), Labrador (Canada), Tochigi, Yamagata (Japan).

Physical/chemical properties: colorless, reddish grey, white, transparent to translucent, vitreous luster, conchoidal to univen fracture, brittle tenacity. Fabrication process of anorthite ceramics suitable for low temperature, sinterable, multilayered substrates is well known. Anorthite ceramics are synthesized from relatively pure kaolin and calcites of varying particle sizes. Mixed powders are uniaxially pressed and fired between 900° and 1200 °C. Firing at 1000 °C yields a dense anorthite ceramic when the finest calcite powder is used. The relative density and water absorption of the anorthite ceramic are 94% and almost zero, respectively. S.G. = 2.74–2.76; H = 6–6.5.

Nepheline [from the Greek "nephele" for "cloud" since it turns cloudy when treated with strong acid] (1800) [47].

$Na_3K(Al_4Si_4O_{16})$—Feldspathoid group—Hexagonal.

Natural occurrence: Campania, (Italy), Saxony (Germany), Kola (Russia), Langesundsfjord (Norway), Sierra de Moncique (Portrugal), Maine, Arkansas, Colorado, New Mexico (USA), Ontario, Quebec (Canada).

Physical/chemical properties: white, grey, yellowish, transparent to nearly opaque, vitreous to greasy luster, subconchoidal fracture, brittle tenacity. Crystallization of the nepheline glass system has been investigated using differential thermal analysis, X-ray diffraction, transmission and scanning electron microscopy. The major crystalline phase was found to be nepheline, $Na_2O{\cdot}Al_2O_3{\cdot}2SiO_2$. The high Vickers micro-hardness and the good thermal expansion coefficients make the nepheline glass ceramics particularly suitable for their application as dental porcelain. S.G. = 2.55–2.66; H = 5.5–6.

4 Carbonates

Carbonates are used in ceramics, glass and glass–ceramic production as source of lithium, sodium, potassium, magnesium, calcium, strontium and barium oxides.

Calcite [Ancient name. Named as a mineral by Latin naturalist Pliny the elder (?–79) from Calx, Latin for Lime] (79) [48].

$CaCO_3$—Calcite group—Trigonal.

Natural occurrence: abundant worldwide. Cumbria, Cornwall (England), Hartz, Saxony (Germany), New York, Misouri, Tennessee, Arizona, Oklahoma (USA), Chihuahua (Mexico), Tsumeb (Namibia), Katanga (Zaire), Primorskiy Krai (Russia).

Physical/chemical properties: white, yellow, red, orange, blue, green, brown, gray etc., transparent to opaque, vitreous to pearly and resinous luster, conchoidal fracture, brittle tenacity. S.G. = 2.71 g/cm^3; H = 3.

Calcite modifies the viscosity and increases the durability of the glass/glaze.

Dolomite [Named in honor of the French mineralogist and geologist, Déodat Dolomieu (1750–1801)] (1791) [49].

$CaMg(CO_3)_2$—Dolomite group—Trigonal.

Natural occurrence: Piedmont (Italy), Tirol (Austria), Saxony (Germany), Binntal (Switzerland), Minas Gerais, Bahia (Brazil), Missouri, Oklahoma, North Carolina, Tri-State, Wayne, New York (USA), Chihuahua (Mexico), Kola (Russia).

Physical/chemical properties: colorless, white, gray, reddish-white, brownish-white, or pink, colorless in transmitted light, transparent to translucent, vitreous to pearly luster, sub-conchoidal fracture, brittle tenacity. S.G. = 2.86 g/cm^3; H = 3.5–4.

One of the most important elements of furnaces, boilers and other heating units is the structure (lining), usually made of silica–alumina, basic or special refractories. The basic refractories are materials that are increasingly in demand and whose manufacturing involves necessarily the use of MgO and CaO, obtained mainly by magnesite ($MgCO_3$) and dolomite [$CaMg(CO_3)_2$] which contribute significantly in industrial ceramics-refractories.

Magnesite [In allusion to the composition, containing mainly magnesium] (1797) [50].

$MgCO_3$—Calcite group—Trigonal.

Natural occurrence: Magnesia (Greece), Piedmont (Italy), Styria, Carinthia (Austria), Snarum (Norway), Bahia (Brazil), Nevada, California, Washington (USA), Ontario (Canada), Liaoning (China).

Physical/chemical properties: colorless, white, greyish-white, yellowish, brown, faintly pink, lilac-rose; colourless in transmitted light, transparent to translucent, vitreous luster, conchoidal fracture, brittle tenacity. S.G. = 3.0; H = 3.5–4.5.

Cordierite ceramics comprises MgO as the main ingredient. Cordierite composition can be prepared with raw magnesite and with waste magnesite generated during the magnesite processing, sintered at various temperatures, between 1250 and 1350 °C. The bulk density, water absorption and porosity properties measured on cordierite composition prepared with waste magnesite and sintered at 1325 and 3 °C/min heating rate, demonstrates its suitability for the cordierite production.

Natron [In allusion to its sodium content (Latin = NATRium)] (1747) [51].

$Na_2CO_3 \cdot 10\ H_2O$—Monoclinic.

Natural occurrence: Debrecen (Hungary), Campania, Sicily (Italy), Tirol (Austria), California, Colorado, British Columbia (USA), Quebec (Canada), Kola (Russia).

Physical/chemical properties: colorless to white, greyish, yellowish, colourless in transmitted light, semi-transparent, vitreous luster, conchoidal fracture, brittle tenacity. S.G. = 1.48 g/cm^3; H = 1–1.5.

Natron deposits, the best known of which being those at Wadi Natrun in Egypt, have been used as the flux in the production of vitreous materials from the early 4th millennium BC onwards. The history of the use of natron as a flux can be traced from its beginnings in the glaze of Badarian steatite beads, through its use in glass production starting in the first millennium BC, until its apparent shortage during the seventh to ninth centuries AD. The possible reasons for the replacement by plant ash as the flux used in glass production during the ninth century AD include the possibility that, because of the massive scale of glass production, the demand for natron exceeded its supply. Further possible reasons can be the effect of climatic changes and the potentially disruptive role of political events in the Wadi Natrun–Delta region.

Trona [Trona is a Swedish term, deriving ultimately from the Arabic "natrum", native salt] (1747) [52].

$Na_3H(CO_3)_2{\cdot}2H_2O$—Monoclinic.

Natural Occurrence: widespread in arid regions. Sahara Desert (Africa), Nile Delta (Egypt), Lake Magadi (Kenya), Lake Chad (Chad), Lake Katwe (Uganda), Lake Nyassa (Tanzania), Otavi (Namibia), California, Wyoming, Utah, Nevada (USA), Merida (Venezuela), Ankara (Turkey), Campania (Italy).

Physical/chemical properties: colorless, gray-white, light yellow, colourless in transmitted light, translucent, vitreous luster, univen to sub-conchoidal fracture, brittle tenacity. S.G. = 2.11; H = 2.5–3.

Lightweight GC material similar to foam glass can be obtained at 700–800 °C directly from alkali-activated silica clay and zeolitized tuff without preliminary glass preparation. This material is characterized by low bulk density of 100–250 kg/m^3 and high pore size homogeneity. Pore formation in both compositions is caused by dehydration of hydrated sodium polysilicates formed during alkali activation.

5 Phosphates

Calcium phosphates, due to their similarity to the inorganic fraction of mineralized tissues, are of great importance in treatment of bone defects.

Fluorapatite [Renamed from the original apatite of Abraham Werner (1749–1817) to emphasize the chemical composition. From the Greek ἀπατάω (apatao), to deceive, as apatite was often confused with other minerals (e.g. beryl). It was added the "Fluor-" prefix in allusion to the dominance of fluorine in the composition] (1860) [53].

$Ca_5(PO_4)_3F$—Apatite group—Hexagonal.

Natural occurrence: Saxony (Germany), Salzburg (Austria), Panasqueira (Portugal), Kunar (Afghanistan), Gilgit (Pakistan), Minas Gerais (Brazil), Potosì (Bolivia), Durango (Mexico), Maine (USA), Quebec, Ontario (Canada), Lake Baikal, Kola (Russia), Transvaal (South Africa), Laverton (South Australia).

Physical/chemical properties: colorless to white when pure, also green, blue, pink, yellow, brown, violet, purple, transparent to translucent, vitreous to sub-resinous luster, conchoidal to uneven fracture, brittle tenacity. S.G. = 3.1–3.25 g/cm^3; H = 5. Melts congruently at 1630 °C.

Apatite-containing GCs are highly biocompatible and can induce bone formation in vivo. Therefore, apatite-based GCs are highly attractive for medical and dental applications. A bioactive material is defined as a material that exhibits a biological response at the interface once in contact with a biological tissue. A bioactive material may induce a biological response through its surface topography or by a controlled release of therapeutic ions. A number of bioactive apatite-containing glass–ceramics have been developed for orthopedic applications, and these can be categorized based on the type of secondary crystal phases present in the GC; apatite–wollastonite (A–W), commercially known as Cerabone®; apatite–fluoromica (A–FM), commercially known as Bioverit®; and apatite–mullite (A–M). Several apatite-containing GCs have also been developed for restorative dentistry applications for the fabrication of dental inlays, crowns, bridges, and veneers. These are namely apatite–leucite (A–L), commercially known as IPS d.SIGN®; and with apatite as the only phase, for example IPS e.max ZirPress® and IPS e.max Ceram®.

Hydroxyapatite [Named from the ἀπατάω (apatao), to deceive as Apatite was often confused with other minerals (e.g. Beryl), plus the "hydro-" prefix for the water-rich (as hydroxyl) nature of the mineral] (1912) [54].

$Ca_5(PO_4)_3(OH)$—Apatite group – Hexagonal.

Natural occurrence: Uri (Switzerland), Baden-Wurttenberg (Germany), Lombardia (Italy), Georgia, Arizona, Missouri (USA).

Physical/chemical properties: white, grey, yellow, green, violet, purple, red or brown, transparent to translucent, vitreous to sub-resinous luster, conchoidal fracture, brittle tenacity. S.G. = 3.14–3.21 g/cm^3; H = 5. $CTE_{(200–700\ °C)} = 14.1 \times 10^{-6}$/K.

Hydroxyapatite (HA) is the hydroxyl endmember of the complex apatite group. The OH^- ion can be replaced by fluoride, chloride or carbonate, producing fluorapatite or chlorapatite. Up to 50% by volume and 70% by weight of human bone is a modified form of hydroxyapatite, known as bone mineral.

Synthetic HA nanoparticles are widely used as bone repairing material due to their excellent biocompatibility and they are called bio ceramic.

References

1. Bernard, J.H., Hirsl, J.: Minerals and Their Localities, 1st edn. Granit, Praha (2004)
2. Klein, C.: Mineral Science, 22nd edn. Wiley, New York (2002)
3. Artioli G., Oberti R.: The Contribution of Mineralogy to Cultural Heritage, 1st edn. E.M.U. and Mineralogy Society of Great Britain & Ireland (2019)
4. Tomalino, M.U.: Una storia della mineralogia. 1st edn. M.A.G.M.A.X, Asti-Italy (2011)
5. Strunz, H., Nickel, E.,H.: Mineralogical Tables, 9th edn. E. Schweizerbart'sche, Stuttgart (1999)
6. Gaines, R.V., Skinner, H.C., Foord, E.E., Mason, B., Rosenzweig, A.: Dana's New Mineralogy, 8th edn. Wiley, New York (1997)
7. Anthony, J.W., Bideaux, R.A., Bladh, K.,W., Nichols, M.C.: Handbook of Mineralogy: Silica, Silicates, 1st edn. Mineral Data Publishing, Tucson (1995)
8. Höland, W., Beall, G.: Glass-Ceramic Technology. The American Ceramic Society, Westerville (2002)
9. Back, M.E.: Fleischer's Glossary of Mineral Species, 12th edn. Mineralogical Record, Tucson (2018)
10. Lee, S., Kim, Y.J., Moon, H.S.: Phase transformation sequence from kaolinite to mullite investigated by an energy-filtering transmission electron microscope. J. Am. Ceram. Soc. **82**, 2841–2848 (1999)
11. Beall, G., Beall, L., Pinckney, L.R.: Nanophase glass-ceramics. J. Am. Ceram. Soc. **82**(1), 5–16 (1999)

12. Sivaranjani, G., Ravikumar, K., Murthy, B.V.S., Basu, B.: Zirconia toughened mica glass ceramics for dental restorations. Dent Mater **34**(3), 36–45 (2018)
13. Wu, J., Li, Z., He, H., Huang, Y., Wu, H.: Preparation of forsterite-based glass ceramics for LTCC. J. Mater. Sci.: Mater. Electron. **24**, 2271–2276 (2013)
14. Tulyaganov, D.U., Tomalino, M.U.: Porcelain from Tableware to Dental Restoration, 1st edn. C.L.U.T., Torino, Italy (2017)
15. Kalantari, E., Naghib, S.M., Iravani, N.J., Esmaeili, J., Naimi-Jamal, R., Mozafari, M.: Biocomposites based on hydroxyapatite matrix reinforced with nanostructured monticellite for biomedical application: synthesis, characterization, and biological studies. Mater. Sci. Eng. **105** (2019)
16. Zaid, M.H.M., Matori, K.A., Hj, S., Aziz, A., Kamari, H.M., Wahab, Z.A., Fen, Y.W., Alibe, I.M.: Synthesis and characterization of low cost willemite based glass–ceramic for opto-electronic applications. J. Mater. Sci. Mater. Electron. **27**, 11158–11167 (2016)
17. Omar, S., Abdel-Hameed, M.: Crystallization of calcium-zinc aluminosilicate glasses. Ceramics – Silikáty **53**(3), 171–179 (2009)
18. Deer, W.A., Howie, R.A., Zussman, J.: Rock Forming Minerals. 1: Ortho and Ring Silicates, 1st edn. Longman, London (1963)
19. Merlini, M., Gemini, M., Giuseppe, C., Gilberto, A.: Phys. Chem. Mineral. **35**(147) (2008)
20. Reddy, A.A., Tulyaganov, D.U., Goel, A., Kapoor, S., Pascual, M.J., Ferreira, J.M.F.: Sintering and devitrification of glass-powder compacts in the akermanite–gehlenite system. J. Mater. Sci. **48**, 4128–4136 (2013)
21. Reddy, A.A., Tulyaganov, D.U., Goel, A., Kapoor, S., Pascual, M.J., Ferreira, J.M.F.: Sintering and devitrification of glass-powder compacts in the akermanite-gehlenite system. J. Mater. Sci. **48**(11), 4128–4136 (2013)
22. Reddy, A.A., Goel, A., Tulyaganov, D.U., Kapoor, S., Kannan, P., Pascual, M.J., Ferreira, J.M.F.: Study of calcium-magnesium-aluminum-silicate (CMAS) glass and glass-ceramic sealant for solid oxide fuel cells. J. Power Sources **231**, 203–212 (2013)
23. Hao, X., Hu, X., Luo, Z., Liu, T., Li, Z., Wu, T., Lu A.: Ceramics Int. Part B **41**(10), 14130–14136 (2015)
24. Van Valkenburg, A., Weir, C.E.: Bull. Geol. Soc. Amer. **68**, 1808 (1957)
25. Sarver, J.F., Hummel, F.A.: Stability relations of magnesium metasilicate polymorphs. J. Am. Ceramic Soc. **45**(4) (1962)
26. Bowen, N.L., Andersen, O.: The binary system MgO-SiO_2. Am. J. Sci. **37**, 487–500 (1914)
27. Beall, G.H.: Refractory glass-ceramics containing enstatite. US Patent 4,687,749 (1987)
28. Baldi, G., Generali, E., Leonelli, C., Manfredini, T., Pellacani, G.C., Siligardi, C.: Effects of nucleating agents on diopside crystallization in new glass-ceramics for tile-glaze application. J. Mater. Sci. **30**, 3251–3255 (1995)
29. Tulyaganov, D.U., Ribeiro, M.J., Labrincha, J.A.: Development of glass-ceramics by sintering and crystallization of fine powders of calcium-magnesium-aluminosilicate glass. Ceramics Int. (2002)
30. Tulyaganov, D.U.: Phase equilibrium in the fluorapatite-anorthite-diopside system. J. Am. Ceram. Soc. **83**(12), 3141–3146 (2000)
31. Soares, V.O., Daguano, J., Lombello, C.B., Bianchin, O.S., Goncalves, L.M.G., Zanotto, E.D.: New sintered wollastonite glass-ceramic for biomedical applications. Ceramics Int. **44**, 20019–20027 (2018)
32. Azarov, M., Maiorova, V., Oborina, M.A., Belyakov, A.V.: Wollastonite raw materials and their applications (a review). Glass Ceram. **52**(9), 237–240 (1995)
33. Knickerbocker, S., Tuzzolo, M.R., Lawhorne, S.: Sinterable β-Spodumene Sinterable β-Spodumene glass-ceramics. J. Am. Ceram. Soc. **72**(10), 1873–1879 (1989)
34. Brindley, G.W., Robinson, K.: The structure of kaolinite. Mineral. Mag. **27**, 242–253 (1946)
35. Newnham, R.E., Brindley, G.W.: The crystal structure of dickite. Acta Crystallogr. A **9**, 759–764 (1956)
36. Uddin, F.: Montmorillonite: An Introduction to Properties and Utilization. Books Google.com (2018)

37. Mukhopadhyay, T.K., Sushovan, G., Maiti, H.S.: Pyrophyllite as raw material for ceramic applications in the perspective of its pyro-chemical properties. Ceram. Int. **36**(3), 909–916 (2010)
38. Baik, D.S., No, K.S., Soung-Soon Chun, J., Yoon, Y.J.: Mechanical properties of mica glass-ceramics. J. Am. Ceramic Soc. **78**(5):1217–1222 (1995)
39. Höland, W., Vogel, W., Naumann, K., Gummel, J.: Interface reactions between machinable bioactive glass-ceramics and bone. J. Biomed. Mater. Res. **19**(3), 303–312 (1985)
40. Twells, R.: The effect of Lepidolite in a high tension electrical porcelain body. J. Am. Ceram. Soc. **11**(8), 644–648 (1928)
41. Partyka, J., Pasiut, K., Jeleń, P., Michałek, J., Kaczmarczyk, K., Kozień, D.: The impact of nano-quartz on the structure of glass-ceramic glazes from the SiO_2–Al_2O_3–CaO–MgO–Na_2O–K_2O system. Ceram. Int. **46**(15), 23888–23894 (2020)
42. Beall, G.H., Kohli, J.T.: Glass-ceramics containing lithium disilicate and tridymite. US Patent 5, 744, 208 (1998)
43. Breneman, R.C., Halloran, J.W.: Effect of cristobalite on the strength of sintered fused silica above and below the cristobalite transformation. J. Am. Ceram. Soc. **98**(5), 1611–1617 (2015)
44. Morsi, M.M., Khater, G.A., Range, K.J.: Glass ceramics in the system diopside–anorthite–orthoclase prepared by using some industrial waste materials. Glass Technol. **42**(6), 160–164 (2001)
45. Fuertes, V., Cabrera, M.J., Seores, J., Muñoz, D., Fernández, J.F., Enríquez, E.: Hierarchical micro-nanostructured albite-based glass-ceramic for high dielectric strength insulators. J. Am. Ceram. Soc. **38**(7), 2759–2766 (2018)
46. Kobayashi, Y., Kato, E.: Low-temperature fabrication of anorthite ceramics. J. Am. Ceram. Soc. **77**(3), 833–834 (1994)
47. Wang, M.-C., Wu, N.-C., Hon, M.-H.: Preparation of nepheline glass-ceramics and their application as dental porcelain. Mater. Chem. Phys. **37**(4), 370–375 (1994)
48. Ke, S., Cheng, X., Wang, Y., Wang, Q., Wang, H.: Dolomite, wollastonite and calcite as different CaO sources in anorthite-based porcelain. Ceram. Int. **39**(5), 4953–4960 (2013)
49. Sadik, C., Moudden, O., El Bouari, A., El Amrani, I.: Review on the elaboration and characterization of ceramics refractories based on magnesite and dolomite. J. Asian Ceramic Soc. **4**(3), 219–233 (2016)
50. Aşkın, A., Tatar, I., Kılınç, S., Tezel, O.: The Utilization of waste magnesite in the production of the cordierite ceramic. Energy Procedia **107**, 137–143 (2017)
51. Shortland, A., Schachner, L., Freestone, I., Tite, M.: Natron as a flux in the early vitreous materials industry: sources, beginnings and reasons for decline. J. Archaeol. Sci. **33**(4), 521–530 (2006)
52. Kazantseva, L.K., Rashchenko, S.V.: Chemical processes during energy-saving preparation of lightweight ceramics. J. Am. Ceram. Soc. **97**(6), 1743–1749 (2014)
53. Jmal, N., Bouaziz, J.: Fluorapatite-glass-ceramics obtained by heat treatment of a gel synthesized by the sol-gel processing method. Mater. Lett. **215**, 280–283 (2018)
54. Chai, C., Ben-Nissan, B., Pyke, S., Evans, L.: Sol-gel derived hydroxylapatite coatings for biomedical applications. Mater. Manuf. Processes **10**(2) (1995)

Traditional Ceramics Manufacturing

Manuel J. Ribeiro and Dilshat Tulyaganov

Abstract Ceramic is associated to clay because earliest ceramic articles were made from naturally occurring materials such as clay minerals. Clays and many of the same raw materials are still serve as the main constituents of traditional ceramics that in broad sense encompass heavy clay products, construction materials, whitewares, refractories, glasses, etc. The aim of this chapter is to provide the basic information about the key technological operations and advances in production of structural clay products, wall and floor tiles, vitreous china sanitaryware, stoneware, majolica pottery and porcelain that would be useful to understand the world of these materials.

Keywords Traditional ceramics · Clay · Shaping · Drying · Firing · Properties · Application

1 Heavy Clay Industry Products

1.1 Composition and Shaping

The heavy clay products in accordance with European DIN-standards include the following categories of ready-made goods: solid bricks and vertical-core bricks, lightweight vertical-core bricks, high-strength bricks and strength clinkers, ceramic clinkers, lightweight horizontal-core bricks and lightweight horizontal-core brick panels, roofing tiles, sewer clinkers, bricks for floor and wall panels, and floor bricks. Brick clays of varying compositions can be used in heavy clay industry (Table 1). They are high in iron content and often contain gross amounts of other impurities.

Natural moisture content of brick clays is around 10–25%, the refractoriness 1150–1250 °C, and the sintering temperature (up to 5% of water absorption) close to

M. J. Ribeiro (✉)
Materials Research and Development Center (UIDM), Polytechnic Institute of Viana do Castelo, Rua Escola Industrial e Comercial de Nun'Álvares, 4900347 Viana do Castelo, Portugal
e-mail: ribeiro@estg.ipvc.pt

D. Tulyaganov
Turin Polytechnic University in Tashkent, 17 Small Ring Street, Tashkent 100095, Uzbekistan

F. Baino et al. (eds.), *Ceramics, Glass and Glass-Ceramics*, PoliTO Springer Series,
https://doi.org/10.1007/978-3-030-85776-9_3

Table 1 Chemical compositions (wt%) and refractoriness (°C) of some brick clays

	SiO_2	TiO_2	Al_2O_3	$Fe_2O_3 + FeO$	CaO	MgO	$K_2O + Na_2O$	Loss of ignition	Refractoriness
Ukraine	51.30	0.63	12.33	2.04	15.82	1.12	2.06	16.40	1140
Russia	52.51	–	15.42	6.77	5.08	2.97	3.09	14.16	1180
	61.50	1.30	19.50	6.00	1.15	2.75	2.75	4.25	1140
Uzbekistan	50.75	0.56	10.92	4.58	12.52	3.07	3.76	13.84	1050
Portugal	51.67	1.37	25.78	7.24	0.13	0.59	1.72	11.5	1180
North	58.53	0.78	19.73	6.45	3.29	2.44	4.44	3.87	1100
Center									
Brazil	59.20	1.50	20.10	6.60	0.20	0.60	1.80	10.00	1200

1100–1150 °C [1]. Because of the high impurity content, fluxes are not necessary and the body can be fired at a relatively low temperature. The quartz sand content in brick clays is up to 50% and in loamy soil—from 50 to 85%. Those clays contain only 10–30% of clay minerals and as a rest fine silica dust called as loam and are very suitable for heavy clay manufacturing since clay minerals introduced here have large fraction of particles less than 1 μm [1]. In general, there are four main steps in heavy clay products manufacturing namely raw materials preparation, ceramic bodies shaping, drying and firing. Figure 1 presents the flowchart for bricks production.

The collective term "preparation" includes all necessary treatments and processes which the raw materials and pastes have to undergo in order to be as desired in the shaping plant which follows, and to be fired to a product with its specific ceramic properties. Raw materials preparation step renders (a) separation of inclusion materials which reduce the plasticity, negatively influence the drying and firing behavior, strength characteristics and appearance of the product; (b) evening out of qualitative and quantitative variations in the composition of the raw materials to allow stable long-term production parameters and to be able to produce products with the same characteristics and the similar appearance; (c) homogenization of the body to make the shaping process easier, improve drying and firing behavior and shorten processing times [2, 3].

The most common shaping methods used in heavy clay products manufacturing are extrusion and semi-dry pressing. Extrusion of plastic body with moisture content

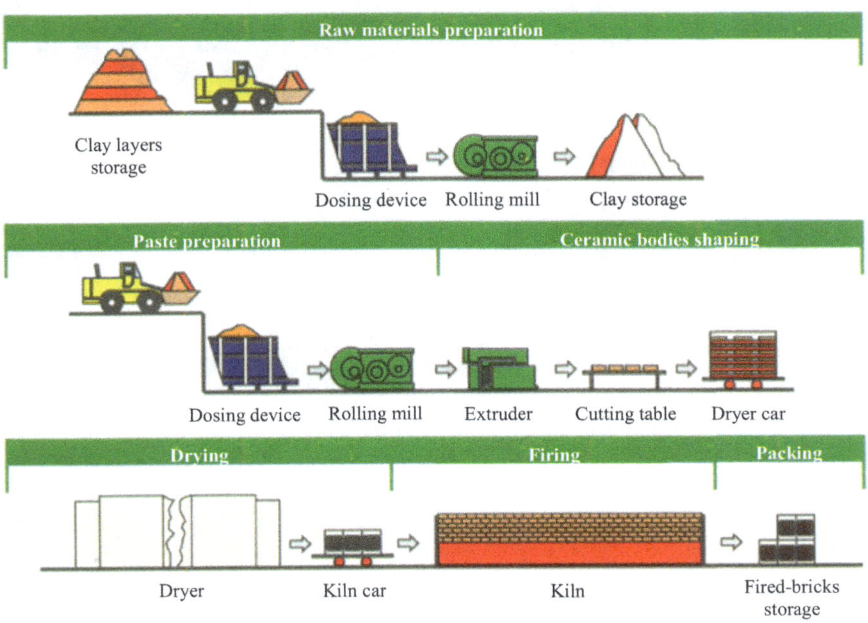

Fig. 1 Flowchart of the bricks production through wet or semi-wet preparation process [2]

of 15–20% is performed in a pugmill (e.g. called also as extruder) under vacuum, so that an air entrapped can be removed before consolidation. This allows improving both plasticity of the clay and strength of the plastic body [4]. After being de-aired in a vacuum chamber, the body chips are compressed and extruded through a die followed by cut into pieces of suitable lengths. For instance, the roof tiles shaping process includes a pre-formed rod extrusion with a geometry that resembles the final tile configuration (Fig. 2a). After this, the extruded plastic paste is cut into a small curved sheet, and automatically pressed in a steel mold by a hydraulic-press. For semi-dry pressing material should possess moisture content of 6–8% and demonstrate plastic properties which enable the feed powder to flow and fill up the die when pressure is applied [3, 4].

Tunnel dryers are commonly used to dry the bricks before sending to firing. Thus, pieces after shaping are stacked on the tunnel kiln cars which pass through the tunnel dryer and then through the tunnel kiln upon firing.

Fig. 2 **a** double rod extrusion of a pre-formed roof tiles, **b** bricks' geometry, **c** dried roof tiles in the tunnel kiln inlet, boxed in sets with 15 pieces, **d** general view of the unfired brick kiln cars area (**a** and **b**, by courtesy of Sotelha Company, Portugal, and **d** by courtesy of SACMI Company)

1.2 Problems Associated with Extrusion and Firing

The differences in the clay-paste velocity through the die (for example, due to a die misalignment), may cause differences in the extrusion pressure and, consequently, in the paste compaction (Fig. 3). In general, a higher outlet velocity corresponds to a higher extrusion pressure and therefore a greater compaction of the extruded brick.

In the regions with lowest paste velocity, the clay particles are subjected to traction and, consequently, the distance between particles will be greater, so that during drying these zones shrink more than those that were extruded at higher pressure. These shrinkage differences increase the traction forces, which are created in the extruded zones with lower pressure, causing brick deformations followed by drying or firing cracks (Fig. 4).

Before the kiln is designed the firing cycle must be determined, taking in account the type of paste. To determine the firing cycle curve, tests are usually made on

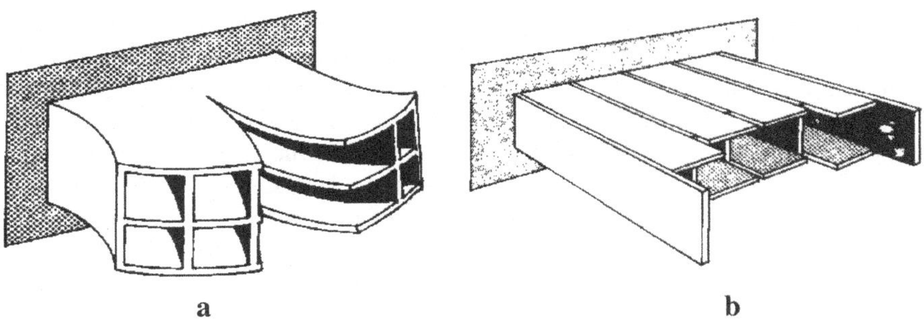

Fig. 3 Bricks shaping through extrusion: **a** central extrusion pressure is higher than lateral one and **b** lateral extrusion pressure is higher than central one

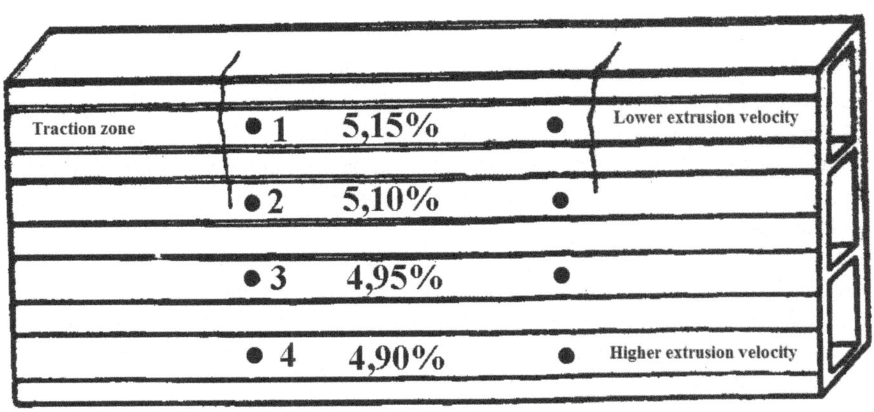

Fig. 4 Schematic view of the shrinkage differences at four different positions of extruded brick emerged due to a die misalignment [5]

the clay and/or clay mixture including differential thermal analysis (DTA), thermogravimetric analysis (TGA), thermal expansion analysis, plus drying, firing and shrinkage tests. With this test data, a firing time/temperature curve can be established. There is another major factor which affects the firing time and that is the load of each kiln-car. For example, with more traditional setting of 14 bricks high in a bung, firing cycle might be 30 h cold to cold. In this case it could be present big temperature differences between the top and bottom of kiln: mainly in the heating zone up to 400 °C. To overcome this problem different kind of stream re-circulation are used. If the paste density in each kiln-car is reduced in size to 8 bricks high in a package the firing cycle time might be reduced to 20 h. However, in the two high concept where the package density is actually two bricks, one set on top of the other, the firing cycle time could be reduced to as six to ten hours cold to cold [6].

A high percentage of structural clay products are burned in the kiln which means that at the end of the firing period, during the early cooling, the kiln atmosphere contain several potential reducing gases, although the necessary fresh air injection, immediately after the firing zone, guarantees an oxidizing atmosphere and the typical red color of this ceramic products. In general, the materials firing can create a whole variety of colors going from red, to garnet, to purple, to brown and black. The complications with firing is connected to obtaining a firing homogeneity throughout the setting density package. The smaller setting density always lowers firing consistency problems, while promoting an increase in specific fuel consumption. The most popular type kilns for face-bricks are large wide tunnel kilns, however, there are three types within this category. These kilns can be top fired index push, side fired index push, or under and over fired with continuous push. However, traditionally the top fired index push kilns have been the most popular in Europe, and in other hand the under fired continuously pushed kilns have been most popular in the United States.

2 Wall and Floor Tiles

The European standard, approved in November 1981, refers "ceramic wall and floor tiles" as building materials, which are generally designed for use as floor and wall covering, both indoors and outdoors, regardless of shape or size. Ceramic tiles also include mosaics, slabs, and pavers for industrial paving, and components for swimming pools, as well as the corresponding accessories (edge, corner and skirting tiles, beads and other pieces) [2].

This European standard provides the generally used definitions, classification, characteristics and marking requirements of ceramic wall and floor tiles. Tiles are prepared from a mixture of ball clays, quartz sand, fluxes, coloring agents and other mineral raw materials, which undergo preliminary treatments including milling, screening, blending, wetting, etc.

The quality of the raw materials determines the individual stages of preparation and shaping. Thus, the proportioning of raw material is performed gravimetrically taking into consideration the moisture of the raw materials. Raw materials in large pieces are pre-crushed in roller mills or auger machines, then milled in periodic or continuous ball mills. A mixture of raw materials (batch) after fine grinding to a particle size < 0.1 mm in the form of the suspension is screened and then stored in stirred storage. The suspension is processed either to a plastic body ('extrusion paste') or to 'dust pressing powder'. The slip casting processing is not common in the manufacture of wall and floor tiles [7].

Ceramic tiles are manufactured in standard ceramic production processes such as dry pressing, extrusion, casting followed by drying and then by firing. Wall and floor tiles may be glazed, unglazed or engobed (engobe is a white or colored clay slip coating applied to a ceramic body to give it decorative color or improved texture). Actually, the glazed wall and floor tiles are normally produced by single-firing (simultaneous firing of ceramic paste and glaze) or, in special tiles, by twice-firing (ceramic paste in the 1st firing—biscuit, and glaze in the 2nd firing). Water absorption E, is expressed as the water absorbed mass in percentage and is the main referred standard tiles classification, ordered as follows: group I (low water absorption) $E \leq 3\%$; group II (medium water absorption) $3\% < E \leq 10\%$ (divided in to different sub-groups, a and b), group III (high water absorption) $E > 10\%$ [2, 3].

2.1 *Shaping Process A—Extruded Floor Tiles*

In production of extruded floor tiles, the suspension is dewatered in filter presses or rotation filters to a water content in the range of 20–25%. A special process to prepare raw materials for plastic shaping is fine grinding in dry mills, mixing and then adding about 20% of water. Organic or inorganic substances such as alginate, dextrin, lignin, methyl cellulose, ethyl cellulose and paraffin are added to the plastic compound to maximize the plasticity.

Extruded tiles are formed by passing the raw materials in the form of a paste through a special tube (die) forming a strip of clay, which is then cut into tiles. Split tiles are formed as double tiles, which, after firing, are split into individual tiles. They may be glazed or unglazed. Split tile recipes and firing temperatures vary according to the locally available raw materials. The following receipt is valid for European conditions, wt%: 25–50 plastic clays, 10–25 chamotte and/or ceramic rejects, 10–25 quartz, 5–20 feldspar, 0–5 talc. From time to time brick clays and secondary raw materials, for example materials recycled from coal fired power stations, are also used. The firing temperatures are around 1200 ± 50 °C. In the USA, a favorite main raw material for colored quarry tiles is shale, an illitic clay rich in Fe_2O_3 which sinters at relatively low temperatures. Fast firing is only partially possible with this raw material due to its tendency to expand the body.

There are also individually extruded tiles such as mosaic tiles that produced individually, after being cut from a length of extruded body. They are subsequently pressed many times and may be glazed or unglazed. The production process for individually extruded tiles can differ from that of split tiles in two ways: shaping and the possibility of single layer firing. Shaping is by means of punches, which may stamp reliefs simultaneously. In contrast of the split tile, individually extruded tiles can also be fired in a single layer e.g. in roller hearth kilns. Individually extruded tiles are used for wall, floor and facade covering. Whereas European standard (EN 87) specifics a frost resistance test, in addition to water absorption, for tiles which may be subjected to frost, in other hand the American Society for Testing and Materials only deal with water absorption (ASTM C 373).

2.2 *Shaping Process B—Dry Pressed Wall and Floor Tiles*

The flowchart of tile production of dry pressed tiles through spray-dried granulation process is shown in the Fig. 5. Wall and floor tiles are made from powdered, fine ceramic paste rounded particle, which are pressed in a steel mold under high pressure.

The prepared suspension slip is processed by means of spray drying to the final granular material with moisture content between 4 and 8%. This granular material is round in shape (Fig. 6) and compared with old, brittle pan mill granules, offers significant advantages: the particle spectrum dimensions is narrow with a lower fine dust content, which presence can result in defective pressing. The rounded granules flow very easily and their moisture content is easily controllable. This good flow

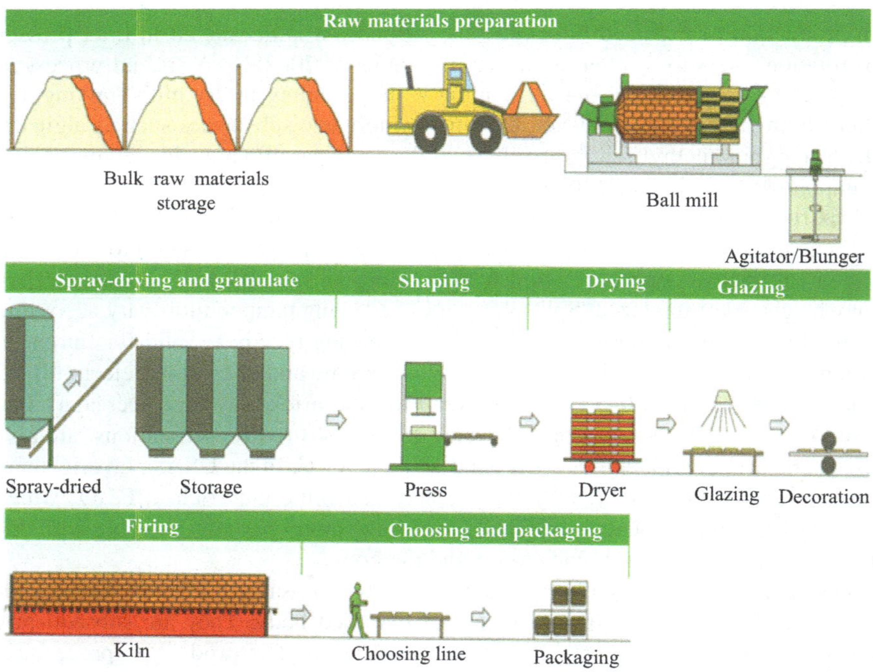

Fig. 5 Flowchart of tile production through spray-dried granulation process [2]

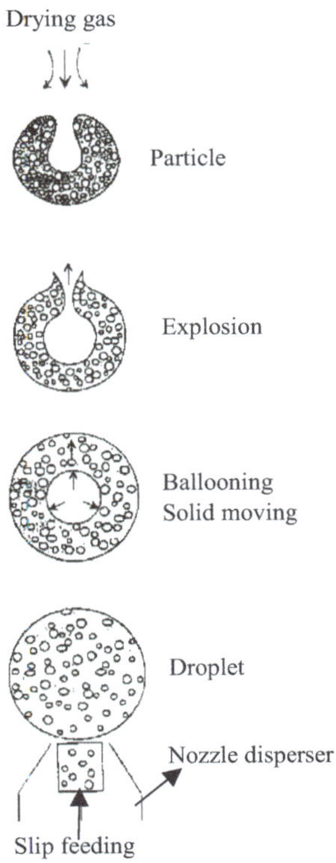

a

b

Fig. 6 Schematic representation of the evolution of spray-dried ceramic particle (**a**), and a photo of granulate batch obtained in a nozzle spray dryer [10] (**b**), amplification ~ 30x

ability permits beneficial pressing mold filling and creates better conditions for de-airing in the initial compression step [8, 9, 10]. A typical particle distribution of granulate produced in a nozzle spray dryer with laminar flow is given in Table 2.

The drying process takes place in a cylindrical spray drying chamber: ceramic slurry is pulverized as a cloud of fine droplets with large exposed surface to meet a

Table 2 Typical sieve analysis for a tile granulate prepared in nozzle type spray dryer with laminar flow

Sieve aperture (μm)	Material retained (wt%)
1000	1.5
600	17.5
355	42.5
250	13.5
150	10.5
< 150	14.5

stream of hot gas and is converted into solid granules, which fall from the gas stream on the conical base of the dryer. The exhausted gas is extracted from the chamber via cyclones and wet scrubbers to the atmosphere. The tiles bodies are made basically from spherical fine particles which are pressed in a steel mold under high pressure. The benefit arising from spray drying represents a significant progress from labor-intensive batch processing to low labor continuous system capable of sophisticated control [8, 9].

2.3 Pressing

There is a wide range of hydraulic presses on the market with pressing forces ranging from 1000 kN to more than 10,000 kN [11, 12]. The choice of press is determined by the size of the tile being produced and by cost effectiveness. Figure 7a shows a SACMI hydraulic press with automatic tile thickness measurement and regulation systems, and Fig. 7b demonstrates a general view of an automatically SACMI exit system from the press mold tiles.

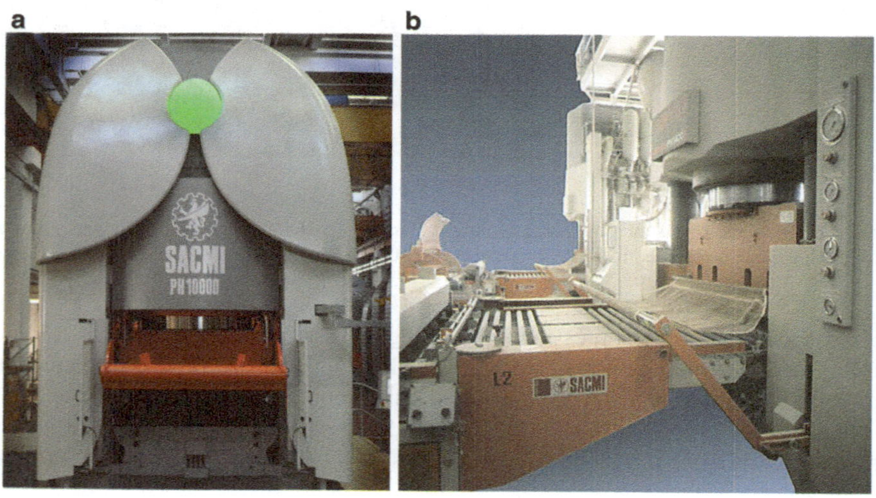

Fig. 7 SACMI hydraulic press PH 10,000 [12], with a pressing force of 10,000 tons which can compact even the very largest sizes at very high specific pressures (**a**) and automatically SACMI exit system from press mold tiles (**b**) (by courtesy of SACMI Company)

2.4 Drying

For single firing, roller dryers are generally used, for the less common twice firing traditional tunnel driers, with stack loading, or vertical dryers for individual tiles. Drying temperatures vary depending on the type of technology, e.g. in tunnel dryers 300–350 ºC, in vertical dryers between 200 and 220 °C while drying time is between one and four hours depending on the water content of the green ware. To avoid fissures and glazing errors in the firing process the required residual moisture content is less than 1% [11].

2.5 Firing and Glazing

Tiles might be prepared as single fired products (glazed or unglazed) at temperature range of 1050–1200 °C or as glazed double or even triple fired products. Double firing process considers that the tiles pass the biscuit firing first that is performed at temperatures between 1050 and 1150 °C in classical tunnel kilns (in 20–50 h) or in modern roller hearth kilns (in 1–2 h). Thereafter the tiles are sorted automatically and transported to the glazing process. Glaze is applied to the tiles by spraying or watering. Both so called “raw glazes” and “frit glazes” are used in tiles production that applied through the various types of glazing procedures. There are the following types of glazing namely press glazing, dry glazing and wet glazing [13, 14].

Press glazing means the pressing of a dry glaze onto the green body. In dry glazing as a rule, the ceramic body has already been pre-glazed using a size solution or sized glaze before the dry glaze is applied. Dry glazing can produce different effects to those of wet glazing. Wet glazing unit include glaze centrifuges, waterfall, bell, dropper unit, and spray guns, which may be either fixed or mobile. This is a system for airless applications, consisting of a pressurizing unit designed to spray any type of ceramic suspension: engobe, glaze or transparent glaze. The technical-construction features of the electro-pump and the pressure control unit make it a reliable user-friendly device, ensuring consistency of applied glaze weight.

For frit glazes the following raw materials are used: borates (boric acid H_3BO_3, anhydrous borax $Na_2O \cdot 2B_2O_3$ and colemanite—$2CaO \cdot 3B_2O_3 \cdot 5H_2O$), sodium carbonate, potash, kaolin, zircon, silica, magnesia, zinc oxide. For single layer fast firing the glaze with following composition is applied, wt%: boric acid—19.5; colemanite—30.4; sodium carbonate—5.0; potash—5.5; kaolin—11.7; zircon—11.3; silica—14.7; magnesia—0.4; zinc oxide—1.5. A glaze may contain one or more frits plus unfired components such as ball clay, bentonite, etc. Fine grinding of frit with admixtures carry out in ball mills during 24–30 h. The ready-made glaze has moisture around 40–45%, density of 1.62–1.68 g/cm^3, while its expense onto the tile with size 150 × 150 mm is 20–24 g; for the curtain flow process, the density of glaze should be 1.55–1.60 g/cm^3 and its expense 18–22 g per tile [13, 14].

2.6 Decoration

Decoration techniques include hand painting, transfers, sprays, screen printing and more recently the inkjet printing. Screen printing (silk screen) is the mainly used for tiles decoration process. Screen printing permits underglaze and onglaze decoration in any sequence, with onglaze decoration being on raw or fired glaze. Depending on the application, either water-insoluble, oil-based screen printing media or polyglycol-based water soluble media are made into a paste with the inorganic pigment (color agents). They are prepared in roller mills or grinding mills. The consistency of rheological properties of the pigment-paste which are of great importance determined by the type of medium, solids content and temperature [15].

Materials used for printing screen include steel ware, nylon and polyester, with mesh size varying between 15 and 140 mesh per centimeter. Flat screen printers can print single or several tiles at the same time, depending on their size, the maximum output with average sized tiles being about 100 tiles/min. In the case of continuous rotary screen printing output is between 130 and 210 tiles/min, but the number of colors is normally limited because of the high operation speeds and slightly lower degree of precision (Fig. 8) [15, 16].

Ceramic inkjet printing technology is basically a "printer" type process in which the terminals of the "printer" have the so-called "heads or dispensers" each of which have millions of pixels distributing the specialty inorganic ink over the body of the tile, before going into the kilns to be fired (Fig. 8). In the early days of this decoration technology process, most factories had equipment with 4 or 6 print-heads and hundreds of pixels [15].

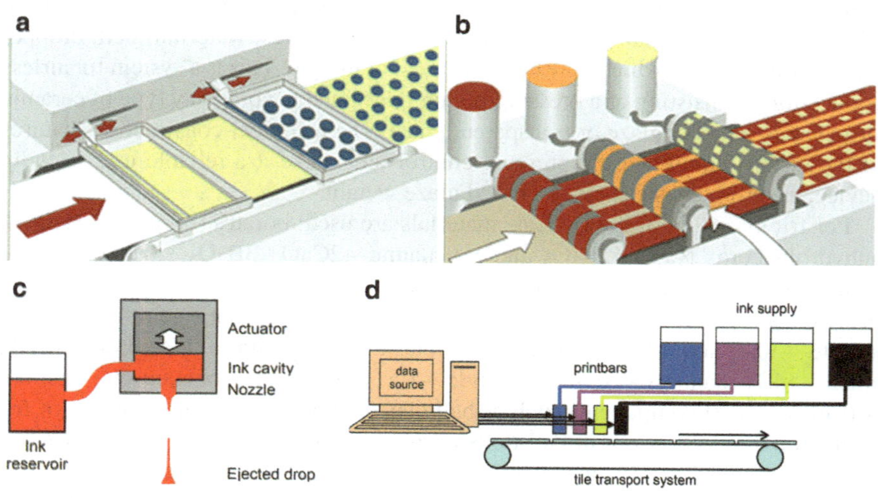

Fig. 8 Schematic presentation of **a** flat screen, **b** rotary screen printing, **c** operation principle of a drop-on-demand in a print-head, **d** a diagram of an in-line, single-pass four-color ink-jet tile printer [15, 16]

2.7 *Single Layer Fast Firing*

The fundamental features of single layer fast firing are the favorable specific heat consumption and the high degree of flexibility in the production line containing this option. Certain basic demands have to be made of the ceramic tiles used for fast firing, for example not too high an organic matter content, and weak thermal reactions. However, even at the shortest possible throughput times, e.g. 35 min for twice firing and 60 min for single firing, it is possible to adapt the firing cycle to specific thermal reactions. The choice of kiln depends on the production process and the demands made of the product. A roller kiln is a kiln that has conveyor to move the product through the kiln instead of kiln cars. The continuous roller kiln operation is similar to a tunnel kiln, and the airflow pattern inside the roller kiln is the same as a tunnel kiln, where a lot of heat is put into the center of the kiln and the exhaust gases are pulled toward the kiln-entrance, in contraflow to the product movement. The cooling zone of a roller kiln works in the same way as in tunnel kiln in that the cooling air is pulled from the exit and toward the center of the kiln, where it is normally pulled out with an air-cool fan. The conveyor itself is made of ceramic rolls or stainless steel rolls depending on the temperature working range. The most popular type of ceramic rolls are mullite, alumina, and recrystallized silicon carbide. Then burners are usually arranged to fire both underneath and above the roll conveyor. The major advantage of a roller kiln over a tunnel kiln is that since there are no kiln cars which must be heated and cooled, and essentially only the ceramic products (the tiles) are moving through the kiln, and due this an extremely fast firing cycles are possible. There are several types of roll drive mechanisms with a chain drive being the least expensive to build and the gear drive being the most precise. Figure 9a shows a close up view of a gear drive system for a roller kiln [11, 12].

This design has the advantage that the ceramic rolls are free to expend and contract and easy to replace without interrupting the overall operation of the conveyor system. The gear drives where the ceramic roll is fixed to a metal end which in turn is

a b

Fig. 9 Image of a gear drive system for a roller kiln movement (**a**), and photo of the exit in SACMI roller kiln single layer, for the ceramic tiles firing (**b**) (by courtesy of SACMI Company) connected to gears is the most popular type of drive today because it is more precise and does not slip [11].

The most successful use of roller kilns has been in the ceramic tile industry where most size tiles can be fired directly on the rolls without the requirement for any kiln furniture. This is extremely important since it eliminates the cost of the kiln furniture and more than halves the weight that has to be heated in order to fire the product. Figure 9b shows a typical roller kiln for firing tiles. Today the majority of wall tile in the world is fired in roller kilns [11].

2.8 Sorting and Packaging

Sorting units may consist of the following individual stages:

(1) Removal of reject tiles through the control of breaking load, cracks and surface evenness; (2) Storage unit for use during breakdowns; (3) Sorting with magnetic marking; (4) Automatic control of plane, parallelism and caliber; (5) Packaging: tile stackers, packaging unit and automatic marking of the packed boxes.

2.9 Problems Associated with the Drying and Firing

Thermal dewatering in spray dryer is expensive in terms of energy costs, and it is essential to minimize these costs. The other important factor governing the cost of spray drying is the solid content of the slip, and the importance of preparing a high solids content. Clays which contain large amounts of minerals of the montmorillonite group present special drying problems which are complicated still further by the water driven out from the interfacial layer, and by the inner crystalline shrinkage associated with this.

The use of very fast firing cycles in the tiles processing increases the probability to the firing problems occurrence, most of them typical of this ceramic products. The most frequent are as follows: (1) geometric defects (warps, poor flatness and lack of squares) due to the small temperature variations along the tile-body caused by the lower temperature uniformity inside the kiln; (2) "black heart" or "explosions" caused by poor degassing of the ceramic body (release of interparticle-air, organic matter burning, and carbonates decomposition) due to the reduced time available in pre-firing region of the kiln; (3) heating or cooling cracks caused by the fast heating or fast cooling of the tiles, respectively.

3 Vitreous China Sanitaryware

According to a modern unified coding system (e.g. it was introduced on January 1, 1988 to harmonize the trading classification systems used worldwide) the product group ceramic sanitary ware consists of (a) ceramic sanitary of porcelain or china (HS 691010) and ceramic sanitary (HS 691090). Examples of ceramic sanitary ware products are ceramic sinks, wash basins, wash basin pedestals, baths, bidets, lavatory pans, flushing cisterns, and similar sanitary fixtures as soap dishes, sponge holders, tooth-brush holders, towel hooks, and toilet paper holders [17, 18].

Historically demand on ceramic sanitary ware increased only towards the end of the 19-th century when Europe and USA went through a period of industrialization and urbanization while industrial production of sanitary ware was based on earthenware and fireclay. The articles were shaped through pressing plastic clay body in plaster molds, then were stuck together in the leather hard condition, dried and biscuit fired, followed by glazing and glaze firing. Crazing of the glaze was common, since the porosity of the ware led to moisture expansion of the body so that crazing of the glaze was common [19]. New technologies were needed to solve several embarrassments substituting expensive plastic pressing for cheaper manufacturing, optimizing body and glaze compositions so that after firing the cooling body compressed the glaze sufficiently to avoid tensile stresses or to offset the subsequent moisture expansion during use. In the early twentieth century the productivity of manufacture was improved by introduction of slip casting in gypsum molds while in the period 30th–50th intermittent coal-fired kilns where replaced by continues muffled tunnel kilns. In continuous kilns, the waste heat from the firing zone was commonly used to preheat the incoming ware and sometimes the heat from the cooling ware was applied to aid combustion or for drying. Further reduction in fuel consumption and the improvement in the uniformity of firing temperature was achieved through adopting open-flame tunnel kilns and efficient modern intermittent kilns.

The development of vitreous china with composition (in wt%) 25–30 china clay, 20–28 ball clay, 25–30 quartz and 14–20 feldspar or nepheline syenite became revolutionary step and allowed to turn to once-firing by spraying raw glaze onto the dried ceramic body (i.e. glost firing was eliminated). Apparently, the new product was impermeable to water and stronger than earthenware. However, adaptation of vitreous china's recipe was accompanied with several changes in the manufacturing process: (a) the mold size had to be enlarged owning to the fact that vitreous china's firing shrinkage was twice higher than of earthenware, (b) eliminating the glaze firing required intensive inspection of ware before spraying, (c) to minimize the extra shrinkage and sag, an unusual techniques of placing the ware on the kiln cars had to be designed [18].

3.1 Modern Manufacturing

In accordance with the British Standard definition: "Sanitary appliances such as wash-basins, water closet, urinals, bidets and others, made from a mixture of clays and other minerals and rendered impervious to water by firing to high temperature. After the ceramic firing treatment the water absorption of pieces should be exceed 0.5%. Vitreous china sanitaryware is coated on all surface exposed to view in normal use with an impervious glaze giving a smooth, shiny, white or colored finish which is durable and easy to clean".

Body formulation. Traditionally sanitary ware manufactures have used a number of china clays and ball clays in body formulation. This was necessary to obtain the required slip casting properties, also to provide some protection against variations in an individual clay that is common practice in many ceramic factories. The choice of the clays has a major influence on casting rate, also on cast properties such as green strength: rise in the proportion of the finer particle size ball clay increases the cast green strength and decreases the casting rate. The choice of ball clay can also affect the quartz content of the body formulation and when, for instance, siliceous ball clay is used, the quartz content can be reduced.

Flint (e.g. microcrystalline quartz) is widely used as an alternative source of silica filler in the body and in this case, the flint pebbles are selected, calcined at 1000–1200 °C, crushed, and ground in ball mill. To give acceptably low iron contents the quartz and feldspars usually require enrichment. Apart of feldspar nepheline syenite might be used in vitreous china body as fluxing agent. Moreover, water quality is essential: demineralized water was adopted to remove high concentrations of sulfates, because their presence have a flocculating effect on the slip-casting process, and the removal of them aids control of the slip rheology.

Slip casting. The crucial state in the manufacturing process is preparing the clay-water slip through the raw materials wet mixing without the intermediate step preparation of filter press cake. The properties of the slip to be considered from the point of view of casting performance are the fluidity and thixotropy. A higher proportion of sodium silicate in the sodium silicate—sodium carbonate deflocculant mixture (typically the deflocculants additions are close to 0.1–0.65%, depending on the rheological clays used properties) allows to minimize the amount of alkali present in the slip casting. The generally accepted solids content of the resulting suspension ranges between 72 and 75% by weight that correspond to slip densities of 1.80–1.85 g/cm^3. To have an opportunity for small corrections to be made if the original deflocculant additions were insufficient the mixed slip has to be stored in a tank for at least (with slow agitation, 14–15 rpm) for at least one day before casting procedure. It helps to reduce pinholes in the sanitary body allowing the fine bubbles introduced during the filling to escape [17, 18].

Slip is normally distributed to the casting-wareroom by pumps through ring mains, which pass through the casting shops and return to the casting tanks. The suspension is poured into the plaster forms or molding in special polymer forms under high pressure. Casting in plaster molds is more traditional, but is already outdated technically and environmentally. One plaster form can be used for production of no more than two hundred pieces (Fig. 10). In the case of the polymer forms more, than twenty-five thousand products might be casted and this technology involves the use of modern technologies and advances in equipment for the production of high quality ceramic sanitaryware (Fig. 11).

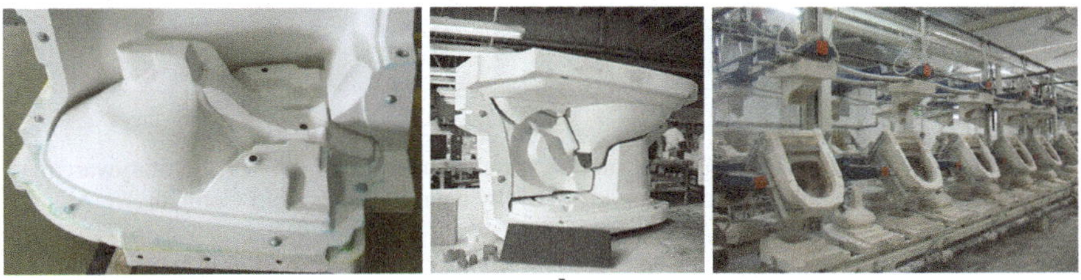

a b c

Fig. 10 WC molds made in gypsum (Paris plaster); **a** mold with 3 parts, **b** mold with 4 parts, **c** production line (by courtesy of Ceracast™ Company) [17]

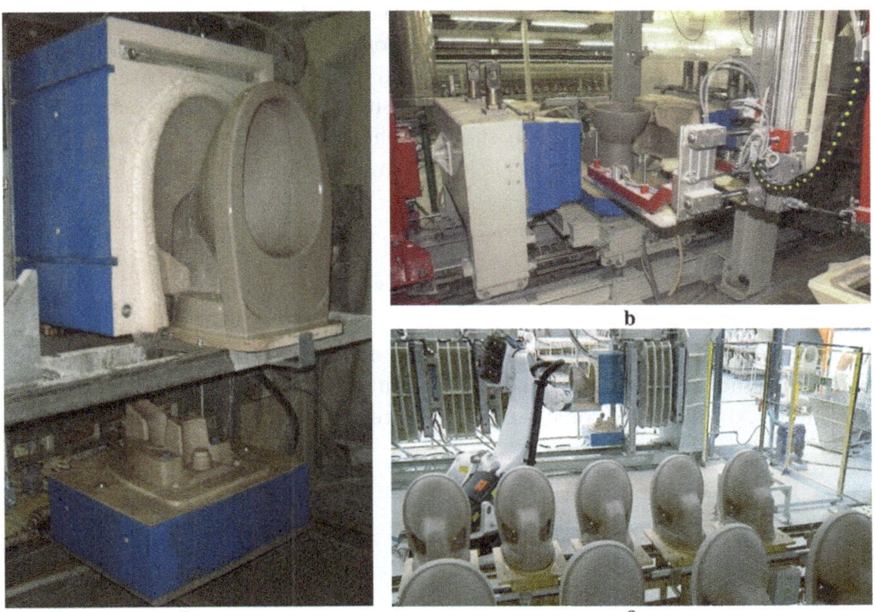

a b c

Fig. 11 WCs production; **a** resin mold, **b** hydraulic manipulator demolding, **c** general view of an automatic WCs line production (by courtesy of SACMI Company)

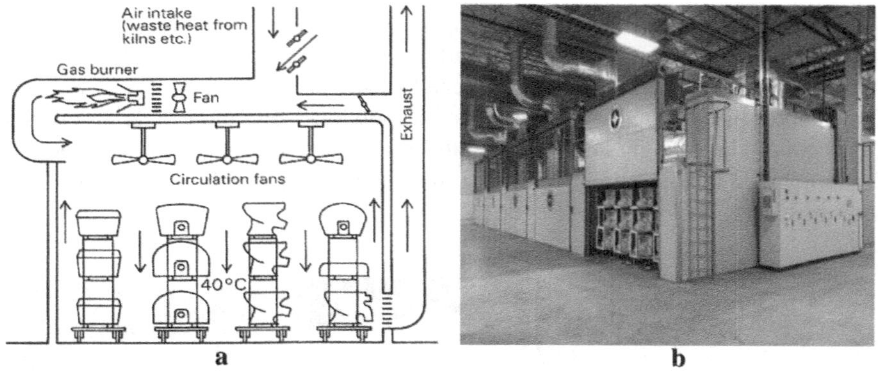

Fig. 12 Schematic view of the enclosed sanitaryware chamber dryer (**a**) [19], sanitaryware dryer with internal environmental control technology (**b**) (by courtesy of Progetti Company)

3.2 Sanitaryware Drying

The open shops (e.g. air is simply heated to 40 °C and about 50% relative humidity and moved around the ware at 0.5–1.5 m/s using ceiling fans) are used to dry of washbasins, pedestals, simple water closets and flood-cast cisterns over one night to reduce moisture content to less than 2%. For drying complex water closets and double cast cisterns a chamber dryer can be used or those can be dried over two nights in the open shop. Figure 12a demonstrates a simplified schematic cross-section of a chamber dryer, while Fig. 12 b shows a photo of sanitary dryer with internal environmental control technology [17].

3.3 Sanitaryware Inspection

Inspection is essential to ensure that the well dried pieces are free from pinholes, cracks and other defects before sending the sanitary ware to glazing and then to firing operations. At this stage, some faults can be rectified while ware with irreparable faults can be easily recycled.

3.4 Glazing

Sufficient raw feldspathic glaze, suspended in water, is sprayed on the dried sanitaryware (Fig. 13). Commonly glaze materials such as tin oxide and/or zirconium silicate, feldspar, quartz, calcium carbonate and silicate, dolomite, china clay, zinc oxide in their pre-ground state are dispersed in water, and after applied on the dried

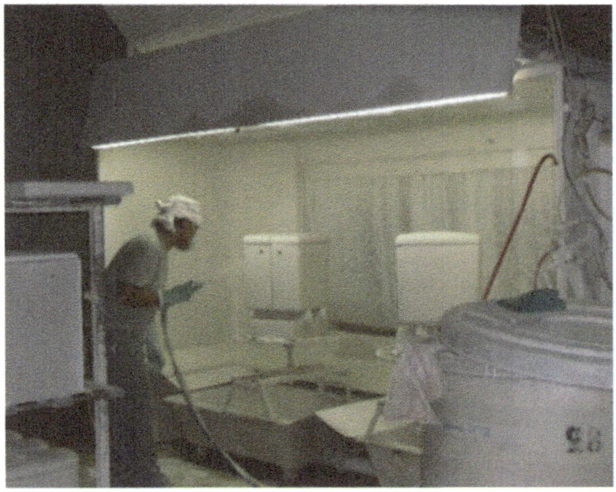

Fig. 13 Manual spray glazing of sanitaryware (by courtesy of Sanindusa Company)

body using compressed air (Fig. 13) to get a thickness of about 300–400 μm after firing.

3.5 *Once Firing*

To avoid damage by knocks and dirt, leading to rejects before and after firing it is important that glazed ware should be handled carefully to the tunnel kiln for firing. The schedule of once firing suggests maximum safe rates of heating and cooling: it can make twelve hours from cold to cold, larger of more complex shapes require up to twice as long. Gas tunnel kilns of small with high temperature uniformity and large cross section, which require longer cycle time (Fig. 14) could be used. A typical modern open flame tunnel kiln can have 90 m long, 1.4 m wide and 0.7 m high. On

Fig. 14 Gas tunnel kiln for sanitaryware **a** kiln inlet, **b** kiln exit (by courtesy of Sanindusa Company)

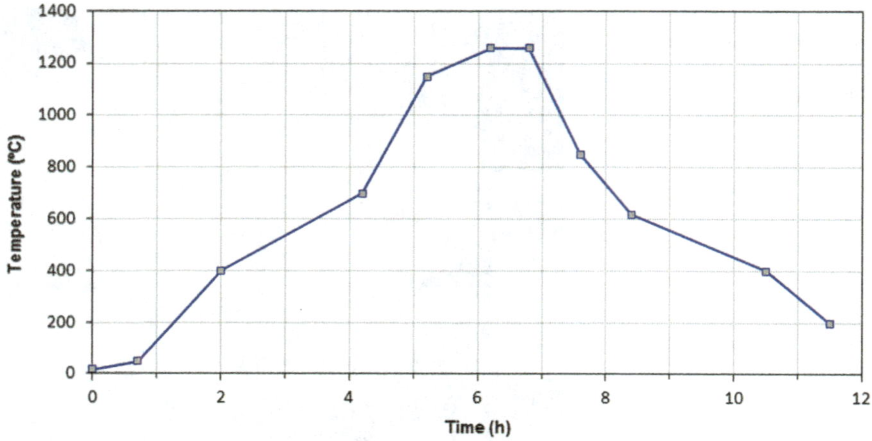

Fig. 15 Typical sanitaryware firing cycle

a 16-h cycle, it once fires 1200 pieces per day at fuel consumption 10,500 kJ/kg of ware. Figure 15 presents a firing cycle commonly used in sanitaryware production.

3.6 Final Inspection

Sorting in general takes place at each stage of the sanitaryware production, but after firing the inspection is carried out especially carefully for cracks, warping, glaze appearance, pinholes, etc. Usually, an inspection marking is done with two colors—one that can be eliminated and it does not affect essentially on the performance of the sanitaryware product, and the other one that cannot be eliminated and has a significant impact. Those marked products that can be corrected, and those that cannot be corrected, are sent to the appropriate departments. For the correct products, the last stage is the finished sanitaryware packaging, where the pieces are packed in a special corrugated cardboard, are supplied with an information sticker.

4 Stoneware

From the Combined Nomenclature of the European Communities, a European industry standard states: "Stoneware, which though dense, impermeable and hard enough to resist scratching by a steel point, differs from porcelain because it is more opaque, and normally only partially vitrified. It may be vitreous or semi-vitreous. It is usually colored grey or brownish because of impurities in the clay used for its manufacture, and is normally glazed". Stoneware is common in craft and studio pottery and covers vast range of modern commercial glazed tableware and kitchenware [3].

Stoneware can be once-fired (more frequently) or twice-fired: maximum firing temperatures can vary significantly, from 1100 to 1300 °C depending on the flux content.

4.1 Body and Glaze Composition

The basic raw material components are quartz, feldspar, china clay and ball clay. Both wet preparation and dry body mixing methods are used. The decision between wet preparation and dry mixing should be made based on how the strength is achieved. The green strength of the body gives the two necessary conditions for the once-firing process: (1) handling strength from the shaping to the filling for firing and (2) non-swelling properties of the body wetted during glazing. Stoneware glazes cover an enormously wide sector from bright to matt glazes, from opaque to transparent, from unstained to colored and crystalline glazes [20].

4.2 Shaping and Drying

Stoneware bodies are shaped using plastic forming (e.g. jiggering and jollying), slip casting, wet and dry pressing techniques. In the jollying process, both surfaces of an item are formed by squeezing the body between a metallic profile tool and a plaster of Paris mold [7, 21, 22]. The profile tool is carried down manually or automatically onto the revolving mold, thus shaping the inside, whilst the outside surface of a hollow article is formed on the mold. The jiggering process is used for the production of flat wares: the mold is rotated in a jigger head and a profile tool is brought down onto the body, to shape the back of the ware (Fig. 16).

Figure 17 presents an automatic line designed for the manufacturing of small-sized ceramic hollowware (e.g. cups) and large-sized pieces (e.g. plates) through the jollying and the jiggering shaping processes, respectively. The automated lines normally integrates de-airing extruder, automatic roller heads, mold drying system, the shaped piece demolding and final finishing of the ceramic body.

Another traditional body shaping used in production of stoneware is the slip casting. Similar to processing of earthenware and porcelain, slip is poured into a porous gypsum mold commonly made from plaster of Paris. The liquid is absorbed into the mold leaving a layer of the required thickness when the mold is inverted and the excess slip is poured away. Then the mold is placed upright again, and the casted ceramic body is allowed to partially dry in the mold [7, 21, 22]. The parts are then removed from the mold due to shrink of the cast (Fig. 18).

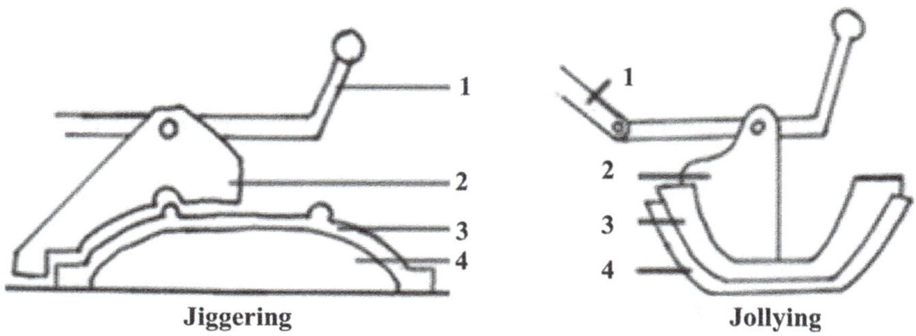

Fig. 16 Schematic view of jollying and jiggering shaping processes: (1) handle, (2) template, (3) clay-paste, (4) plaster of Paris mold

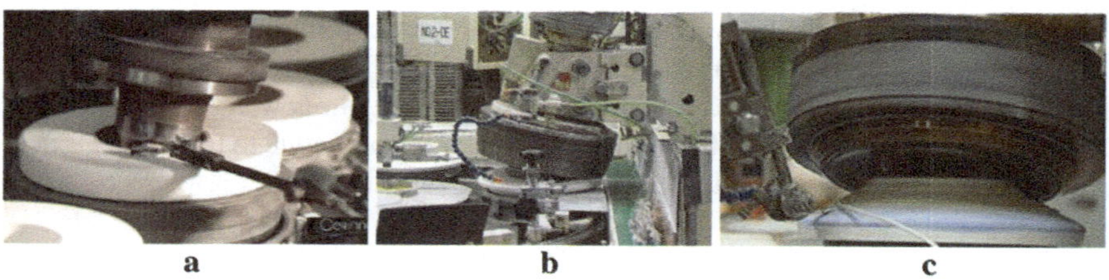

Fig. 17 The automatic line for shaping through jollying (**a**) and jiggering (**b**, **c**) (by courtesy of Cerinnov Company)

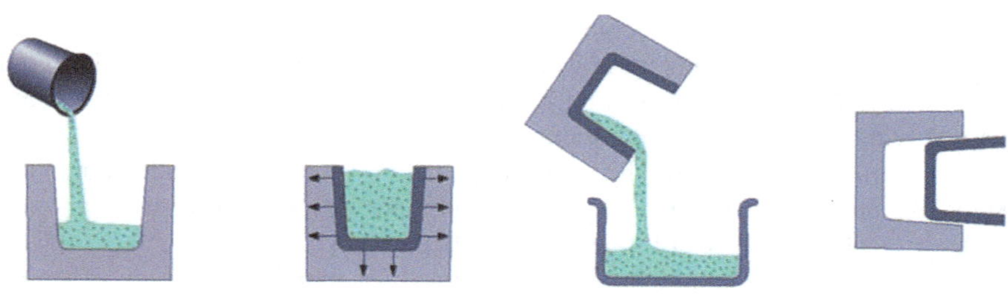

Fig. 18 Schematic representation to the drain slip casting shaping process, using a porous gypsum mold: **a** fill mold with slip, **b** mold extracts liquid, forms compact along mold walls, **c** excess slip drained, **d** casting removed after partial drying [21]

New approaches that allow to increase casting rate include pressure casting, vacuum-assisted casting and centrifugal casting techniques. Thus, Fig. 19a illustrates the five stages making up a whole working cycle of the automatic pressure casting machine with mold made of porous polymer material. The automatic high-pressure casting system is designed in a simple and compact manner and can involve robot demolding and finishing operations as well. Alternatively, demolding and finishing operations might be performed manually (Fig. 19b).

Fig. 19 Schematic view of pressure casting. **a** 1-closed porous resin-mold, 2-slip entry, 3-slip entry and water outlet, 4-vacuum and compressed air injection, 5-demolding step, with help of compressed air injection; **b** Manual demolding (by courtesy of Loucania Company)

After the shaping process, the ware is sent to drying so that to avoid cracking or shattering during firing.

4.3 *Decoration*

In comparison to decorating techniques applied to twice-fired ware the application of colors to green ware is somewhat limited. Hand painting by brush, manual or machine banding, steel screen pad printing transfers onto the unfired glazed ware are among the techniques applied [23]. The latter technology is used to transfer a two-dimension (2-D) image onto a three-dimension (3-D) object and opens up new possibilities, which were not available with traditional printing processes. In particular, Fig. 20 shows the details of the pad printing process where an image from a printing pre-fired plate (Fig. 20a) is transferred onto the object by soft silicone pad (Fig. 20b).

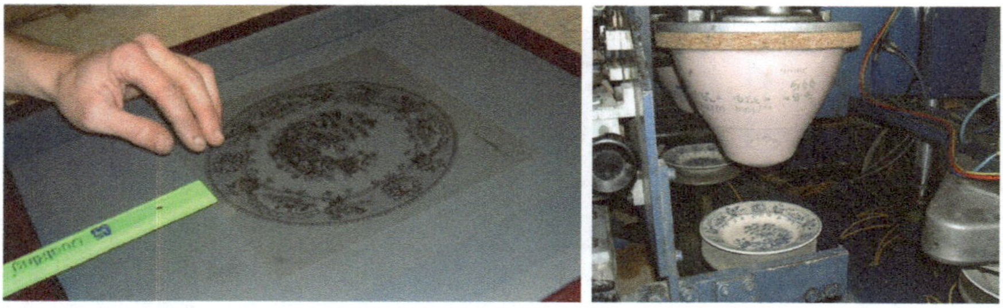

a b

Fig. 20 Details of the pad printing process: **a** Steel screen ink impression step, **b** the silicone pad in the position after image printing on the ware (by courtesy of Faianças Primagera Company)

4.4 Glazing

Raw glazes are mostly used in stoneware production whilst fritted components are of minor importance. Basically, glaze for once-fired stoneware have a high percentage of clay and can be applied to a green body by spraying or dipping. In spray glazing (Fig. 21a) that is most suitable for the once-fired ceramics the solid content is kept at a higher level since lower water content speeds glaze drying. Dipping of clayware is a very convenient for glaze application, but it is more seldom especially if pieces are large. The glaze defect known as crawling where the molten glaze withdraws into "islands" leaving bare clay patches is common for once-fired ware (Fig. 21b). This is explained by difficulties to achieve and maintain a good mechanical glaze-body bond while the chemistry of glaze may be such that the surface tension of the melt encourages crawling [24].

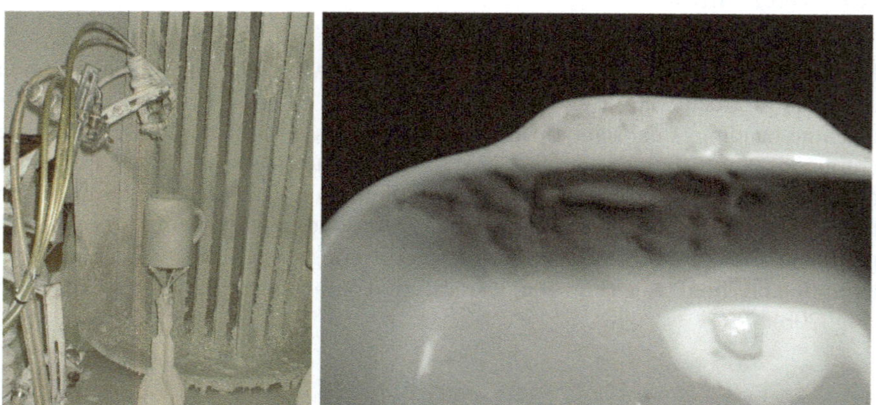

a b

Fig. 21 **a** Automatic spray glazing: an excess glaze can be collected from the filter and the inside of the booth and reused. **b** Crawling defect of the glaze (by courtesy of Vianagrés Company)

4.5 Once-Firing

After glaze application, pieces are allowed to dry and then carefully handled to the kiln avoiding fingerprints or glaze that is knocked off during handling. Importantly that reactions taking place at biscuit and glaze firing occur simultaneously during once-firing. Therefore, firing is done more slowly than usual because the volatile components resulting from the raw materials decomposition of the ceramic paste may be harmful to the glaze layer if the real temperature gradients are too steep [25]. As the results, a more controlled manner of once-firing allows combustible impurities completely burn out before the glaze layer becomes impermeable (Fig. 22).

This in turn ensure high quality of the sensitive glaze layer and helps to overcome crawling. A once-firing curve, typically used in stoneware ceramic factories and respecting all these requirements, is shown in the Fig. 23.

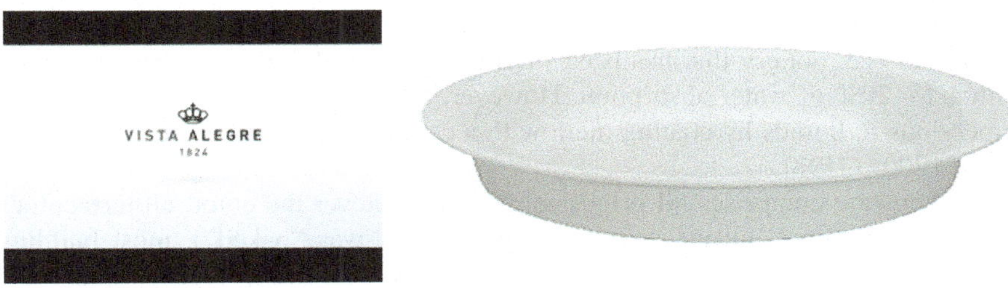

Fig. 22 Vista Alegre 020104 fusion stoneware Casserole (from Internet. https://chefstoys.com/39220/vista-alegre-020104-fusion-stoneware-casserole-dish-hite-22-oz)

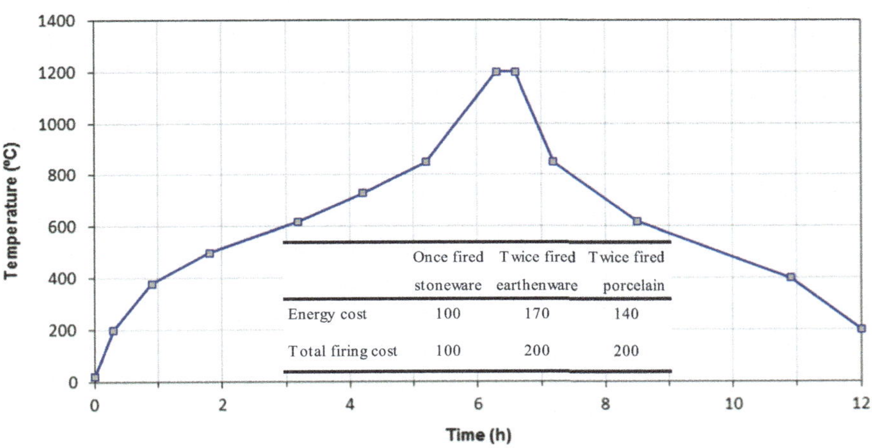

Fig. 23 Typical stoneware one-firing curve and comparison of the relative costs for once and twice fired ware

The special glazing technique applied on the once-fired body by adding salting mixture of sodium chloride and water to the kiln at high temperatures is called salt glazing. Salt glazing gained popularity since 17-th century and normally was done on stoneware at temperatures above 1100 °C. A glassy coating is formed owning to the reaction between sodium oxide from the salt and silica from the clay body according to the following reactions: (1) $2NaCl + 2H_2O \rightarrow 2NaOH + 2HCl$, (2) $2NaOH \rightarrow Na_2O + H_2O$, (3) $Na_2O + SiO_2 = Na_2SiO_3$. It was promoted for studio pottery during 20-th century, but environmental considerations (e.g. negative effect of hydrogen chloride release to the atmosphere) led to the demise of widespread use of salt glazing [26, 27].

5 Earthenware: Terracotta—Majolica—Faience

Earthenware is a group of ceramic fired products that can absorbs water because the biscuit (i.e. pottery that has been fired but not yet glazed) is porous, typically with a 15–20% of water absorption. However, earthenware products can be made impervious to liquids by coating them with a ceramic glaze and normally are fired below 1200 °C [23].

Earthenware comprises all primitive pottery whatever the color, all terracotta's (from Italian "terra" means "earth" + "cotta" that was "baked"), most building bricks, most of the wares of Egypt, Persia and the near East; Greek, Roman and Mediterranean, and some of the Chinese, nearly all European pottery up to the seventeenth century and the fine earthenware which forms the greater part of our tableware today [23].

In general, the term "earthenware" can be applied to the group of porous ceramic materials made from clays, which might be mixed with other typical raw-materials such as quartz, feldspar, calcite and/or kaolin. In this sense, terracotta, majolica and faience are merely different types of "earthenware". Thus, terracotta a hard red-brown unglazed earthenware, used for pottery and building construction. Earthenware products coated with transparent lead glazes applied directly to a biscuit ceramic body has been called as "majolica". In the majolica products engobing is an operation that serves to conceal (to mask) the color body and give a light tone to the surface of the piece (Fig. 24): the engobe consists of a porous clay-based mix that is applied to the raw piece before glazing by transparent lead glazes [23].

"Faience" is the conventional name in English for a tin-glazed earthenware made in France, Germany, Spain, Portugal and in the East (e.g. the oriental potters had discovered and widely used the excellent opacity of tin oxide in tin-glazes much earlier when in 9-th century under Harun al Rashid Mesopotamian potters came into contact with Chinese porcelain for the first time and consequently great efforts were made to imitate these ceramic pieces) [23]. The addition of a tin oxide to the slip of a lead glaze led to the invention of a white, shiny and opaque pottery glaze suitable for painted decoration (Fig. 25). In Italy, this kind of tin-glazed earthenware is still

Fig. 24 Typical majolica articles: decoration was applied above the white engobe layer followed by thin transparent lead glazing

Fig. 25 Hand painted faience, Primavera LCS Lavorazione Ceramiche Sestesi, Italy (private collection, by courtesy of Mrs. Farida Kaxarovna Kadirova)

Table 3 Comparison between compositions used for terracotta, faience, stoneware and porcelain

Ceramic body	Ball clay	Kaolin	Quartz	Feldspar	Calcite	Talc	Red clay
Terracotta and majolica	–	–	5–10	–	–	–	90–95 (*)
Faience	20	32–38	26–30	–	8–14	–	–
Stoneware	25	25	24–28	18–22	–	0–10	–
Porcelain	5–10	45–50	25	25	–	–	–

(*) It is usually a mixture of two red clays with different plasticity levels, with illite as the main clay mineral

called in that case as "maiolica", and this distinguished name can sometimes cause some confusion.

Table 3 gives the typical compositions for terracotta, majolica, faience, and for comparison, stoneware, and porcelain.

5.1 *Mold Preparation*

Porous molds for ceramic pieces shaping and in particular for earthenware are produced from plaster of Paris (the name came from the fact that it was first of all made by heating gypsum which was mainly found in Paris). To produce plaster of Paris natural gypsum ($CaSO_4 \cdot 2H_2O$) is heated up so that it loses 75% of its chemically bonded water. Calcium sulphate hemihydrate ($CaSO_4 \cdot 0.5H_2O$) is then milled to the necessary fineness and maintains the property of picking up water again and known as plaster of Paris [3].

Let us describe a mold making sequence to produce a block model for a pot to be shaped by slip casting (Fig. 26). Firstly, model in clay is prepared (I) and then marked on its middle line vertically (II). The model is placed upon a flat plate made of plastic clay and pushed downwards to the middle line (III) while mouth and bottom are closed by clay. Then the flat plate of clay is covered by thin layer of plaster of Paris that is stained by anti-sticking liquid. The construction is fixed into a wooden cage made up of four frames and the suspension of plaster of Paris is poured (IV). The cage is opened when the plaster is set, the flat supporting plate made of clay is removed and incisions are made on the half model obtained (V). The same operations are repeated to produce the other side of the model. Then to produce the bottom in plaster, the clay of the bottom is removed, anti-sticking liquid is stained and plaster suspension is poured out into the cavity. Finally, the three parts of the finished mold are detached (VI) [3]. All the mold making sequence it is schematically described in Fig. 26.

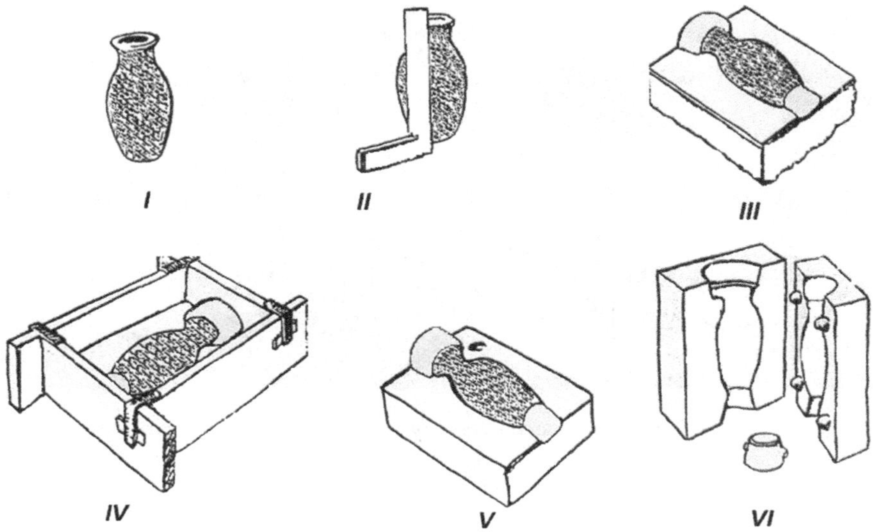

Fig. 26 Sequence of plaster mold preparation for a pot to be shaped by slip casting

5.2 Earthenware Manufacturing

The main shaping process are normally the same as used in the stoneware manufacturing, and reported in the previous section. Majolica's paste can also be formed by pressing or turning. In the last years, mixtures of majolica clay with white earthenware clay have been adopted for slip casting. Traditionally the dried raw pieces are fired first at temperatures ranging between 1040 and 1080 °C (biscuit firing), whereas the glazed body is fired between 980 and 1020 °C (glost firing), using chamber or tunnel kilns. Actually, the most of the glazes of earthenware are fritted since glost firing temperature does not normally exceed 1050 °C (note that raw glazes are usually applied at glost firing temperatures higher than 1050 °C).

Fritting ensures chemical reactions between individual components so that soluble and/or toxic components transform to insoluble and/or nontoxic. Additional advantages include reduction of the melting point of the frit compared to the individual components and possibility for faster glost firing cycle versus that of biscuit firing.

Lead glazes facilitate melting, are easy to use, add color and provide a smooth bright finish and therefore have traditionally been used on ceramic earthenware products (Fig. 27).

The transparent glazed earthenware coated with a thin lead glazing and intended for kitchen use frequently shows a fine network of cracks right after firing or after some weeks of use. This is explained by a higher thermal expansion of the glaze than that of the body or to the latter moisture expansion of the body. Another disadvantage is that lead release in glazed ware is a critical thing to consider: the use of lead compounds causes poisoning of the workers; moreover, the organic acids food always

Fig. 27 Lead glazed majolica (private collection, by courtesy of Mrs. Farida Kaxarovna Kadirova)

contains attack the leaded glaze of the finished pieces, which results in the formation of poisonous lead salts. Over the past decades, the entire ceramic industry has been forced to progressively move away from leaded glazes and colors over the past decade. For the production of domestic faience pieces (i.e. cups and plates) the glazes should be lead free.

Traditionally the earthenware ceramic bodies are fired at least twice (it's believed the two-step process was first tried in Europe when tin oxide was introduced as a glaze component: tin oxide is not compatible with single fired work as too much damage was caused to the original pieces during the firing process, so a second firing was added). Before the first treatment called biscuit firing the shaped earthenware undergo a drying process and then loaded into the kiln. During the initial period of biscuit firing, the temperature is slowly raised, firstly eliminating the residual water still present in the objects after drying. Further, in the interval 450–550 °C the clay mineral starts to decompose and hydroxyl groups present in the clay structure are driven off as water in the reaction, which is known as dihydroxylation: $Al_2O_3 \cdot 2SiO_2 \cdot 2H_2O \rightarrow Al_2O_3 \cdot 2SiO_2 + 2H_2O$. At any temperature between about 300 and 700 °C or even higher, organic matter in the body may burn off: the temperature and ease of removal depending on the type of organic material present and the rate of heating. Sintering reactions related to the real strengthening begin only above 300–400 °C and continuing up to the maximum temperature ~ 1100 °C [23].

Strength in the fired body is in these cases mostly by complex solid-state reactions starting with destroying the actual crystalline lattice, followed by decomposition of molecules and emission of glasses and ending with recombination of residual oxides. A liquid phase might be formed in a negligible amount due to low temperature eutectics formed between silica and alkali oxides. This first treatment renders a body with particles become fewer and larger and porosity is generally reduced. It is resistant to water and a mechanical strength sufficient withstands glaze operation. Before the second firing, called glost firing or glaze firing, a glaze slip is applied on to the porous biscuit: after drying of glazed ware, in some cases, a decoration

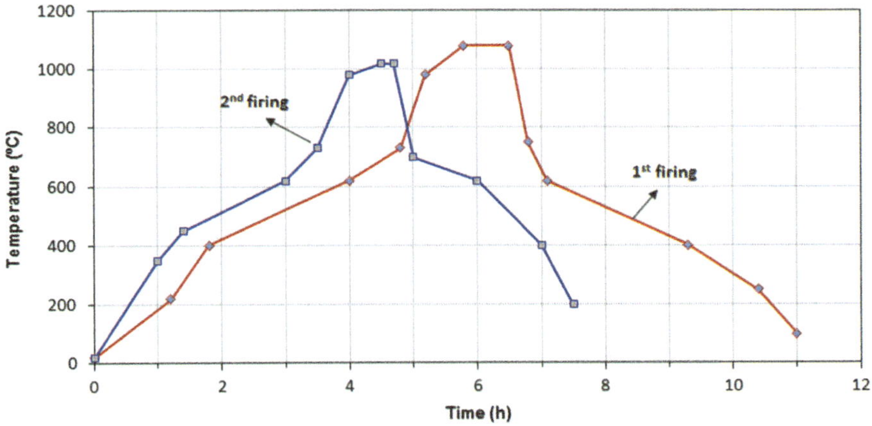

Fig. 28 Typical faience two-firing curves

is applied above the glaze layer. Glazed articles loaded into the kiln and fired at a temperature lower than that of the biscuit fire (Fig. 28).

A third firing is needed in the cases when special colors or glazes are applied by a second decoration treatment at temperatures as low as 850 °C.

5.3 *Typical Defects of Earthenware*

Incorrect firing schedules might result in cracks during heating, black core, too high porosity and too soft a biscuit, deformation and overfired parts, cooling cracks. The most part of those cracks can appear during cooling and might be attributed to the silica content since phase transition paths between silica polymorphs gives a rapid volume change at 573 °C (in that case the β-phase to α-phase, in cooling). Similar to stoneware, during the glaze firing the earthenware products may form non-uniform surface along with crazing and peeling of the glaze due to crawling and poor CTE adjustment, respectively [19].

6 Porcelain

Porcelains are actually the fired product of ternary mixture of fine–grained clay (commonly kaolin), flux (usually feldspar) and filler (commonly quartz). Porcelain is usually valued more than other commodities because of the dense, white, translucent quality, smoothness of its surface and normally classified according to firing temperature. So called "hard porcelains" with the standard batch formulation of 50% kaolin, 25% feldspar and 25% quartz have been usually fired at 1380–1460

°C and "soft porcelains" (e.g. feature high content of fluxes) at lower temperatures 1200–1250 °C. Apparently, soft porcelains have lower production cost due to the lower energy consumption during firing, and can be decorated with a wider range of colors. According to the use, there are tableware, sanitary, chemical-technical, art and electrical porcelains, all made of quartz, feldspar and kaolin. Moreover, there are special types of porcelains, such as "frit porcelains" which is manufactured mainly in France and made of kaolin, quartz and alkali rich frit, or "bone china" which is produced mainly in England and consists of 20–30% kaolin, 25–30% pegmatite (e.g. Cornish Stone) and 25–50% bone ash [24].

The production of porcelain became very popular in China under the Thang dynasty (618–907), so that the poet Tu-Mu was enhancing its beauty by saying that "*The porcelain is so light but solid and the wonderful white cups are even more shining than the pure crystals of snow*". Such a better production of porcelain in terms of quantity and quality was mainly due to the discovery and exploitation of the ore located at Kau-ling or Kao-ling (meaning high mountain) or Gaoling according to the Pinyin grammar, located in the province of Kiangsi in the South-Eastern China. The Kau-ling mine, next to Kaulin village, gave the name to the exploited kaolin rock, whose hard and no-plastic nature should be more similar to a feldspar-rock rather than the kaolin-clay properly named in Europe and recognized to be softer and with a less content in silica. The European discovery of porcelain is ascribed to Johann Friedrich Böttger and Ehrenfried Walther von Tschirnhaus who were working for August the Strong, Elector of Saxony and King of Poland while a scale-up production of porcelain begins in the Dresden laboratories in 1709. A large-scale manufacturing industry properly occurred in Europe during the eighteenth century and later (Fig. 29) when the potteries from a craft institution evolved towards important manufacturing sites like Meissen (1710) and Höchst (1746) in Germany (Fig. 1), Wien in Austria (1718), Sevres in France (1770), Staffordshire in England (1744–1750), Imperial porcelain factory in St. Petersburg in Russia (1744), Vista Alegre in Portugal (1824), etc.

6.1 Body Preparation

In the manufacturing kaolin is washed before being used to eliminate the coarse particles, which consist mainly of quartz, however very fine silica and mica can remain in the washed kaolin to a certain extent.

Ball clay may be added (not more than 5–10%) to improve the plasticity of the porcelain body while with regard to the necessary white color of the fired product only few ball clays with a low iron content can be taken into consideration. Quartzite or quartz sand are added to the batch formulation: to avoid thermal shock faults due to the inversion behavior of quartz (e.g. α to β quartz inversion is accompanied by a linear expansion of 0.45%) cooling rates are not exceeding 50 °C/h. The most used type of feldspar in porcelain bodies is potassium feldspar whilst sodium feldspar is used only occasionally. Potassium feldspar (that is preferably introduced through

Fig. 29 The first row—Höchst Porcelain—Museum of Frankfurt am Main (by courtesy of Mrs. Patricia Stahl); the second row—gold painted porcelain, Vista Alegre company, first half of the twentieth century (private collection, by courtesy of Mrs. Maria Fernanda Ribeiro)

the pegmatite) demonstrates good stability against distortion of the porcelain during the firing process due to formation of a melt of great viscosity, which decreases only slightly with increasing temperature. The most important properties of all these raw materials are sufficient purity, especially low iron content. Determining the percentage of moisture in each raw material is performed carefully so as to compound the body on a dry weight basis [24].

Wet processing is common for the production of porcelain ensuring high quality of the final product. In this processing route, kaolin and ball clay can be either wet milled, or slip can be prepared from them. When quartz and feldspar are ground to the desired fineness, they are mixed with the clayey components in the ball mill to achieve 56 μm sieve residue in the range 0.5–1%. Then the body slip is pumped under pressure onto cloth or nylon sheet of the filter press, in which dewatering process takes place: the water passes through, the solid material is left behind and forms a filter cake of plastic with a water content of about 22–23% [7, 24]. However, the moisture distribution in filter press cakes is not uniform throughout the cross-section and they are still full of fine air bubbles, and with unwanted orientations of the paste particles. To achieve a greater degree of homogeneity filter cakes are fed to vacuum extruder that develops uniformity to a high degree, improves plasticity of the extruded body that is then cut to rods.

Several methods of shaping of the porcelain body are applied such as slip casting, plastic making (jollying or jiggering), pressure casting and isostatic pressing that depends mostly on the shape of the item to be manufactured and the lot dimension (the number of pieces to be produced).

Plastic making is done using different methods and all of them based on the development of plasticity in the body being shaped. Clay content in porcelain formulation and the amount of water in porcelain's body determine its plasticity that along with the low moisture content ensure easy shaping and safe drying. The mechanism behind plasticity that determine the foundation of many present shaping methods as pressure casting and isostatic pressing as well is still not wholly understood. Nevertheless, it is thoroughly discussed that clay minerals form thin sheets extended in length and width, and thus clay particles create a great surface area attracting water molecules to the surface. This acquired water layer might serve as lubricant facilitating movement of one clay plate over another and resulting in plasticity of clays, and consequently controlling the plasticity level of the all clay containing formulation. As for a mechanism of water layer formation, it should be considered the existence of intermolecular interactions acting between water and clay particles in the form of non-bonding dipole-ion forces and hydrogen bonding due to some attraction between hydrogen in the polar water molecule and oxygen in the clay's tetrahedral layer, and between oxygen in the water molecules and hydrogens in the clay's octahedral layer respectively. In this regard, the presence of ball clays and the addition of small amounts of montmorillonites are essential to improve plasticity. Aging of filter cakes and the pugging process greatly contribute to a uniform distribution of water throughout the body and eventually helps to obtain consistent working properties, including plasticity [24].

Isostatic dry pressing for the production of porcelain tableware, flatware, bowls, salad bowls, platters, casseroles, pans and similar items, is nowadays the most suitable and cost-effective method to produce most shapes, either in small or large amounts. Isostatic dry presses are most suitable for the production of oval, rectangular and square flatware by using ceramic spray dried granulates (Fig. 30). The pressing technique replaces increasingly the plastic porcelain tableware production because staff, energy and production floor area are saved thus achieving a better quality and maintaining good production flexibility [25].

6.2 Drying

Drying is optimized considering the sensibility of the body to drying, its moisture content, as well as the shape of the articles and their distribution in the dryer. For instance, in case of porcelain wares molded by traditional plastic making the whole process is divided into two parts: the low-temperature drying of molded ware on the plaster molds, and the finish drying of the articles that have already been withdrawn

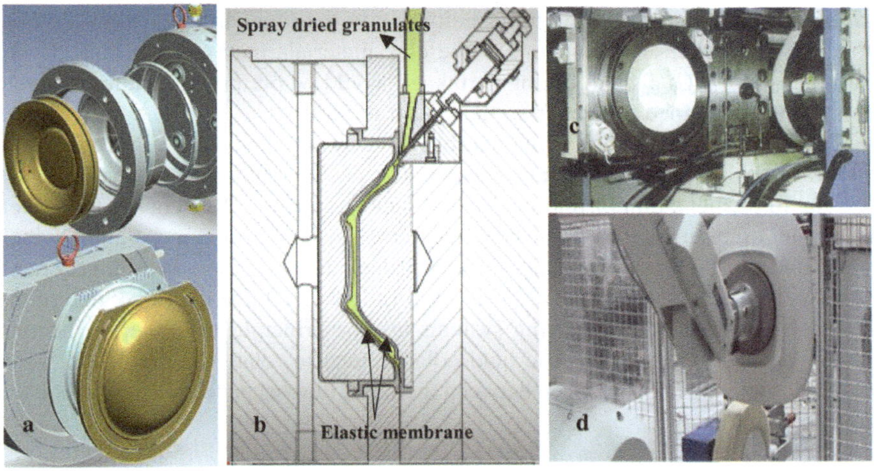

Fig. 30 Isostatic pressing: **a** elastic membranes for a plate production, **b** schematic view of the spray-dried granulated feed to the mold (space between the elastic membranes), **c** isostatic press, **d** automatic finishing of a porcelain plate by robot (by courtesy of SAMA Company) [25]

from the plaster molds. Importantly that the second part of the drying proceeds can be much faster than the first one. This is because moisture has already reached the required value called as *critical moisture content* when no more shrinkage will further occur since water remained only in the free spaces between particles touching each other, and that allows to proceed at a faster rate, without risks of cracking. The typical moisture content of porcelain wares molded by traditional plastic making and slip casting, before and after the drying process is shown in the Table 4.

Drying results in the porcelain body possessing a certain green strength required for further smoothing of edges and the removing of seam scratches before glazing (in case of single firing) and the installing of the ware for firing. A high green strength

Table 4 Moisture content (in %) of porcelain wares molded by traditional plastic making, slip casting and isostatic pressing

	Moisture content		
Shaping process	After molding	After withdrawal from plaster molds	Final
Hollow pieces molded by plastic making	23–27	16–18	1.5–2
Flat pieces molded by plastic making	20–21	12–14	1.5–2
Hollow pieces molded by slip casting	31–32	18–20	1.5–2
Flat pieces molded by isostatic pressing	3	–	< 3

is normally correlated with high plasticity due to the fine particle size of the clay materials. However, extra amount of the clay materials may cause a high drying shrinkage, although a small drying shrinkage would be preferable to allow an easy removal of the shaped articles from the plaster mold.

The modern globalization of markets has recently enforced porcelain producers to find solutions for a manufacturing cost reduction. As an important step in this regard was the implementation of microwave and radio-frequency (RF) energy to shorten the drying time. In contrast to the heat transferred by convection or radiation, which is provided to the body from the outside in these new systems, heat is generated directly inside the body and distributed uniformly through the entire volume of the article. Microwave and RF drying systems use the principles of dielectric heating when change in external electric field polarity tend to rotate the dipoles of water molecules and the friction arisen between the water dipoles generates heat inside the green ceramic body.

6.3 Glazing

Whilst lower-fired counterparts such as majolica, faience, and sanitary ware are generally covered by fritted glazes, raw-glazes are preferably used to glaze tableware porcelain to bring gloss, whiteness and translucence. The feldspar-based raw glazes with the molecular formula $(R_2O + RO) \cdot (0.5–1.4)\ Al_2O_3 \cdot (6–12)\ SiO_2$ (where R_2O and RO are alkali and alkaline earth oxides, respectively) are used in the hard porcelain manufacturing with preferable ratios of oxides $K_2O/Na_2O \geq 2$, $CaO/MgO \geq 1$ and $\Sigma RO/\Sigma R_2O \geq 3$ and spreading temperatures of 1300–1450 °C. For instance, the batch composition of a glaze with a firing temperature of 1380 °C is as follows, in wt. %: 36 feldspar, 27 quartz sand, 14 dolomite, 7 kaolin, 16 scrap of fired porcelain [24].

A decrease in the Al_2O_3/SiO_2 ratio lowers refractoriness and the new range of glaze compositions having the molecular formula $(R_2O + RO) \cdot (0.3–0.6) \cdot Al_2O_3 \cdot (3–4)SiO_2$ and with the firing temperatures 1200–1280 °C becomes suitable for soft porcelain [24].

During glazing the utmost care must be addressed to each single action, in order to ensure an overall success before loading the kiln: the articles already passed through the drying stage or bisque wares are handled with disposable gloves, to avoid resist spots where the glaze adheres unevenly or not at all, all the dust must be removed before glazing, special caution should be taken when cleaning single fired articles, bearing in mind their low mechanical strength.

Glazing procedure is performed by means of conventional techniques such as dipping, spraying and pouring techniques. Unfortunately, raw glazes decrease whiteness of the final product by 5–7% even when pure raw materials are used while increase the wares' compressive strength by 40–50%, their tensile strength by 30–60% and their bending strength by 30–40% that depend also on the glaze's thickness and on the body's composition. In this regard, a perfect conformity between the thermal expansion values of the glaze and the body is determinative. Considering that fact that the glaze, as other brittle materials withstand compressive forces much better than tension forces the glaze should possess lower a coefficient of thermal expansion (CTE) than the body [24].

6.4 Firing

Porcelain undergo a controlled high temperature firing that determines set of its unique properties. The most porcelain wares were traditionally fired twice with bisque firing (1st firing) required to give the pieces enough strength before glazing followed by the glost firing (2nd firing) that brought to maturity the glaze and the body. Traditionally additional firing (3rd firing) is required when decoration is applied above the surface of a glazed ware [24].

For the first (900–1000 °C) and the third firing (650–850 °C) the batch and the continuous roller hearth kilns are generally used. The major advantage of a roller hearth kiln over batch and continuous tunnel kilns is absence of kiln cars that enables a fast firing cycle. In the tunnel kilns glazed ware is placed either directly on loading platforms or inside refractory boxlike containers—saggars—to protect ceramic articles from the flame, guaranteeing in this way a good temperature homogeneity throughout the porcelain ware and preventing the warping (Fig. 31).

The glost (2nd firing) firing cycle of porcelain tableware includes a stage with kiln oxidizing atmosphere up to 950–1050 °C, followed by a reduction period until the reaching of soaking temperature (1350–1400 °C), and then again, a kiln oxidizing atmosphere for cooling. The reduction period starts out with a very strong reducing atmosphere of 4–5% CO and then is changed at about 1300 °C to 1–1.5% CO, until maximum firing temperature [24, 26].

In the tunnel kilns, the car travels from left to right, while the exhaust gases flow in counterflow direction to the firing porcelain wares, from the firing zone through the preheating zone to the kiln entrance, and transfer their enthalpy on to the firing

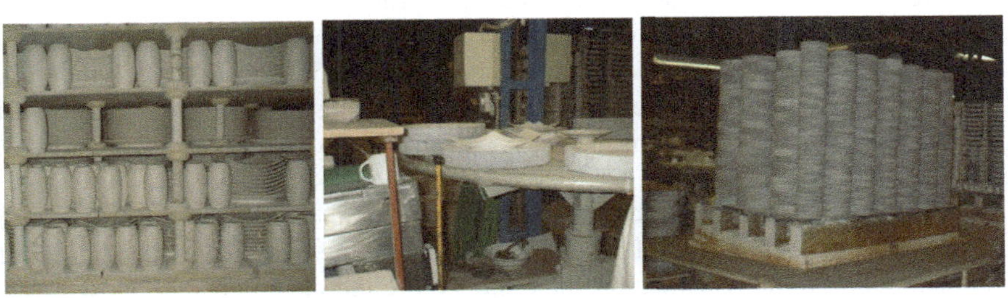

a b c

Fig. 31 Preparation for firing: **a** dried cups and plates are stacked before the 1st firing, **b** glazed ware placed in the refractory containers (saggar) before the 2nd firing; **c** a kiln car with stacked saggars prepared for the 2nd firing (by courtesy of Porcel Company)

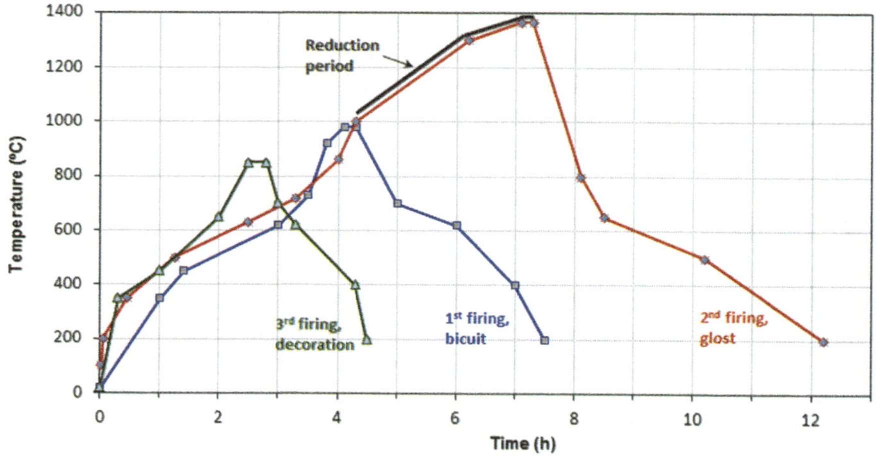

Fig. 32 Typical hard porcelain firing curves, with the indication of 1st firing (biscuit firing), 2nd firing (glost firing), and 3rd firing (decoration)

goods, while cooling off themselves. In contrary, cooling air flows into the cooling zone in counterflow direction to the goods towards the firing zone, cooling the firing products while heating up itself. This heat is extracted by the so-called intermediate exhaust immediately after the firing zone, and is applied for drying or preheated air. The reduction is not influenced by the cooling air since it is achieved in the burner zone between ~ 1150 °C and the maximum temperature, by adjusting the gas/air ratio of the burners [24, 26].

The glost firing process often begins by warming up the product from 20 to 500 °C for a few hours (Fig. 32), in an oxidizing atmosphere considering that quite a large proportion of the heat is absorbed by the kiln's walls, as well as by the hearth of kiln cars, loading refractory platforms and saggars (Fig. 31a and c). Considering that in industrial kilns, clay minerals decompose in the interval 550–850 °C (although, according to the thermal analysis, decomposition occurs at 550–750 °C) after preheating phase at 600–700 °C, the temperature is raised to about 1000 °C with a higher heating rate. The oxidizing period covers the following transformation periods including clay minerals decompose in the interval 550–850 °C, a polymorphic transition from α to β quartz, decomposition of carbonates in glaze and normally last until a temperature of 1050 °C is reached, with an air factor excess of 1.20–1.25 [24, 26].

During glost firing the carbon monoxide present in the fuel will transform to carbon and carbon dioxide and thus carbon settled in the body have to be burnt out completely before the vitrification stage, i.e. before the open pores are closed by a liquid phase. There are some recommendations to maintain the CO content around 3–4% and 6–8% at 1000–1100 °C and 1100–1250 °C, respectively, while keeping the CO/CO_2 ratio higher than 0.3 (Fig. 32). In case of carbon that is not burned completely before glaze reaches its maturation, it causes defects called as pinholes and crater-like dimples. Indeed, during reduction period a significant amount (e.g. 30–35% by volume) of vitreous phase is formed, due to the melting of feldspar, then meta-kaolinite and quartz are gradually dissolved in the feldspar's melt, while undissolved meta-kaolinite turns into primary mullite and amorphous silica [24, 26].

Intense shrinkage is usually observed during the reduction process, between 1050 and 1250 °C, when the capillaries and pores in the ware are progressively being closed, and the diffusion of gases in the core gradually fades out, preventing the reduction process from reaching an end. The following reactions (1) $2Fe_2O_3 \rightarrow 4FeO + O_2$, (2) $FeO + SiO_2 \rightarrow FeO{\cdot}SiO_2$, (3) $FeS_2 + CO + 3O_2 \rightarrow 2SO_2 + CO_2 + FeO$, (4) $CaSO_4 + CO \rightarrow CaO + SO_2 + CO_2$ take place with increasing CO content in the burning gases owing to reduction conditions. Tremendously that whiteness is improved due to the formation of fayalite $FeO{\cdot}SiO_2$ and another colorless FeO-containing alumino-silicates while iron oxide favors the vitreous phase formation and widens the sintering temperature range to a certain extent. The vitreous phase, in turn, promotes a desirable mullite phase formation [24, 26]. Importantly that the primary mullite is originated from a solid phase and it forms in the region of clay relicts as scaly aggregates at ~1100 °C (Fig. 33 a and b) and interpreted as a pseudomorph of mullite after kaolinite. The liquid phase of the eutectic composition can dissolve silica and mullite. The liquid content increases with further potassium feldspar melting at about 1150 °C followed by alkali diffuse out of feldspar and nucleation/growth of secondary mullite. Indeed, the loss of small amount of alkali in the initial liquid phase will shift the composition of the liquid phase to the field of mullite crystallization at the phase diagram K_2O–Al_2O_3–SiO_2. Newly formed mullite crystals in the shape of idiomorphic and predominantly thin needles are called as secondary mullite (Fig. 33b) [24].

For the firing technique of porcelain the dissolution of quartz is of great importance since the decrease in viscosity at elevated temperatures is compensated by a progressive dissolution of silica (Fig. 33v) that ensures sufficient stability against distortion. Generally, the cracking severity is governed by the particle size of residual quartz crystals and by the cooling rate, essentially in the 2nd and 3rd firings. Apparently, relaxation within the glassy phase impedes the development of residual stresses at

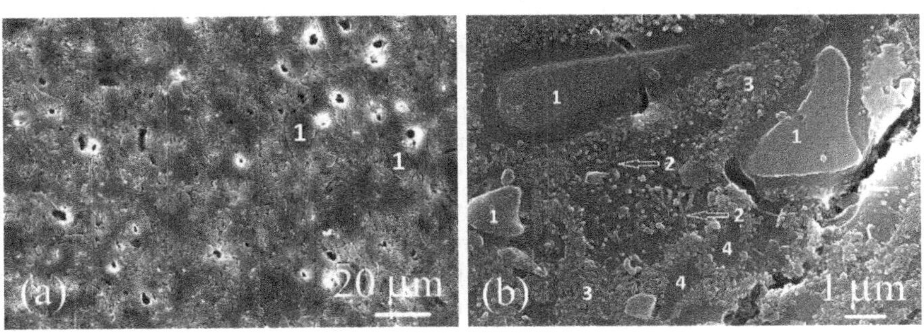

Fig. 33 SEM images of Li-containing porcelain body fired at 1150 (**a**) and 1250 °C (**b**): 1-α-quartz, 2-secondary mullite, 3-primary mullite, 4-glassy matrix [24]

temperatures above the glass transition temperature. However, on further cooling in the temperature range 20–750 °C residual stresses are developed owning to the significant difference between the linear coefficient of thermal expansion of the glassy phase (CTE ~ 3×10^{-6}/K) and the crystalline α-quartz (CTE ~ 23×10^{-6}/K) that may causes the formation of cracks within the quartz grains and surrounding glass phase (Fig. 33b). Moreover, additional stresses within a glassy matrix might appear during the cooling through β- to α-quartz and β- to α-cristobalite inversions that result in a volume decrease. In standard triaxial porcelain bodies, the principal crystalline phases are mullite and α-quartz, while bodies with high quartz content may also contain cristobalite (Fig. 33b) [24, 27].

The structure of the porcelain after glost firing is quite heterogeneous, as it comprises both crystalline phases and pores embedded in the glassy matrix (Fig. 33b) [24, 28].

Unlike conventional porcelain, bone china is fired first at high temperatures (ranging from 1200 to 1300 °C) in an oxidizing atmosphere, to obtain a durable, white and translucent body. Yet, when considering a short firing range (i.e. a specific temperature range at which the properties of porcelain fully satisfy the required commercial standards), the firing schedule must be carefully controlled, in order to avoid deformation and cracking. To prevent this, the pieces are generally placed on refractory setters, called *cranks* ("delta" or "Y" cranks), to let them follow the profiles of the setters. The irregular shaped pieces that are not suitable for on-crank firing are set in saggars, instead. The glaze, consisting of a mixture of frit and kaolin, is applied by spraying and then glazed pieces are fired at around 1100–1200 °C in oxidizing atmosphere [27, 28].

6.5 Properties of Porcelain

Porcelain possess the unique set of properties when compared to other ceramic products (Table 5). Porcelain is following a general characteristic, found in all ceramics and demonstrates far higher compressive strength than its tensile strength. As long as the coefficient of thermal expansion of glaze is correctly adjusted to that of the body the glazed items a tensile might increase by 30–60 MPa and flexural strength might increase by 30–40% than unglazed counterparts. Pelling of glaze that is the opposite of crazing caused by compressive forces in the glaze, which exceed the bonding strength at the glaze/body interface [29]. Application of raw glazes and the specific microstructure with > 60 vol% of high-viscosity glass gives porcelain another important feature as translucency while grain boundaries and pores will scatter the light and therefore somewhat reduce the translucency. Therefore, porcelain becomes more translucent at elevated temperatures as the number and size of the pores are reduced, while the crystalline components are dissolved and the amount of glassy phase increases. Moreover, the formation of secondary mullite causes a drop in translucency, while cristobalite improves the translucency of porcelain [28, 29].

Table 5 Final properties for different ceramic products

Ceramic body	Water absorption (%)	Firing temperature (°C)	Glaze	Glassy phase content (vol%)	Bending strength (MPa)
Terracotta/majolica	14–18	950–1050	Opaque/translucent/transparent	5–8	20–35
Faience	14–18	1st firing, biscuit: 1050–1080 2nd firing, glost: 980–1020 3rd firing, decorative: 650–850	Opaque	5–10	35–45
Stoneware	3–6	1180–1280	Opaque	30–40	50–60
Hard Porcelain	< 0.1	1st firing, biscuit: 900–1000 2nd firing, glost: 1350–1400 3rd firing, decorative: 650–850	Translucent/transparent	~ 60	80–100

Another important characteristic that strictly distinguishes porcelain from all the other white ware bodies is whiteness that depends mainly on the purity of the raw materials and the kiln atmosphere during the glost firing. Thus, a reducing atmosphere contributes to the formation of colorless FeO containing aluminosilicates, while the addition of a very small quantity of cobalt oxide might be beneficial, by compensating the yellow shades in the ceramics. Whiteness is determined by comparing the diffuse reflected light of the sample with the one of the reflectance standard, namely barium sulphate. At temperatures between 20 and 600 °C the linear-CTE of a porcelain body varies from 4×10^{-6}/K to 6×10^{-6}/K: this range of CTE values is in general comparable with the linear-CTE of mullite (5.4×10^{-6}/K) and desirable with regard to resistance to thermal shock. A series of test methods for thermal shock and thermal shock endurance of several brittle materials, including ceramic items that are used in ovens or as tableware is specified by the European Standard EN 1183:1997. In case of porcelain, thermal shock resistance is improved as quartz gradually dissolves in the glassy phase: the overall CTE of the body decreases due to both the diminishing high thermal expansion of crystalline quartz content and the enrichment of the glassy phase with pure silica glass, having a very low CTE ($\sim 0.5 \times 10^{-6}$/K). The bodies with a quartz content of 10–15 wt% are used for the production of the chemical-technical porcelains, which require a high resistance to thermal shock [28, 29].

6.6 Decoration of Porcelain

The decorations of porcelain, with their colors, enliven the whiteness of the material and create a vivid contrast with the plain white shape, providing a most pleasant appearance to many articles for everyday use.

So called *underglaze decoration* applied on the biscuit surface (Fig. 34a) must withstand the high temperatures of glaze firing [30, 31]. The layer of glaze will protect such decoration from mechanical and chemical damage when in use. The lower the temperature that the coloring material will have to withstand, the wider the choice of colors available and vice versa.

So called *on-glaze* or *overglaze decoration* applied on top of the matured glaze requires further firing to mature the paint and fix it firmly into the glaze surface [30, 31]. Therefore, pieces are fired at 650–850 °C again, 3rd firing, to make the pigments fuse into the matured glaze. Since the final firing requires rather low temperatures, a wider range of colors can be used, including reds and oranges that would not withstand intense heating. It is worth to note that patterns applied on the glaze are not as smooth as the glaze itself, and normally they are detectable to the touch.

The transfer method that is widely used to apply pre-printed images to porcelain tableware, ornamental ware, etc. is the decal method when designs are created and screen-printed onto water slide decals. The printed transfer is soaked until the design can be slid off onto the porcelain and positioned. This process is followed by firing in a kiln. During firing, the transfer medium evaporates and the glaze fuses onto the porcelain just as it would if it had been painted directly and fired.

Porcelain, bone china, stoneware and earthenware pieces are generally decorated with on-glaze decals, which are applied to the ceramic object's glazed surface. During the firing, the colors fuse onto the glaze and the decoration becomes durable. In case

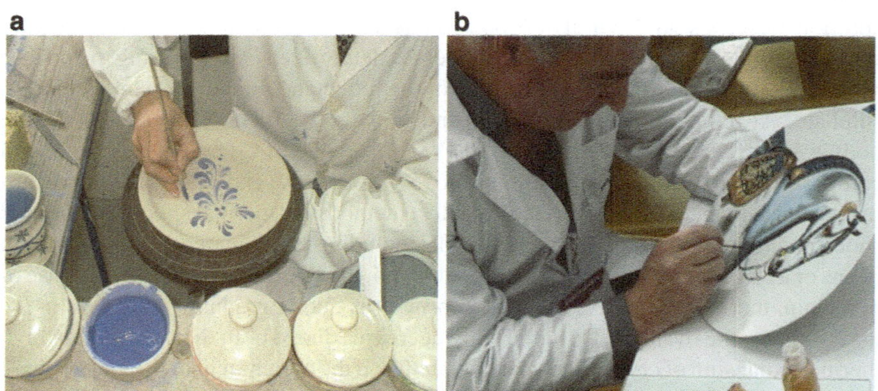

Fig. 34 Two types of porcelain ware decorations: **a** underglaze cobalt decoration applied on the biscuit surface, **b** overglaze hand-paint decoration, the 3rd firing at 650–850 °C (by courtesy of Porcel and Vista Alegre Companies)

of in-glaze decals, they are also applied to a glazed surface, but then the piece is fired to a higher temperature, so that the decal sinks into the glaze, and this provides greater durability. A full range of precious metals are also available, including bright, matt and burnished gold or platinum pastes, while colored lusters give a metallic iridescent look to wares.

6.7 Final Inspection

Quality control and removal of defective products is usually performed at the end of the process, but also at the intermediate phase: after the ware shaping process, the first firing, and the glazing phase. The very high quality requirements of the customers for the porcelain wares are obliging the companies working in this industry to deliver only first class quality products. Simultaneously, the product inspection criteria have become very diverse, the number of products inspected has increased and also the complexity of the control tasks.

Nowadays, the final porcelain products inspection criteria is very diverse. Normally, it depends on the number of products inspected, the customer requirement level, and on the complexity of control. Porcelain defects can be classified into several categories such as: ware asymmetries, curves, deformed edges and warps, glazing application traces, flaking, fissures, cracks, indenture, scratches, degraded color such as "black colored" zones or "yellowish" zones (due to the excess of reduction or lack of reduction during the 2nd firing, respectively), etc.

References

1. Karaman, S., Ersahin, S., Gunal, H.: Firing temperature and firing time influence on mechanical and physical properties of clay bricks. J. Sci. Ind. Res. **65**, 153–159 (2006)
2. Manual, B.M. (ed.): APICER. Coimbra, Portugal (2000).(in Portuguese)
3. Augustinik, A.I.: Ceramics, 2nd edn, Stroiizdat, Leningrad, Russia (1975)
4. Ryan, W.: Properties of Ceramic Raw Materials, 2nd edn. Pergamon Press, Oxford (1978)
5. Ribeiro, M.J., Ferreira, A.L., Labrincha, J.A.: Fundamental aspects of the extrusion of red ceramic pastes. Cerâmica Industrial 8(1) (Jan/Feb, 2003) (in Portuguese)
6. Wieck, R., Duailibi, J.: Red Ceramic extrusion: basic principles, problems and solutions. Cerâmica Industrial **18**(3) May/June (2013) (in Portuguese)
7. Kingery, W.D., Bowen, H.K., Uhlmann, D.R.: Introduction to Ceramics, 2nd edn. Wiley—Wiley series on the science and technology of materials, New York (1976)
8. Masters, K.: Spray Drying Handbook, 4th edn. Wiley, New York (1990)
9. Lukasiewicz, S.J.: Spray-Drying Ceramic Powders. J. Am. Ceram. Soc. **72**(4), 617–624 (1989)
10. Cao, X.Q., Vassen, R., Schwartz, S., Jungen, W., Tietz, F., Stöever, D.: Spray-Drying of Ceramics for Plasma-Spray Coating. JECS **20**, 2433–2439 (2000)
11. "Tecnologia de la fabricacion de azulejos", SACMI Imola, Asociacion Tecnicos Ceramicos (1990)
12. https://sacmi.com/en-US/Ceramics/Tiles/pressing (Oct 2020)
13. Facincani, E.: Tecnología Cerámica los Ladrillos. Faenza Editrice Iberica S.L., IMPIVA, Castellón (1993)

14. Ceramic tiles application manual. APICER, Coimbra, Portugal 1998 (in Portuguese)
15. Hutchings, I.: Ink-Jet Printing for the Decoration of Ceramic Tiles: Technology and Opportunities. Qualicer'10. 11th World Congress (2010)
16. http://eippcb.jrc.es
17. https://www.usgboral.com/en_ip/products/ceramics/sanitary-ware/ceracast.html
18. Fortuna, A., Fortuna, D.M., Martini, E.: An Industrial Approach to Ceramics: Sanitaryware. Plinius **43**, 138–145 (2017)
19. Fraser, H.: Ceramic Faults and Their Remedies. A & C Black Publishers Ltd, 35 Bedford Row, London (1986)
20. Mitchell, B.S.: An Introduction to Materials Engineering and Science. Wiley, New Jersey (2004)
21. Askeland, D.R., Phulé, P.P.: The Science and Engineering of Materials, 4th edn. Global Engineering Publisher, USA (2003)
22. Callister, W.D.: Materials Science and Engineering: an introduction. 6ª ed. Wiley, University of Utah (2003)
23. Dictionary of Ceramics. Arthur Dodd & David Murfin, 3rd edn. The Institute of Minerals
24. Tulyaganov, D.U., Tomalino, M.U.: Porcelain from Tableware to Dental Restoration. CLUT publisher, Politechnico di Torino, Italy (2017)
25. https://www.sacmi.com/en-us/ceramics/tableware/Isostatic-pressing
26. Birlutiu, A., Burlacu, A., Kadar, M., Onita, D.: Defect Detection in Porcelain Industry based on Deep Learning Techniques. Computer Science. In: 19th International Symposium on Symbolic and Numeric Algorithms for Scientific Computing (SYNASC), pp. 263–270 (2017)
27. Eppler, R.A., Eppler, D.R.: Glazes and Glass Coatings. The American Ceramic Society, Westerville, Ohio (2000)
28. Iqbal, Y., Lee, W.E.: Fired Porcelain Microstructures Revisited. J. Am. Ceram. Soc. **82**(12), 3584–3590 (1999)
29. Carty, W.M., Senapati, U.: Porcelain-Raw Materials, Phase Evaluation and Mechanical Behavior. J. Am. Ceram. Soc. **81**(1), 3–20 (1998)
30. Ceramic Monographs—Handbook of Ceramics, Verlag Schmid GmbH Freiburg i. Brg (1980).
31. Ceramic dictionary. http://ceramicdictionary.com/en/t/612/temple

Silicate Glasses and Glass–Ceramics: Types, Role of Composition and Processing Methods

Dilshat Tulyaganov and Francesco Baino

Abstract Glass has been a versatile and fascinating material since the early stages of civilization. The aesthetic and functional properties of glasses are mainly dictated by the composition, which in most cases is a mixture of inorganic oxides and can be properly designed according to the end use. Glass–ceramics are polycrystalline materials produced by the controlled crystallization of certain parent glasses and contain one or more crystalline phases embedded in a residual amorphous matrix. The distinct chemical nature and microstructural features of these phases have led to various combinations of properties and applications in many industrial, medical and high-tech fields. This chapter introduces the reader into the "mystery of glass", providing a picture of the structural theories, formation criteria and main processing methods for glass and glass–ceramic products with focus on a selected set of silicate materials..

Keyword Glass · Glass–ceramic · Melt-quenching · Sol–gel · Crystallization · Sintering

1 Definition of Glass: The Attempts to Disclose Its Secret

Glass is indeed one of the most "universal" and fascinating materials made by mankind. Use and evolution of glass products have often been associated to key steps from technological, historical and societal viewpoints over the long path of human progress. Generally speaking, the term "glass" refers to a fragile and transparent matter having a dramatically broad range of current applications in arts, building, industry, medicine, optics and optoelectronics, and everyday life [1–14]. Although the most traditional glasses are inorganic and non-metallic, the term "glass" has been

D. Tulyaganov (✉)
Department of Natural-Mathematical Sciences, Turin Polytechnic University in Tashkent, Tashkent, Uzbekistan

F. Baino
Institute of Materials Physics and Engineering, Department of Applied Science and Technology, Politecnico di Torino, Turin, Italy

F. Baino et al. (eds.), *Ceramics, Glass and Glass-Ceramics*, PoliTO Springer Series,
https://doi.org/10.1007/978-3-030-85776-9_4

gradually extended to include optically-transparent high-quality polymers [the so-called organic glasses, e.g. poly(methyl methacrylate)] and amorphous metals (also called metallic glasses).

According to the ASTM definition of glass, as initially proposed by Committee C-14 in the tentative standard about glass in 1941 and still issued in ASTM C162 in 1992, glass is an inorganic product of melting which has been cooled to a rigid condition without undergoing crystallization [15, 16]. This definition is valid for most commercial glasses but does not take into account some "modern" alternative methods to produce glasses, such as vapor deposition, sol–gel process and neutron irradiation of crystalline materials.

From a general viewpoint, glass is typically hard and brittle, and exhibits a conchoidal fracture. It may be colourless or coloured, and its optical transparency to visible light can vary depending on the presence of crystalline phases embedded in the glassy matrix. These glass-crystal composites are actually called glass–ceramics; devitrification (i.e. the nucleation and growth of crystalline phases inside glass) make the material partially or totally opaque.

Many other definitions for glass have been proposed over the last decades, including the attempt to distinguish between glass and amorphous solids (not all amorphous solids can be considered as glasses: for example, wood and amorphous silicon are actually amorphous but exhibit no glass transition temperature) and the concept of supercooled liquid (however, glass does not exhibit any viscous flow at room temperature) [1, 6, 8, 13]. Perhaps, the most fascinating definition of glass was provided by Prof. Patrick Charbonneau in a press release of Duke University in 2014: "glass is a mystery". He also added: "There have been beautiful mathematical models, but with sometimes tenuous connection to real, structural glasses. Now we have a model that is much closer to real glasses". In fact, Prof. Charbonneau used advanced mathematical theories to analyze how glass molecules behave and concluded that glasses seem to obey fractal models.

Finally, a practical definition of glass can be provided on a phenomenological basis: glasses have no long-range, periodic atomic arrangement, and exhibit a time-dependent glass transformation behaviour. This behaviour occurs over a temperature range known as the glass transformation region. Hence, any material (inorganic, organic, or metallic) formed by any technique, which exhibits glass transformation behaviour, is a glass [5].

Most common glasses are formed via rapid cooling of melts in order to avoid crystallization, since little time is allowed for the atomic ordering processes. The glass transformation behaviour is usually analysed and discussed by plotting the volume (or enthalpy) vs. temperature (Fig. 1) [5, 14, 17]. Most liquids upon cooling abruptly solidify at a fixed temperature, T_m (the melting point or liquidus temperature), with a significant change in volume following the line b-c in the diagram shown in Fig. 1. This is due to the formation of a long-range, periodic atomic arrangement. Continued cooling of the crystal will result in a further decrease in volume due to the heat capacity of the crystal. If the rate of *thermal* energy removal is *faster* than the rate of *crystallization*, the latter will not occur and the material can be considered as a supercooled liquid, while the volume follows the line b–e in Fig. 1. The

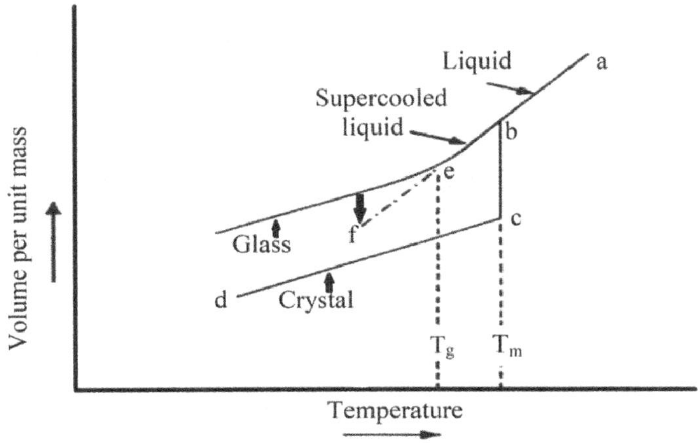

Fig. 1 Relation between the glassy, liquid and solid states: T_m—melting point or liquidus temperature, T_g—glass transition temperature. Dotted line e–f is the extrapolated line for the equilibrium liquid

viscosity of the supercooled melt continues to increase as the temperature is reduced until a temperature is reached, called glass transition temperature (T_g), around which the volume-temperature plot undergoes a significant change in slope ("knee point"). Below T_g, the material can be considered rigid and solid for most practical applications [3, 5]. This temperature range around T_g, the limits of which correspond to the enthalpies of equilibrium liquid and frozen solid, is known as the transition (or transformation) range; here, the viscosity of glass is about 10^{12} Pa s [3, 5]. Glass transition corresponds to the transition from the liquid state to the glassy state (or the reverse one); in a disordered liquid, each structural unit can thermally diffuse over the whole size of the system while, in the glassy state, such a thermal motion is restricted within a cage of nearest neighbors due to ultrahigh viscosity (internal attrition). However, only below T_g the material descended from the frozen melt may be correctly referred to as a glass [5, 14]. At room temperature, the glass exhibits a structure similar to that which was frozen in the melt at T_g. Below T_g a glass, like supercooled liquid, has a higher free energy than a crystalline phase due to the fact that the energy absorbed during fusion of crystals was not completely given up during glassy solidification [7]. Hence, glasses are thermodynamically metastable solids and tend to spontaneously evolve towards the crystalline state (devitrification). However, since structural rearrangements can occur just extremely slowly at temperatures well below T_g, glasses can be considered stable for practical purposes.

The definition and understanding of the glassy state are fundamental to glass science and technology. A concise picture of the complexity of the glassy state, which combines features of both liquids and solids and also brings along its own peculiar characteristics, was effectively provided in 2017 by Zanotto and Mauro [17] as follows: (i) glass is non-crystalline and thus is absent of long-range atomic order which is characteristic of most solid materials; (ii) the structure of glass is very similar to that of its parent supercooled liquid; (iii) glass is a non-equilibrium solid, meaning that the glassy state cannot be described using equilibrium thermodynamics or statistical mechanics and the macroscopic properties of a glass depend on composition and thermal history; (iv) a glass is frozen by quenching from the liquid state, but over longer times it indeed flows and relaxes toward the supercooled liquid state; (v) glasses relax and then crystallize upon continuous heating at any temperature above absolute zero. As a result, an alternative definition for glass was then proposed: "Glass is a non-equilibrium, non-crystalline condensed state of matter that exhibits a glass transition. The structure of glasses is similar to that of their parent supercooled liquids (SCL), and they spontaneously relax toward the SCL state. Their ultimate fate is to solidify, i.e., crystallize" [17]. The last part of this definition was criticized by Schmelzer and Tropin [18] who suggested that "even if this statement would be true, it seems to us not to be reasonable to include such a statement into the definition, as it does not supply any additional information as to what glasses are. In addition, if at all, crystallization proceeds at a perceptible rate for states below the glass transition range also only at time scales exceeding the limits of human history". Another critique was addressed to the flow behaviour of glasses as follows: "Because everything flows on such historical time scales, this feature is not a specific property of glasses and cannot be used to distinguish it from any other states of matter".

As regards crystallization of glasses, it was pointed out [18] that intensive nucleation occurs in a relatively small range of temperatures (with a maximum close to T_g) while the kinetic factor is correlated with viscosity, which significantly increases as temperature decreases. Therefore, perceptible crystal nucleation (i.e., devitrification) does not occur for a variety of glasses under normal conditions on the Earth. Assuming that local nucleation processes do proceed, this does not necessarily lead to full crystallization of the material, since the maximum growth rate is typically located at much higher temperatures [18].

Thermal analysis methods, including differential thermal analysis (DTA), differential scanning calorimetry (DSC) and dilatometry, are commonly used to evaluate T_g since rearrangements that occur at glass transition lead to characteristic jumps of derivative thermodynamic parameters such as the specific heat or the coefficient of thermal expansion [19]. However, the values of T_g obtained by different techniques (e.g. from thermal analysis plots or thermal expansion curves) are not identical. Furthermore, the value of T_g depends on the heating rate applied to obtain those curves. Therefore, T_g is interpreted as the approximate temperature at which the supercooled liquid converts to a solid upon cooling, or, conversely, at which the glass starts behaving as a viscoelastic solid upon heating [5, 6, 11].

2 Theories of Glass Formation

Different cooling rates may be needed to form glasses in different compositional systems. Such experimental observation is behind several attempts to produce an atomic theory of glass formation based on the nature of the chemical bonds and the shape of the structural units involved. Although proposing a structural theory might seem contradictory for a material that is characterized by no long-range, periodic atomic order, some forms of short-range order allow reproducing the same glass from a nominal composition and reliably controlling the overall properties.

Two main approaches [5] were developed to explain glass formation: (i) the structural theories of glass formation, based on structural considerations such as geometry, nature of bond forces, etc., and (ii) kinetic approach, focused on controlling glass formation by changes in processing.

Inorganic glasses are readily formed from a wide variety of precursors, mainly oxides, chalcogenides, halides, salts, and their relevant combinations. There have been many attempts to relate the glass-forming tendency of a material to its molecular level structure. One of the earliest structural concepts was proposed in the 1920s by Goldschmidt [20] who suggested that the ability of an oxide to form a glass might be related to the way in which the oxygen ions were arranged around the cation to form the unit cell of the crystal structure. It can be shown from geometrical considerations that, for an oxide M_xO_y, the coordination number of the cation M is 4 if the ratio of ionic radii R_M/R_O lies approximately within 0.2–0.4. Goldschmidt observed that some glass-forming oxides (e.g. SiO_2, GeO_2 and P_2O_5) typically exhibit a tetrahedral arrangement in the crystalline state and suggested that this might be a criterion of glass-forming ability. However, this theory was later found to be incomplete, being unable to adequately explain a variety of systems. In this regard, although all ionic glass formers satisfy Goldschmidt's rule, there are many systems that satisfy it but are not glass formers (e.g., BeO and most halides).

In the early 1930s, Zachariasen [21] formulated the famous random network theory, according to which glass formers are cations that have high valences (≥ 3) and can create three-dimensional networks of polyhedra. For instance, in silicate glasses, oxygen networks are formed by polymerization of polyhedra. By postulating that the oxygen polyhedra found in the oxide crystals would also be present in the glasses, Zachariasen introduced the concept of a continuous random network structure for a glass, where periodic structural arrangement is prevented by random orientations. He proposed that the structure of glass was similar to that of a crystal but with a larger lattice energy resulting from the disordered arrangements of polyhedral units, yielding a random network that lacks long-range periodicity (as shown schematically in Fig. 2). This was also demonstrated by the absence of sharp X-ray diffraction lines for glasses, the pattern of which is typically characterized by a broad halo. According to his experiments, Zachariasen suggested four rules for a structure to favour glass formation:

1. no oxygen atom must be linked to more than two cations;
2. the number of oxygen atoms surrounding any given cation must be small (typically 3 or 4);
3. oxygen polyhedra can share only corners, not edges or faces;
4. at least three corners of each oxygen polyhedron must be shared with other polyhedral.

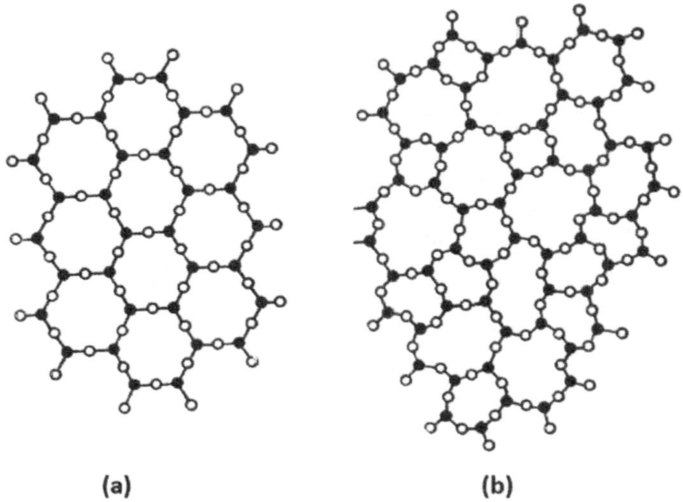

Fig. 2 Schematic representation of a two dimensional structure for **a** crystalline silica and **b** vitreous (glassy) silica (legend for both pictures: silicon – ● and oxygen – ○). A fourth oxygen would be located above each cation in the three-dimensional structures

The condition no. 4 was specifically introduced to ensure that the network would be three-dimensional (although certain glasses can exist in structures describable in fewer dimensions).

These four rules lead to the open structures that can accommodate a distribution of inter-polyhedral bond angles which are associated with the loss of long-range structural order when a crystal forms a glass. Zachariasen concluded that oxides such as SiO_2, B_2O_3, GeO_2, P_2O_5 and As_2O_3 are glass formers because oxygen coordination number for them is 3 or 4, which indicates that they can easily build up a characteristic network structure consisting of triangular or tetrahedral units. Diffraction studies made by Warren [22] and later by Wright [23] confirmed Zachariasen's prediction that glasses and crystals have similar short-range polyhedral structures but different long-range polyhedral arrangements. The Dietzel's field strength consideration (i.e. Z/a^2 ratio, where Z is the charge of the cations and a is the cation-oxygen distance) represented an extension of the Zachariasen's network theory [7].

Along with the development of the random network hypothesis, the Lebedev's crystallite theory was also developed and adapted by a great number of glass researchers. According to Lebedev, silicate glasses are composed of assemblages of microcrystalline clusters of SiO_2, called crystallites [7].

The basic concepts of glass structural theories proposed by Zachariasen and Lebedev have been subjected to certain modifications which brought them closer to each other, but they still remained basically different [7].

Jiang and Zhang proposed a phase diagram model to explain various glass structures [24] and pointed out that neither the "random network" theory nor the "crystallite" hypothesis is a universal structural model for glass. Based on measurements from infrared spectroscopy, Raman and nuclear magnetic resonance (NMR), as well as the physical properties of relevant compounds in phase diagrams (e.g. density and refractive index), Jiang and Zhang concluded that glasses and crystalline congruent compounds exhibit similar NMR data and spectral features in a phase diagram. In their phase diagram structural approach, binary glass is considered to be a mixture of the melts of the two nearest congruent compounds in a binary phase diagram; similarly, ternary glass is composed of a mixture of the three nearest congruent compounds in a ternary phase diagram. The structures and properties of the resulting glass can be predicted and calculated from the relevant characteristics of either two or three congruent compounds [24].

Other theories have also been proposed for predicting the easiness of glass formation based on some known data, such as the bonding strength between the constituent elements, the electronegativity values and the existence of deep eutectics in the phase diagrams [5]. Specifically, Shelby [5] proposed the fundamental law of structural models as a further guide to the testing of any complete structural model for glasses, which should take into account: (i) coordination number of all network cations, (ii) distribution of bond angles and rotations, (iii) connectivity of all network units, (iv) dimensionality of the network, (v) nature of any intermediate range order, (vi) morphology, (vii) bonding characteristics (field strength, bond strength, site specific bonding), (viii) nature of the interstitial or free volume, (ix) role of minor constituents, impurities and defects. However, a limitation of structural approaches is that they are unable to take into account the thermal history of the melt [20, 21].

Especially in glass industry, special attention is addressed to the purification of the glass compositions from impurities and mineralisers which might promote crystal nucleation and, thus, unwanted devitrification. Multiple melting cycles, superheating and centrifugation are typically applied to avoid the effect of impurities on nucleation, in order to achieve solidification of liquid phases without undergoing crystallization at a quite considerable degree of supercooling [7].

Glass forming ability is defined in terms of resistance to crystallization of a melt upon cooling and varies with the variation in composition and size of the melt. Hence, the difference in glass forming ability of two closely related compositions can be provided by the kinetic approach to glass formation as discussed by Shelby [5]. A similar concept to glass forming ability is glass stability, which can also be thought in terms of glass resistance to crystallization. Glass stability is often quantified by the difference in temperature between the onset of crystallization (T_c) and T_g for a sample heated at a specific linear rate (i.e., $T_c - T_g$). Alternatively, the temperature of the first maximum or peak (T_p) in the DTA or DSC plot can be considered instead of T_c to calculate this temperature difference [5].

In the 1970s, Hrubÿ [25, 26] proposed another parameter to measure glass stability, defined as:

$$K_H = (T_c - T_g)/(T_m - T_c) \tag{1}$$

According to Hrubÿ, the larger K_H of a certain glass, the greater its stability against crystallization upon heating.

Turnbull and Cohen [27] proposed the determination of kinetic stability on cooling experiments through the steady-state nucleation rate. Gutzow et al. [2] related the glass stability to the non-steady-state time lag, τ. These kinetic approaches assume that one of the three considered parameters, i.e. crystal nucleation rate, crystal growth rate or τ, is dominant over the other two, the contributions of which can be accordingly neglected [28]. Uhlmann et al. [29] considered crystal nucleation and growth rate simultaneously, formulating a kinetic criterion for vitrification; the kinetic theory of glass formation was extended to include non-steady state effects and heterogeneous nucleation. Weinberg et al. [30, 31] found that the volume fractions transformed and the resulting critical cooling rates (R_c) are quite sensitive to the method of calculation. For instance, the nose method based on isothermal Time–Temperature-Transformation (TTT) curves overestimates R_c by up to one order of magnitude. It was also demonstrated that R_c is highly sensitive to the main physical properties dictating the nucleation and growth kinetics, i.e. crystal liquid surface energy, thermodynamic driving force and viscosity [32]. Weinberg [33] also integrated the equation of overall crystallization kinetics to estimate and compare the criteria for vitrification on cooling and glass stability against crystallization on heating, and apparently concluded that glass forming ability and glass stability are, surprisingly, ill-related concepts.

Using experimental values of crystal nucleation and growth rates for four glasses that nucleate in the bulk, Cabral et al. [34] calculated the critical cooling rates for glass formation (R_c) by the TTT method. It was shown a correlation between the Hrubÿ parameter of glass stability (K_H) and glass forming ability. Avramov et al. [28] repeated and extended the calculations initially done by Weinberg [33] and demonstrated that glass forming ability and glass stability are indeed directly related quantities, as already suggested by the experimental results reported by Cabral et al. [34].

Glass stability against crystallization deserves to be taken into account not only during the initial melting procedure of the material but also when massive glass products need to be fabricated, for example, by applying high-temperature treatments on glass powder compacts well above T_g (sintering). It is well known that viscous flow sintering of glass particles can effectively occur when the surface tension is high and the viscosity is low, i.e. when the glass shows its lower viscosity, without undergoing any crystallization process [35, 36]. Lara et al. [37] introduced a parameter that quantifies the sinterability of glass, defined as:

$$S_c = T_c - T_{MS} \tag{2}$$

where T_{MS} is the temperature of maximum shrinkage determined from hot-stage microscopy (HSM).

This parameters measures the competition between glass sintering and crystallization that concurrently occur during heating (sinter-crystallization). The temperature difference between onset of crystallization and maximum shrinkage (i.e., densification) is a measure of the ability of sintering versus crystallization: in other words, the greater S_c, the more independent the kinetics of the two processes. A general rule can be proposed for the interpretation of S_c: if $S_c < 0$, only partial densification is achieved before crystallization begins; otherwise, if $S_c \geq 0$, full densification occurs prior to crystallization [38]. Therefore, higher values of S_c are related to higher final densities, which indicate a better sintering behaviour leading to a more efficient densification of the final glass sample. Other details on these topics are provided in the Sect. 7.

3 Methods for Glass Production: An Overview

The glasses used over most of our history have been silicate glasses traditionally formed by cooling from a melt. Glass processing by melting comprises four main steps: (i) batch preparation, (ii) melting, (iii) fining and (iv) homogenization. Batch preparation involves the selection of raw materials, batch calculation (i.e. assessing the amounts of precursors to achieve a given glass composition), weighing and mixing these materials to get homogeneous starting batch. A series of transformations occur during the conversion of a batch, usually composed by solid powders of oxides, carbonates and other salts, to a melt upon heating, including the release of physically-adsorbed and chemically-bonded water, solid state reactions with formation of double carbonates such as $Na_2Ca(CO_3)_2$ and $CaMg(CO_3)_2$, thermal decomposition of carbonates, sulphates and nitrates, liquid phase formation (e.g. due to melting of eutectic mixtures or melting of some substances). Further heating leads to the increase in volume fraction of a liquid phase, synthesis of new compounds such as silicates (e.g. Na_2SiO_3, Ca_2SiO_4, $Na_2O \cdot 3CaO \cdot 6SiO_2$) and aluminates (e.g. $Na_2Al_2O_4$, $CaAl_2O_4$), dissolution of refractory components such as silica and alumina in the liquid phase. Removal of gas bubbles (i.e. fining) is usually required in order to reduce the heterogeneity of the melt; furthermore, these gaseous inclusions are unwanted in commercial products due to both antiesthetic and functional reasons (internal pores reduce mechanical performance and can be the trigger for crack propagation). Then, additional time for diffusion processes and creation of convection flow via stirring in the melt is usually needed to convert the melt to a homogeneous liquid; gas bubbling through the melt can be performed at this stage if obtaining glass foams is the goal.

The melting temperature varies depending on the glass composition: common silica-soda-lime glasses typically require above 1400 °C to be melted, but phosphate glasses can be produced at significantly lower temperatures (even below 1000 °C in some cases).

In order to obtain glass products of desired shape and size, the melt can be poured into molds replicating the negative of the object. Alternatively, the melt can be quenched into water to obtain a "frit", which is composed by small chunks of glass that are thermally-shocked and reduced in fine pieces; this approach is very useful in the production of glass powder for further processing (e.g. milling followed by pressing or coating and final sintering). Glass fibers can also be drawn starting from a melt-derived glass cylinder ("preform") heated within a specific temperature range above T_g by using a proper equipment (drawing tower).

Melting is indeed the most common, easy and relatively quick method to produce glass and is widely used for commercial mass-production. However, if high purity of the final product is a major goal, glass can also be synthesized via a "wet" method, i.e. the sol–gel process. Sol–gel silicate glasses are inherently nanoporous (or, more specifically, mesoporous) materials (pore size within 2–50 nm) exhibiting a larger specific surface area (well above 50 m^2/g) as compared to melt-derived glass of analogous composition (specific surface area less than 1 m^2/g) [39, 40]. The sol–gel method is classified as a chemistry-based synthesis route in which the polymerization reactions in a colloidal suspension (sol) containing appropriate precursors of glass oxides leads to sol gelation at room temperature. Sol–gel glasses are usually produced by hydrolysis and poly-condensation of alkoxide precursors [e.g. tetraethyl orthosilicate (TEOS) or tetramethyl orthosilicate (TMOS)] followed by aging and drying under ambient atmosphere. The precursors of other oxides (e.g. CaO) are introduced in the sol by adding and solubilizing appropriate salts (e.g. calcium nitrate decahydrate). The overall sol–gel process can be divided in 5 major steps. During step 1, liquid and solid reagents are mixed together at room temperature to induce the formation of covalent bonds between the elements. Hydrolysis and poly-condensation occur simultaneously while the sol undergoes homogenization. Then, during gelation at room temperature (step 2), viscosity increases due to the formation of a three-dimensional interconnected network. Poly-condensation continues during aging at around 60 °C (step 3) and is accompanied by decrease in porosity and improvement in mechanical strength. Both these aspects are fundamental to avoid cracking during drying (step 4), which is typically performed at 120–150 °C to eliminate the liquid phase from the pores. The dried gel is then stabilized by high-temperature thermal treatment in the range of 500–700 °C (step 5), which is lower than the melting temperature typically required to produce melt-derived silicate glasses. The major limitations of sol–gel approach compared to melting route include the long time required for the whole synthesis (several days), the poor mechanical properties of the final glass monolith due to inherent nanoporosity, and the need for a careful control on the synthesis parameters (e.g. pH, environmental temperature) which can deeply affect the properties of the final product (homogeneity, phase segregation during the synthesis, crystallization).

At present, inorganic glasses can be produced starting from many compositions based on silica, phosphorus oxide, boron oxides, halides or chalcogenides. However, silicate glasses are the most important regarding commercial impacts because they typically have excellent transparency, good chemical durability and can be produced by using inexpensive natural precursors.

4 The Role of Oxides on Glass Structure, Properties and Glassmaking

According to the random network theory, Zachariasen [21] introduced the basis for the most commonly-used models for glass structures. Depending on the role in glass production, oxides are usually classified in three groups: (i) network formers, such as SiO_2, B_2O_3, P_2O_5, having oxygen coordination numbers equal to 3 or 4 and tending to produce the basic cross-linked "polymeric" glass structure; (ii) network modifiers, which typically are alkaline and alkaline-earth oxides such as Na_2O, K_2O and CaO, having coordination number equal to 6 or more and generally tending to reduce the degree of polymerization and viscosity; and (iii) intermediate oxides, with metallic cations such as Al^{3+}, Zn^{2+}, Pb^{2+} and Ti^{4+}, having intermediate coordination of 4–6 and acting either as network formers or modifiers, depending on the glass composition [5, 12, 41].

Oxides with high coordination numbers and relatively weak bonds (i.e., the network modifiers) alter the glass-forming network by replacing stronger bridging oxygen (BO) bonds between glass-forming polyhedra with weaker, non-bridging oxygen (NBO) bonds, thereby modifying the polyhedral structure [5]. Figure 3 shows a schematic 2D representation of the random network of an alkali-silicate glass. The network modifiers are important constituents in most technological glasses because they are useful to decrease the melting temperature, with an obvious impact on energy consumption and related costs, and control many properties, such as chemical stability. Modifiers are commonly used to facilitate the glass production at lower temperatures because they promote the viscosity decrease by disrupting the network of the glass melt [12]. Silica glass (without any modifiers) is difficult to process

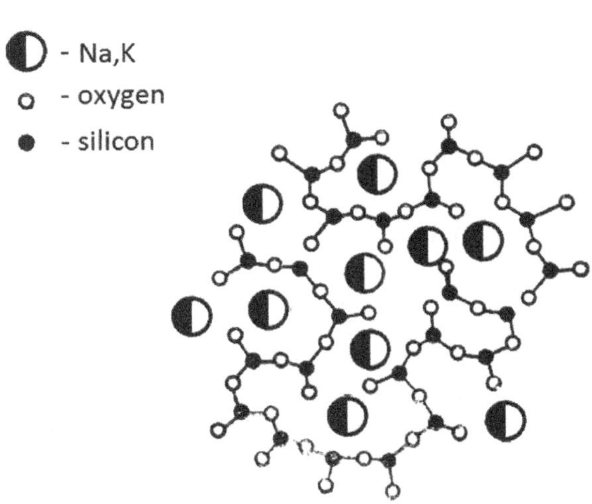

Fig. 3 Schematic representation of atomic arrangement in an alkali-silicate glass. A fourth oxygen would be located above each cation in the three-dimensional structure

because its melting temperature is about 1713 °C, which corresponds to cristobalite-liquid equilibrium. If 25 wt% of Na_2O is added to SiO_2, the liquidus temperature is lowered to only about 793 °C, which is a significant advantage from a technological viewpoint [42]. Alkaline metal ions (e.g. Na^+) are monovalent, mobile and allow ion migration while alkaline-earth ions, being typically bivalent (e.g. Ca^{2+}) and thus electrically compensated by two NBOs, are relatively immobile and can hinder the diffusion of other ions, especially the alkaline ones, thus improving the chemical resistance of the glass [12]. For this reason, most of important commercial glasses are based on silica-soda-lime compositions comprising SiO_2 (network former), Na_2O and CaO (alkaline and alkaline-earth modifiers, respectively).

Intermediates such as Al_2O_3, WO_3, TiO_2, ZrO_2 and SeO_2 can link the continuous network or locally replace cations of formers (e.g. Si^{4+} with Al^{3+}), but are unable to form a glass network by themselves. Introducing Al_2O_3 in a silicate glass can yield better mechanical properties (hardness, strength and elastic modulus) as well as higher chemical stability and durability of the material.

The relative concentration of BO and NBO plays a key role on the structure and properties of glasses [43]. According to the number of BO in a tetrahedral unit of silicate glasses, the following scenarios can be considered:

1. BO = 4 (i.e. [O]/[Si] = 2): each BO is shared by two silicon atoms and the network is three-dimensional with all four corners bridging;
2. BO = 3 (i.e. [O]/[Si] = 2.5), the network is two-dimensional with three corners bridging (note that some tetrahedra may be linked to four others and some therefore to less than three, the said number being the average value over the network);
3. BO = 2 (i.e. [O]/[Si] = 3), the network is formed by one-dimensional chains with one corner bridging;
4. BO < 2 (i.e. [O]/[Si] > 3), the network is composed of individual SiO_4 tetrahedral units, some of them being bound together.

The rigidity of the glass network gradually decreases by replacing BO with NBO until only individual isolated tetrahedra remain. The notation Q^n is usually preferred among glass scientists, where n is the number of BO in a tetrahedron; accordingly, it is possible to refer to (1) Q^4, (2) Q^3, (3) Q^2 and (4) Q^1 or Q^0 structures, respectively.

Considering an oxide glass of general composition $(A_2O)_x(SiO_2)_{1-x}$, where A is the alkaline metal, if x = 0 (i.e. pure silica) only BO exist (Q^4); if the alkali concentration increases, a dramatic decrease of BO and an increase of NBO occur accordingly [12]. The intermediate oxides have coordination numbers and bond strengths between the network formers and network modifiers and tend to have an intermediate effect on glass properties [44].

In the case of oxide glasses, the short-range structure can be extremely well-defined in terms of the coordination polyhedra of the network-forming cation such as silicon. These glasses are characterized by predominantly heteropolar bonding between network-forming cations and oxygens. Bond lengths and angles in the first coordination shell of oxygens around these cations vary only over a narrow range. Glass-forming cation-oxygen polyhedra, like SiO_4 units, are usually corner-linked through BO and form a three-dimensional connected network. The properties of glasses mainly rely on its structure which, in the case of silica glass, consists of well-defined SiO_4 tetrahedra connected to another neighbouring tetrahedron through each corner (Fig. 2b). Neutron diffraction studies revealed that the Si–O distance in the tetrahedron is about 0.16 nm and the shortest O–O distance is about 0.26 nm, which are the same dimensions as found in crystalline silica (quartz). The inter-tetrahedral (Si–O–Si) bond angle distribution is centred around 143° but is much broader than that found in crystalline silica, thus producing the loss in long-range order already shown in Fig. 2b [23, 45].

Similarly to pure-silica glass, the structure of alkali-silicate glasses also consists of a network of SiO_4 tetrahedra, but some of the corners are now occupied by NBO that are linked to the modified polyhedra (Fig. 3). Increasing the concentration of modifiers leads to increasing the relative fraction of NBO associated with the glass network, which leads to reducing T_g and melt viscosity as well as increasing the thermal expansion coefficient and ionic conductivity [46].

The changes in the silicate glass network, and hence the compositional dependence of many of the glass properties, can be described by taking into account the relative fractions of BO and NBO or, alternatively, the concentrations of the different Q^n units [12]. The rigidity of the network decreases gradually by replacing BO by NBO until only individual isolated tetrahedra remain (Q^0). Glasses containing less than 10 mol% of alkaline oxides are considerably more difficult to melt due to high viscosity [5]. Moreover, alkali-deficient glasses are prone to phase separation and devitrification on a scale of 0.1–1 μm [46]. Modifiers disrupt the network and are used to lower the viscosity of the glass melt and, hence, to facilitate glass production at lower temperatures.

The viscosity-temperature relationship is one of the key point in determining the easiness of glass formation of any melt. Glass formation is favored when crystallization is discouraged by the kinetic barrier to atomic arrangement, provided that (i) the viscosity is very high at the melting temperature of the crystalline phase which would form from the melt and/or (ii) the viscosity increases very rapidly as temperature decreases [5]. In commercial glasses, melting usually takes place at a viscosity of $\leq$ 10 Pa s and the viscosity at working point (i.e. when a glass object is delivered to processing/shaping) is about 10^3 Pa s. The softening point corresponds to the value of viscosity that is sufficiently high to prevent deformation under glass' own weight, while the temperature range between the working point and softening point is referred as the working range: melts that demonstrated a wide working range are called as "long glasses" and those with a short working range are known as "short glasses". As soon as a glass object is formed, it is subjected to annealing to release internal stresses: the annealing point typically corresponds to a viscosity range of 10^{12}–$10^{12.4}$ Pa s and is defined as the temperature at which the stress is substantially relieved in a few minutes [5].

It cannot be ignored that some glass formulations that are being cooled to a temperature suitable for forming may partially devitrify, thus developing small crystals embedded in the amorphous matrix. In order to avoid this unwanted phenomenon during glass manufacturing processes, the glass composition should be designed so that the temperature interval of crystallization is narrow, the rate of crystal growth is low and the T_p value is significantly higher than glass-working temperature.

In many systems, the region of glass formation coincides with a region of the phase diagram where the liquidus temperature is low. The silica-soda-lime (SiO_2–Na_2O–CaO) system, which is very common in glass technology, might serve as an instructive example [47]. Morey and Bowen [48] studied the phase relations of the pseudo-binary system $CaO \cdot SiO_2$–$Na_2O \cdot SiO_2$. Later Morey [49] investigated the phase equilibria in the silica-rich region (>50 wt% of SiO_2) of the $CaO \cdot SiO_2$–$Na_2O \cdot SiO_2$ system in greater detail. Morey and Bowen found four ternary compounds: $Na_2O \cdot 3CaO \cdot 6SiO_2$ (devitrite), $Na_2O \cdot 2CaO \cdot 3SiO_2$, $2Na_2O \cdot CaO \cdot 3SiO_2$ and $Na_2O \cdot CaO \cdot SiO_2$. Figure 4 shows the silica-rich region of the silica-soda-lime phase diagram according to Morey and Bowen [48, 49].

New compounds, i.e. $4Na_2O \cdot 3CaO \cdot 5SiO_2$ and $Na_2O \cdot 2CaO \cdot 2SiO_2$, were reported by Segnit [50] who studied the (CaO, Na_2O)-rich part of the system and showed that Na_2O forms solid solutions with dicalcium silicate. Toropov and Arakelyan investigated two compounds, $2Na_2O \cdot 8CaO \cdot 5SiO$ and $2Na_2O \cdot 4CaO \cdot 3SiO_2$, and demonstrated that they are solid solutions of alkali-silicates in calcium orthosilicate [51]. Table 1 collects the invariant points of the system according to Morey and Bowen [48, 49].

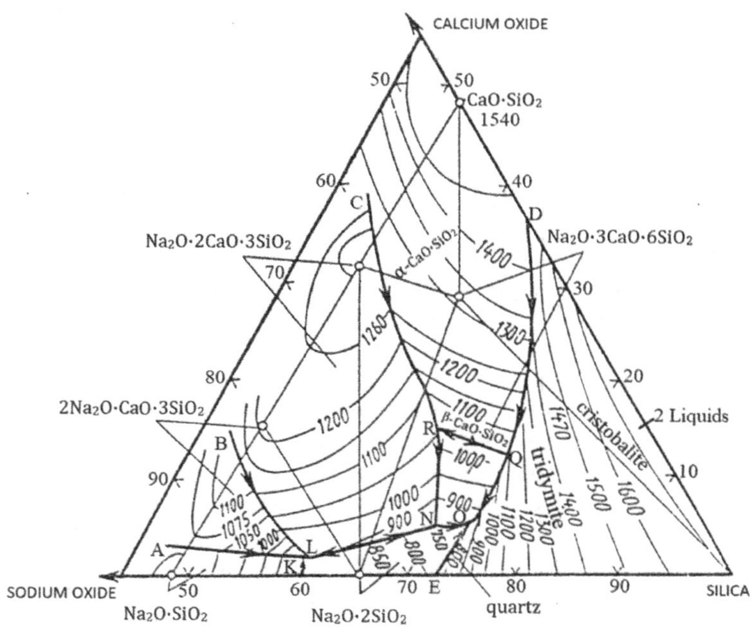

Fig. 4 Silica-rich region of the silica-soda-lime phase diagram (in wt%) [48, 49]

Table 1 Invariant points in the SiO_2–Na_2O–CaO system (see Fig. 4)

Relevant points in the diagram	Equilibrium phases	Process	Composition (wt.%) Na_2O	CaO	SiO_2	Temperature (°C)
(K)	$Na_2O \cdot 2SiO_2 + Na_2O \cdot SiO_2 + 2Na_2O \cdot CaO \cdot 3SiO_2 + L$	Eutectic	37.5	1.8	60.7	821
(L)	$2Na_2O \cdot CaO \cdot 3SiO_2 + Na_2O \cdot 2CaO \cdot 3SiO_2 + Na_2O \cdot 2SiO_2 + L$	Tributary reaction point	36.6	2.0	61.4	827
(N)	$Na_2O \cdot 2SiO_2 + Na_2O \cdot 2CaO \cdot 3SiO_2 + Na_2O \cdot 3CaO \cdot 6 SiO_2 + L$	Tributary reaction point	24.1	5.2	70.7	740
(O)	$Na_2O \cdot 3CaO \cdot 6SiO_2 + Na_2O \cdot 2SiO_2 + SiO_2 + L$	Eutectic	21.3	5.2	73.5	725
(Q)	$SiO_2 + CaO \cdot SiO_2 + Na_2O \cdot 3CaO \cdot 6SiO_2 + L$	Tributary reaction point	13.7	12.9	73.4	1035
(R)	$Na_2O \cdot 3CaO \cdot 6SiO_2 + Na_2O \cdot 2CaO \cdot 3SiO_2 + CaO \cdot SiO_2 + L$	Tributary reaction point	19.0	14.5	66.5	1030

The ternary eutectic point at which quartz coexists with $Na_2O \cdot 3CaO \cdot 6SiO_2$, $Na_2O \cdot 2SiO_2$ and liquid was established to be at 725 °C. The ternary SiO_2–Na_2O–CaO diagram is the basic phase diagram for commercial glass compositions with SiO_2, Na_2O and CaO as major constituents and can also be used to describe, under certain approximations, multicomponent glass systems with some additional oxides (> 3) as minor constituents. Additional modifiers and intermediate may be required to finely modulate the glass forming ability and final properties. For example, addition of Al_2O_3 increases the chemical resistance of the glass as well as the glass stability against crystallization due to the increase network connectivity. Essentially, introduction of Al_2O_3 in a silicate network yields the substitution of Si^{4+} with Al^{3+} cations so that AlO_4 tetrahedra join the network formed by SiO_4 tetrahedra, provided that an additional Na^+ cation occupies an interstitial hole in order to maintain charge neutrality.

Many of commercial glass compositions (Table 2) are located in the region of primary phase formation of devitrite and wollastonite, i.e. inside the triangle with

Table 2 Representative commercial glass compositions (wt%) [47]

Type of glass	SiO_2	CaO	MgO	Na_2O	K_2O	Al_2O_3
Window	71	10	3	14	1	1
Food container	72	3	7.5	15	1	1.5
Plate	71	11	4	13	–	1

the vertexes corresponding to the compounds $Na_2O \cdot 2SiO_2$, $Na_2O \cdot 3CaO \cdot 6SiO_2$ and SiO_2. All the melts with the compositions located in that triangle will solidify in the ternary eutectic with composition 21.3 wt% Na_2O, 5.2 wt% CaO and 73.5 wt% SiO_2 at 725 °C (i.e. in the point "O" in Fig. 4).

Resistance to crystallization can also be improved when CaO is partially substituted by MgO in SiO_2–Na_2O–CaO glasses [52]. This is demonstrated by the data reported in Table 3: for example, looking at the glasses containing 18 wt% of Na_2O (compositions from no. 19–24), one can see that the decrease in CaO content from

Table 3 Compositional effect on resistance to crystallization [52]

Composition No.	Na_2O/CaO weight ratio	Crystallization range (°C)	Maximum crystal growth rate, (μm/min)	Crystalline phases formed
1	12/6	800–1350	90	Quartz
2	12/8	800–1300	60	Quartz
3	12/10	900–1250	40	Quartz
4	12/12	750–1000	30	Devitrite
5	12/14	850–1150	15	Devitrite, Quartz
6	12/16	900–1100	38	Devitrite, Quartz
7	14/6	900–1200	50	Quartz
8	14/8	900–1200	40	Quartz
9	14/10	850–1200	30	Devitrite, Quartz
10	14/12	900–1050	10	Devitrite
11	14/14	800–1100	37	Devitrite, Quartz
12	14/16	800–1100	70	Devitrite, Quartz, Wollastonite
13	16/6	950–1150	35	Quartz
14	16/8	900–1100	10	Quartz
15	16/10	800–1000	10	Quartz
16	16/12	800–1000	20	Devitrite, Quartz
17	16/14	800–1000	75	Devitrite
18	16/16	800–1000	70	Devitrite, pseudo-Wollastonite
19	18/6	800–1000	10	Wollastonite
20	18/8	800–950	5	Quartz
21	18/10	850–950	5	Quartz
22	18/12	800–1000	30	Devitrite, Quartz
23	18/14	800–1100	100	Devitrite
24	18/16	800–1100	200	Devitrite, pseudo-Wollastonite

16 to 8 wt%, which is accompanied with an equivalent increment in MgO, leads to a significant decrease in maximum crystal growth rate from 200 to 5 μm/min. Similar trends can also be observed for the other compositions listed in Table 3.

Chemical composition of glasses is determined by the type of raw materials used to prepare a batch. Glasses are either produced from high-quality, chemically pure components or from a mixture of less pure materials [5]. While glasses for research and high-tech applications are typically produced from expensive, high-purity chemicals, bulk commercial glasses are produced mostly from natural materials. Specifically, alkaline, alkaline-earth oxides and silica are usually introduced through using soda ash, potash, nepheline syenite, calcite, dolomite, magnesite, and silica sand, which may contain some impurities such as Fe_2O_3, TiO_2, Mn_2O_3, MnO_2 and Cr_2O_3. Therefore, a complete chemical analysis of the glass could reveal not only major (i.e. SiO_2, Na_2O, CaO) and minor constituents (e.g. K_2O, MgO and Al_2O_3), but non-negligible traces of impurities. In order to approximate the composition to a ternary system, the major constituents might be arranged into 3 groups, i.e. alkali oxides (from monovalent cations), alkaline-earth oxides (from bivalent cations) and glass formers (silica and others) [47]: compositions having 72–74 wt% of glass formers and 26–28 wt% of alkaline/alkaline-earth oxides which are located near the upper area of the boundary OQ in Fig. 4 exhibit improved glass stability and high chemical resistance [52].

Most glasses for both common and high-tech applications are based on silicate systems, which may also incorporate small amounts of other formers. However, glasses having other major formers have shown great suitability for special applications, such as borate glasses in biomedicine and phosphate glasses in optics; these topics will be discussed in dedicated chapters of this book.

5 From Glasses to Glass–Ceramics: A Short Historical Overview

It has been known for a long time that glasses can be crystallized to form polycrystalline ceramics. One of the early evidences of this knowledge, initially relying on an empirical basis than on a clear theoretical background, dates back to the third century AD. Chemical and microstructural analyses carried out on mosaic tesserae of the Roman Age revealed that the white colour of these archaeological materials was due to the presence of crystalline phases nucleated and grown in the glass matrix as a consequence of a thermal treatment. It was supposed that Roman craftsmen bought relatively cheap and transparent silica-soda-lime glass and, in a second manufacturing step, deliberately heat-treated this colourless material to convert it to a glass–ceramic to be used in the final form of opaque white tesserae [53].

Apparently, this knowledge was lost during the Medieval Age and only after 14 centuries partially-crystallized glasses were produced again according to a controlled treatment. In the first half of the eighteenth century, René-Antoine Ferchault de

Réaumur tested certain glass compositions such as cut glass, porcelain glazes, mirror and window glass in his series of novel experiments on porcelain production. In particular, hollow glass articles, filled with a mixture of silica sand and calcined gypsum and embedded in the same mixture of materials, were heat-treated at temperatures corresponding to red heat for above 24 h [54, 55]. In the course of the annealing process, the originally transparent glass changed into a white stone-like product, which was visually very similar to porcelain. However, unlike conventional porcelain bodies, the new product possessed a high resistance to thermal shocks. The fracture surface of Réaumur's porcelain exhibited thin needle-like crystals grown from its periphery towards the center and embedded in a glass matrix [54]. Réaumur's porcelain was the first commercial material produced in the Modern Age by applying a roughly-standardized heat treatment on glass; it paved the way to future glass–ceramic production, although the crystallization process had not been fully controlled yet [10, 11]. Two centuries later (around 1930), Becker [54] proposed the crystallization of $4Na_2O \cdot 3CaO \cdot 10SiO_2$ glass to produce refractory vessels for glass and metal industries. A breakthrough in this field of research came out after the pioneering work of Donald Stookey at the Corning Glass Works (USA) when the theory of glass phase separation was proposed [56–58]. Glass–ceramics (GCs) were "formally" discovered in 1953—and this was actually the third discovery, after the accidental discoveries made by ancient Roman craftsmen and Réaumur. However, also in this case the re-discovery of GCs was somewhat accidental. A sample of photosensitive lithium disilicate glass with precipitated silver particles was placed inside a furnace for undergoing heat treatment at 600 °C; however, a problem in the controller let the temperature to increase to 900 °C. Stookey expected to obtain a blob of melted glass and a ruined furnace, but he discovered that the lithium disilicate glass had transformed into a milky white plate. Another accident led to an extraordinary observation: when Stookey tried to remove it from the supporting substrate, the sample slipped from the tongs and bounced on the floor instead of shattering [59]. This phenomenon was explained further by formation of crystals in a matrix of glass, resulting in a GC material possessing properties of both glass and ceramics. Crystallization is the process by which well-ordered or regular periodic crystalline structures are produced from a non-periodic structure, like that of common glass [60, 61]. In its simplest form, crystallization is observed when a melt of a single pure substance or compound is cooled. According to Tammann [60], the crystallization process is actually the sequence of two independent processes: (i) nucleation, which corresponds to the formation of crystallization centers and (ii) crystal growth from such centers. The theory of nucleation and crystal growth was expanded by Stookey to develop glass–ceramics in a controlled way [10]. He postulated that (i) almost all known glasses are metastable at room temperature as compared to the more thermodynamically-stable crystalline state and (ii), at least theoretically, they are prone to be crystallized provided that appropriate catalyst crystals are introduced into the melt [58]. Stookey literally said: "I am most proud of opening up a whole new field of science—the nucleation of crystallization of glass—that produced all kinds of new crystalline products with so many different useful properties" [56].

6 Production of Glass–Ceramics

Nucleation is the key factor for controlling crystallization in GCs. Crystal nuclei must be present in the glass matrix so that the crystal growth can start. Nucleation involves the formation of long-range atomic order regions (also called seeds or embryos), which are normally present in melted materials or in supercooled liquids. The embryos turn into nuclei when they attain a critical minimum size that makes them capable of spontaneously developing into large particles of the stable phase. This was initially observed in photosensitive glasses where noble metal crystallites serve as catalysts for the crystallization of lithium metasilicate crystals [58].

Two distinct types of nucleation can occur: (i) homogeneous nucleation or (ii) heterogeneous nucleation [5, 10]. In the process of homogeneous nucleation, the first small seeds have the same composition as the final crystals which will grow upon them, while in heterogeneous nucleation the nuclei are chemically different from the formed crystals. Homogeneous nucleation is believed to occur due to local fluctuations of density and kinetic energy in the absence of any foreign boundaries [10]. On the contrary, heterogeneous nucleation involves the presence of foreign boundaries such as substrates (external boundaries, for example in coatings) and grain boundaries (internal boundaries). Heterogeneous nucleation is also called catalyzed nucleation and is used for the development of most glass–ceramic systems. The theory of nucleation involves a thermodynamic parameter, known as Gibbs free energy (G), which depends on other thermodynamic parameters: the enthalpy (H), i.e. the internal energy of the system, and the entropy (S), i.e. a measurement of the disorder of atoms or molecules in the system [5]. It is worth underlining that, from a thermodynamic viewpoint, a transformation occurs spontaneously only if $G < 0$.

Silicate GCs are the product of controlled crystallization of a SiO_2-based liquid between T_g and the melting point, T_m, of the major crystalline phase [61, 62]. GCs are usually obtained via two steps: first, a glass is formed by a conventional glass-manufacturing process; then, the glass product is shaped, cooled and reheated above T_g. The second step is sometimes repeated as a third step. A very fine-grain microstructure of GCs can be achieved by carefully selecting the crystal nucleation and growth conditions. Relatively coarse-grained GCs are produced during the direct cooling path of a melted viscous liquid; however, this method is seldom applied to induce and control internal crystallization.

Stookey mentioned that the efficiency of a given catalyst crystal depends on a number of factors, including the similarity between its crystal structure and that of the crystal phase to be nucleated [58]. He succeeded in achieving volume crystallization of the Li_2O–K_2O–Al_2O_3–SiO_2 parent glass composition [10] by incorporating metal ions, such as Ag^+, acting as nucleation agents. The typical composition of the parent glass was 80 SiO_2, 4 Al_2O_3, 10.5 Li_2O, 5.5 K_2O, 0.02 CeO_2, 0.04 AgCl (wt%) [10]. Under exposure to UV light, Ce^{3+} was oxidized to Ce^{4+} while metal ions were reduced to the atomic state (e.g. $Ce^{3+} + Au^+ \rightarrow Ce^{4+} + Au^0$), thus forming heterogeneous nuclei for crystallization of lithium metasilicate during subsequent heat treatment. Lithium metasilicate easily dissolves in dilute HF, thus allowing high-precision patterned GCs to be prepared: this was the case of commercial Fotoform®, which is useful in fluidic devices, display screens, lens arrays, magnetic recording head pads, charged plates for inkjet printing and other high-tech devices [63]. Fotoceram®, having $Li_2Si_2O_5$ and quartz as crystalline phases, was the commercial evolution of Fotoform®. A more comprehensive overview of the physics of nucleation can be found in Höland and Beall's book on glass–ceramics [10] as well as in Borrelli's book on photosensative glasses and glass–ceramics [64].

The fundamental research carried out by Stookey stimulated further advances related to GC development while nucleating agents such as fluorides, ZrO_2, TiO_2, Cr_2O_3, Fe_2O_3, P_2O_5 were then commonly added to the parent silicate glass composition to initiate the nucleation process [10, 11, 54, 59, 61–63].

Cr_2O_3 can remarkably increase the crystallization rate mostly for Fe-containing glass compositions as the corresponding mechanism includes the formation of spinel which, in turn, actively catalyses the formation of pyroxene phases [65–73]. Karamanov et al. [65] demonstrated that the Cr_2O_3 addition, in percentages as high as 0.7 wt%, enhances the spinel formation which, accordingly, increases the number of nuclei available for crystallisation; as a result, a higher degree of crystallisation and a finer structure of the glass–ceramic was achieved. The effect of Cr_2O_3 on glass crystallization can be studied by comparing the position of the exothermic peaks in the DTA thermograms for different compositions (Fig. 5): as the Cr_2O_3 content

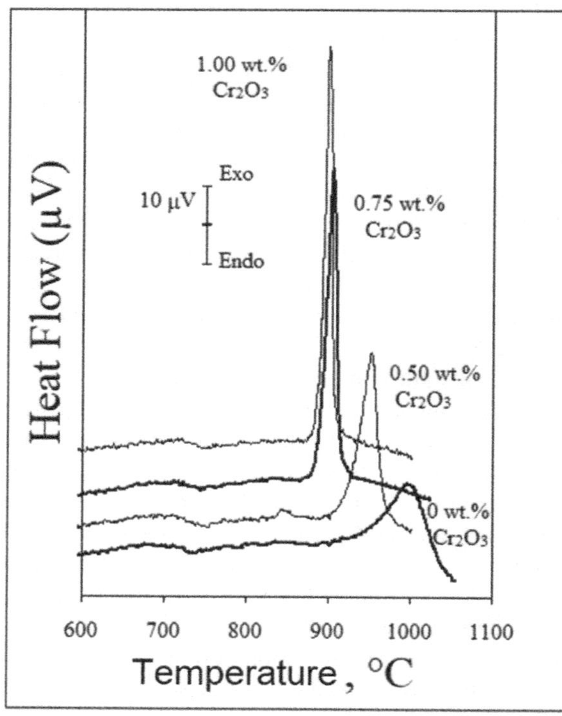

Fig. 5 DTA plots of glass powders with different Cr_2O_3 content [72, 73]

increases, there is a sharpening effect on the shape of the peak along with the shift of T_c and T_p towards lower values.

The crystallization of a major phase may be initiated by a liquid–liquid phase separation without the need for a crystalline nucleation catalyst. Such separation may result in a fine-scale dispersion of second-phase liquid droplets that are very poor glass formers and, therefore, crystallize very readily [63]. In turn, the addition of some nucleation agents such as ZrO_2, TiO_2, P_2O_5, Ta_2O_5, WO_3, Fe_2O_3 and F was reported to cause the formation of primary crystals in the systems Li_2O–Al_2O_3–SiO_2 and MgO–Al_2O_3–SiO_2; these nucleation agents were also shown to accumulate in a specific microphase of phase-separated base glass or, in general, to promote phase separation. Höland and Beall [10] pointed out that "for most commercially significant glass–ceramics, heterogeneous nucleation of the base glass has been achieved with targered development and utilization of microimmiscibility". For instance, P_2O_5 as nucleating agent caused microphase separation inducing the formation of the transition phase Li_3PO_4 prior to the crystallization of lithium disilicate [10]. Likewise, it is worth mentioning that, in tetrasilicic mica glasses, Mg–, K– and F–rich droplets can precipitate in a silicate matrix phase [74, 75]. Furthermore, as regards the behaviour of glasses from tetrasilicic mica ($KMg_{2.5}Si_4O_{10}F_2$)–fluorapatite ($Ca_5(PO_4)_3F$)–diopside ($CaMgSi_2O_6$) system (Fig. 6), it was revealed that mica crystallites were formed from fine-scaled glass-in-glass phase separation containing quasi-spherical mica grains [75]. In this regard, Fig. 7a shows the microstructure of glass corresponding to tetrasilicic mica composition in which round-shaped particles

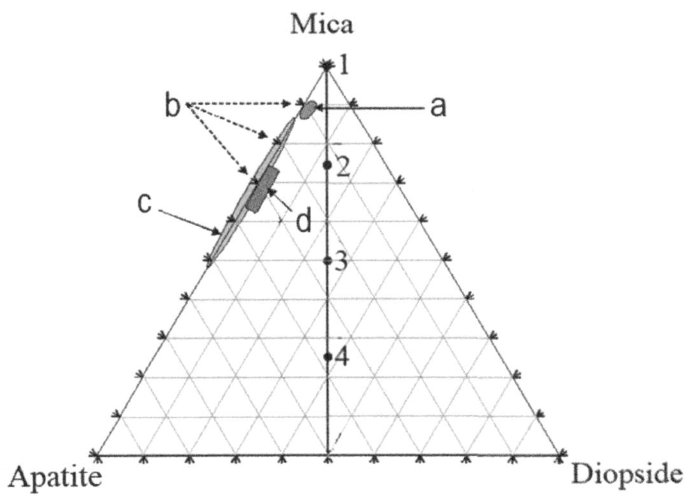

Fig. 6 Position of four different glasses in the tertrasilicic mica (M)—fluorapatite (FA)—diopside (D) system. **1**: glass of tetrasilicic mica composition (M), **2**: 75.0% M, 12.5% FA, 12.5% D, **3**: 50.0% M, 25.0% FA, 25.0% D, **4**: 25.0% M, 37.5% FA, 37.5% D. For comparison, some compositions that have been already reported in the references [76–79] are labelled in the diagram as a, b, c and d, respectively

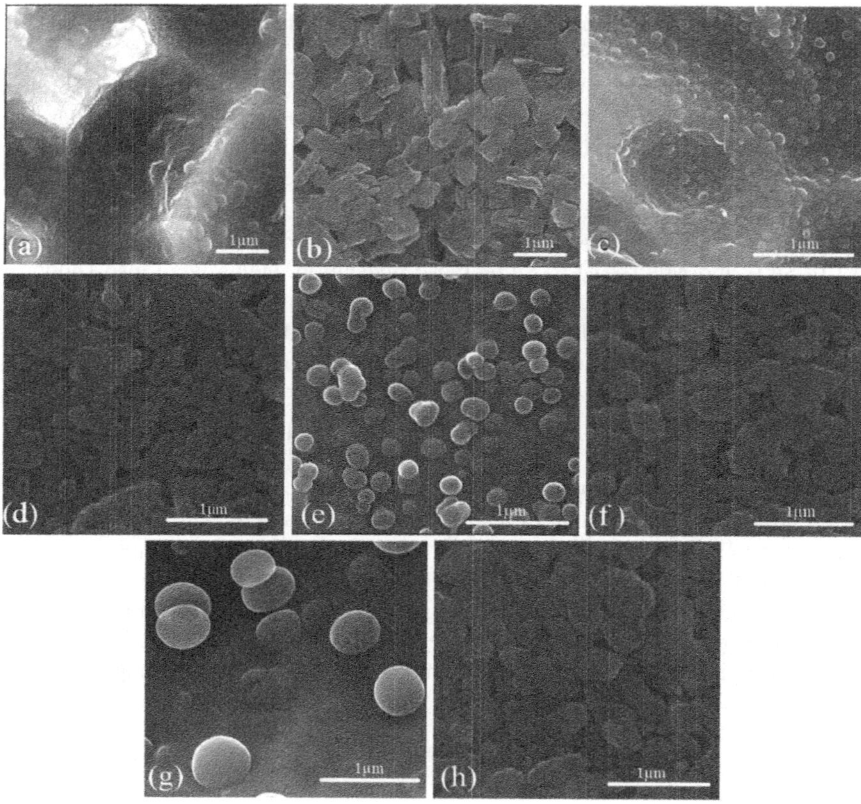

Fig. 7 Microstructure of annealed glasses (position of the compositions are shown in the ternary diagram of Fig. 6) **a** 1, **c** 2, **e** 3, **g** 4, and corresponding glass–ceramics **b** 1 (1075 °C, 5 h), **d** 2 (1000 °C, 1 h), **f** 3 (1000 °C, 2 h), **h** 4 (1050 °C, 1 h) in tetrasilicic mica ($KMg_{2.5}Si_4O_{10}F_2$)–fluorapatite ($Ca_5(PO_4)_3F$)–diopside ($CaMgSi_2O_6$) system [75]

(diameter 0.6–1.0 μm) were embedded in glassy matrix. Interestingly, the nucleation of apatite and diopside might be obtained simultaneously to the formation of tetrasilicic mica in the tetrasilicic mica–fluorapatite–diopside system. It is likely that nucleation is achieved within the droplet glass phases, and phase separation is further enhanced with the addition of CaO and P_2O_5 introduced through apatite and diopside The tendency of glasses to favour phase separation as well as the formation of large spherical droplets yield important optical effects, such as high opacity (light scattering) resulting in the milky appearance of glasses containing apatite and diopside (Fig. 8). The glass compositions mentioned above are preferably crystallized in bulk form within 700–1075 °C [75]. Looking at the diagram in Fig. 6, tetrasilicic mica is crystallized from glass 1: for example, a typical "house of card" structure of randomly-oriented block-like mica crystals (1.5 − 3.0 μm) is formed when the base glass 1 is thermally treated at 1075 °C for 5 h (Fig. 7b). The other intermediate

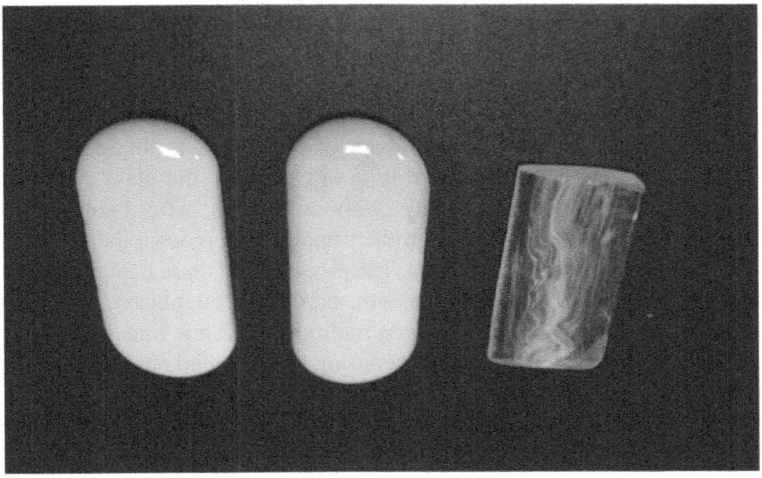

Fig. 8 From the right to the left: pattern of the annealed glass 2 (righttmost), annealed glass 3 (in the center of the figure) and pattern of the glass 3 heat treated at 1000 °C for 2 h (see also Fig. 6)

glasses 2, 3 and 4 (see Fig. 6) can crystallize yielding a combination among mica, apatite and diopside.

7 Monitoring the Crystallization in Glass–Ceramics

Glass–ceramics are fine-grained polycrystalline materials formed when glasses of suitable compositions are heat-treated and undergo controlled crystallization (devitrification). While glasses, in principle, can be prepared from a variety of substances including inorganic materials, organics, metals and even polymers. the term glass–ceramics is typically used for inorganic, non-metallic materials [80].

The key feature of the processing of glass–ceramics is that the crystallization must be controlled. One or more crystalline phases may form during the heat treatment, and both the composition of these phases and the composition of the residual glass are normally different from that of the parent glass [5, 10, 11, 54, 61, 81, 82]. The content of crystalline phase(s) may vary between 0.5 and 99.5%, but in most applications the crystalline volume fraction is in the range of 30–70% [59]. The broader range was achieved by Deubener et al. [80], who reported volume fractions of crystalline phase in the range of few ppm to almost 100%.

Glass–ceramics are highly appealing for several end applications, such as the thermal, chemical, biological and dielectric ones, because the properties of these materials (e.g. transparency, mechanical strength, resistance to abrasion, coefficient of thermal expansion etc.) can be finely modulated by properly selecting the parent glass composition and the heat-treatment parameters (temperature, time, oxidizing vs. reductant atmosphere), which can control the extent of crystallization, crystal morphology, crystal size and aspect ratio.

The ease of fabrication techniques and relatively low production cost offer obvious industrial advantages [5, 10, 11, 54, 61, 82]. Initially, the glass batch is heated to form a homogeneous melt. Then, the desired shape of the product is obtained by applying processes such as pressing, blowing, rolling or casting to the glass at the working point, which corresponds to a viscosity of about 10^3 Pa·s. After being annealed to eliminate internal stresses, the glass product finally undergoes a thermal treatment that converts it into a glass–ceramic [83]. The properties of these materials, especially the mechanical ones (e.g. bending strength, hardness, toughness), are superior to those of most conventional glass and even traditional ceramic materials.

Glass–ceramics are often produced according to controlled nucleation and crystal growth in the volume of the parent glass. However, if the specific surface area of the system is high, crystallization can start from the surface: this is the typical case of glass–ceramics produced by concurrent sinter-crystallization of glass powder compacts (greens). In this case, nucleating agents are not required because the particle surfaces provide nucleation sites and crystallization starts at the interfaces among glass particles [59]. DTA thermographs combined with HSM may serve as useful, complementary instruments for monitoring—and then optimizing—the sintering and crystallization processes. In particular, the glass stability against crystallization can be measured by using the thermal parameters already defined in the Sect. 2 (e.g. K_H and S_c). In general, glasses with a large temperature interval between onset of crystallization and T_g (also called "sintering window") can possibly be well sintered [84]. When the onset of crystallization occurs before the glass is fully densified, further densification will be impeded by the formation of crystalline phase(s) that increase(s) the matrix viscosity [85].

HSM allows the sintering processes to be studied in a "real-time" mode, since it permits glass scientists to follow continuously the contraction process without any external load or friction. Small greens with cylindrical geometry are commonly used in HSM-based sintering studies, so that the axial and radial contractions of the powder compact can be simultaneously recorded, and the anisotropy during sintering can be determined. The HSM projects the silhouette of the sample through a quartz window and onto the recording device; an image analysis system automatically records and analyses the geometrical changes of the sample (e.g. shrinkage or blowing) during heating. The image analyser takes into account the thermal expansion of the alumina substrate while measuring the height of the sample during controlled heating, with the base as a spatial reference. Complementary studies using DTA and HSM under the same heating conditions are therefore very useful for investigating the effect of glass composition on sintering and devitrification phenomena. In general, the desired order of events in glass-powder densification processes occurs when sintering precedes crystallization. This is key to obtain a final high-density material with very low porosity like, for instance, a gas-tight glass–ceramic seal [86] (Fig. 9). On the contrary, if crystallization occurs before sintering is completed (Fig. 10), the viscosity increases sharply and sintering is highly inhibited or even stops.

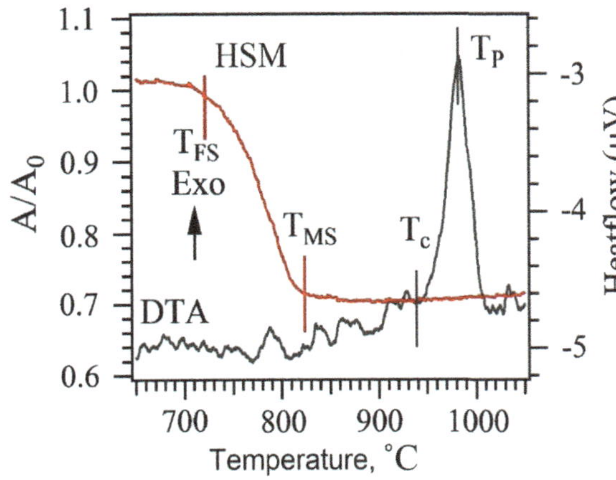

Fig. 9 An example illustrating the relationships between the HSM an$_c$d DTA data for diopside glass system [87] (T_{FS}: first shrinkage, T_{MS}: maximum shrinkage, T: onset of crystallization, and T_P: peak crystallization temperature)

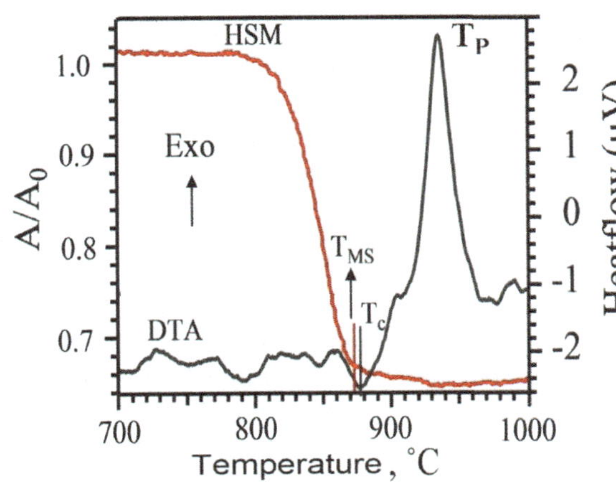

Fig. 10 An example illustrating the relationships between the HSM and DTA data for melilite glass [88] system (T_{FS}: first shrinkage, T_{MS}: maximum shrinkage, T_c: onset of crystallization, and T_p: peak crystallization temperature)

Figure 11 displays another interesting scenario in which glass powder sintering was partially impeded by crystallization. HSM investigation on glass powder compact revealed the presence of two densification steps (i.e. T_{FS1}—T_{MS1} and T_{FS2}—T_{MS2}) that are separated by the temperature range in which crystallization occurs and temporarily stops sintering. At the end of the second densification step, a total variation of A/A_0 close to 0.60 was achieved, corresponding to volume shrinkage of

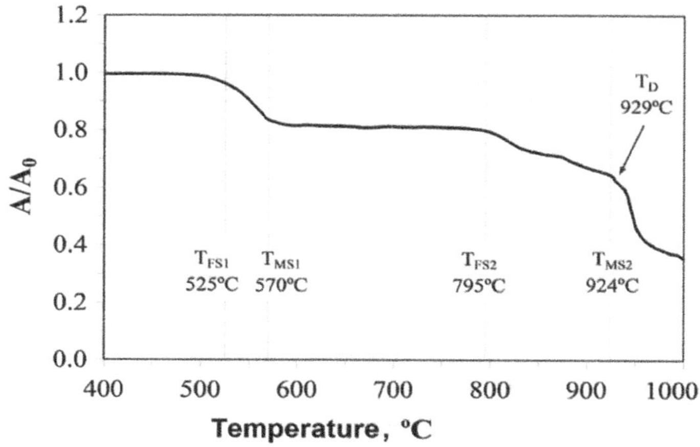

Fig. 11 Variation in relative area of powder sample's silhouette (A/A_0: A_0 is the initial area at room temperature, A is the area at defined temperature) during the HSM measurement

about 40% (under the hypothesis of isotropic shrinkage) and to a practically full densification of the sample.

It was also demonstrated that the glass particle size and the heating rate have an effect on crystallization. Specifically, the finer the particles for a given glass composition, the lower T_g and T_p, provided that all the other experimental conditions are kept constant [89, 90]. This means that a glass powder compact composed by small particles will undergo concurrent crystallization during sintering at lower temperatures as compared to a similar sample of coarser particles. In the same studies, T_g and T_p were also observed to increase as the heating rate increased (Fig. 12).

DTA or DSC also are quite popular methods for studying the kinetics of crystallization processes in glasses. The crystallization kinetics based on these data are usually interpreted according to the Avrami nucleation-growth model and the Johnson–Mehl–Avrami equation [91], which describes the time (t) dependence of the crystallized fraction α as follows:

$$(-\ln(1-\alpha))^{1/n} = kt \tag{3}$$

where k and n (Avrami coefficient) are constant.

The value of n can be related to another parameter, called m, which is the dimensionality of growth of the crystalline phase [92]. The values of m range from 1 (1-dimensional growth corresponding to rod-like crystals) to 3 (3-dimensional growth corresponding to sphere-like crystals); if surface crystallization is dominant, there is the condition n = m = 1. Kinetic aspects of crystallization in glasses and amorphous solids have been comprehensively reviewed and discussed by Malek [93, 94].

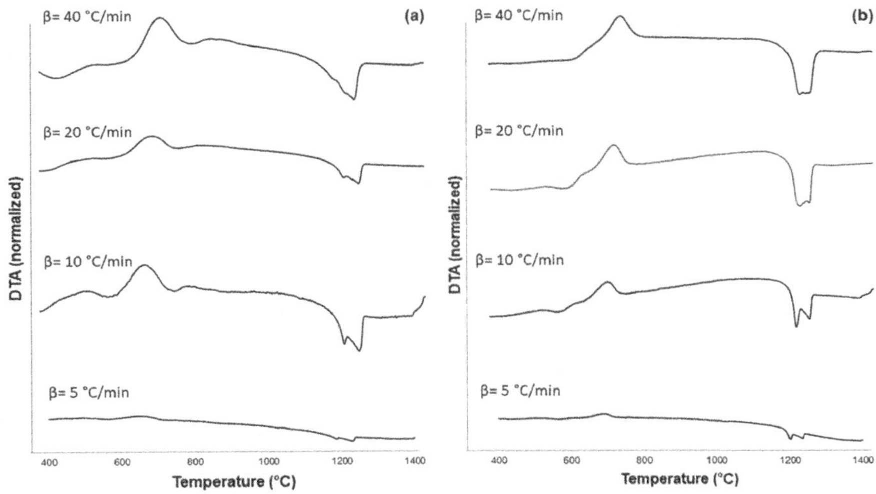

Fig. 12 DTA thermographs of Bioglass® powders (composition $45SiO_2$–24.5CaO–$24.5Na_2O$–$6P_2O_5$ wt%) having different size—**a** less than 5 μm and **b** less than 32 μm–and collected at different heating rates (β) [90]

8 Glass–Ceramic Types and Applications

Over many years, silicate systems were extensively used for developing glass–ceramics for commercialization purposes. Although many other systems are now subject of intense research [80], silicate systems will be inevitably elaborated yet more considering their high flexibility in compositional design and versatility for myriads of applications, ease of processing due to their appropriate glass forming ability, abundance of precursors in the Earth crust, etc. The design of glass–ceramic products takes into account multiple factors, including requirements for external contours, color and aesthetics, surface structure and internal microstructure in relation to the functional properties [10]. The microstructure of glass–ceramics coupled with the phase assemblage (i.e. type of crystalline phase(s), volume fraction of crystalline and glassy constituents) dictate the final functional properties of the product (e.g. resistance to wear, toughness etc.) while the parent glass composition is key to determine the degree of glass workability and to foresee whether internal or surface nucleation can be achieved [10]. Too high rates of crystal growth are to be avoided; materials in which the crystal growth rate is too high do not develop the fine-grained microstructure, necessary for the achievement of high mechanical strength. The major components of any glass–ceramic composition are selected to ensure precipitation of crystalline phase(s) that will impart desired properties but also bearing in mind the need for producing a glass melt with proper viscosity-temperature characteristics to be shaped by using the available techniques.

Just to cite a few examples, the beta-quartz and keatite solid solutions, spodumene or beta-eucryptite are predominantly precipitated in glass–ceramics under tightly controlled conditions in order to achieve zero thermal expansion and make useful materials for transparent cookware, optics, telescope mirror blanks, infrared transmitting electric range tops, wood-stove windows and fire door glazing where the dimensional stability and the ability to resist thermal shock are necessary [95, 96]. High-strength glass–ceramics may contain spinel, mullite and corundum, while glass–ceramics for electrical applications include cordierite, anorthite, diopside and wollastonite [54, 61].

Commercial glass–ceramics can be divided in several categories according to their applications, which may be addressed to both high-tech and more common uses, including glass–ceramics for thermal protection, machinable glass–ceramics, high-strength glass–ceramics, glass–ceramics for biomedical applications, glass–ceramics for electrical and electronic applications, optical glass–ceramics, glass–ceramics used as construction materials., etc. [10, 54, 59, 61]. Commercial glass–ceramics for broad consumer and specialized markets are currently produced by several companies all around the world such as Corning, Schott AG, St. Gobain, Nippon Electric Glass, Ohara, Ivoclar etc. One of the most comprehensive analyses of glass–ceramic commercialization was provided by Maziar et al. in 2015 [97], who performed a detailed patent search and reported that glass–ceramics with specific properties, such as thermal (e.g., low thermal expansion, insulating, high thermal stability, etc.), electrical, (e.g., high ionic conductivity), or optical (e.g., high transparency, high luminescence efficiency) ones, have mainly attracted the attention from companies over the past decade. The range of applications was quite broad (Fig. 13), spanning from more "traditional" fields, such as chemistry, engineering, energy, materials science and medicine, to unusual areas, such as polymer science and food technology (also food packaging). More specifically, above 550 patents on various glass–ceramics for making electronic components, wiring board substrates, cooktop panels, insulators, sealants and heat reflector substrates were found. Some other patents have been granted for glass–ceramics addressed to architectural, biomedical, magnetic, armor, energy, nuclear, and waste immobilization applications. Overall, two opposite trends were found in patent applications for glass–ceramics, i.e. (i) a decrease of electrical, electronic and magnetic applications and (ii) an increment of dental, biomedical, optical, energy, chemical, waste management and refractory. These trends are actually in line with current demands of new high-tech products, thus suggesting prospects for industrial growth of glass–ceramic market in these areas.

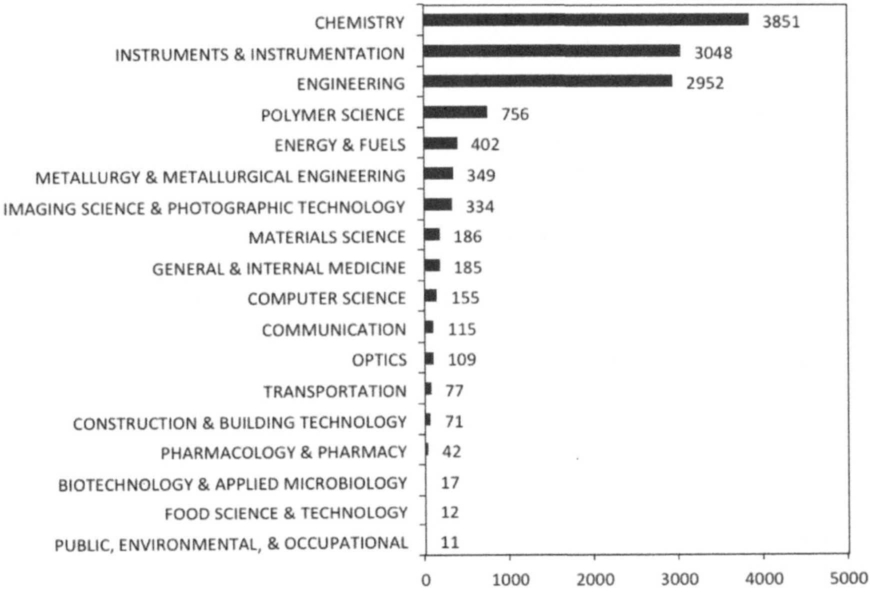

Fig. 13 Breakdown from the Derwent World Patent Index (DWPI) database about the number of patents granted in various fields from 1968 to 2014 by using the keywords "glass–ceramic*" or "glass ceramic*" [97]

9 A Forecast for the Future: The Potential of Modelling and Computational Approaches in the "Glass and Glass–Ceramic World"

Although being an ancient, traditional material, glass—along with its glass–ceramic derivatives—does not stop yet proving to be suitable in a number of new, high-tech fields. Many of these latest applications are described in the next chapters of this book, including the uses of glass and glass–ceramics in medicine, optics and optoelectronics, energy storage and circular economy.

The relatively recent success of many glass–ceramic products relies on achieving unique combinations of attributes, including appropriate optical, thermal, mechanical, and biological properties, which often cannot be achieved by using an amorphous glass. Machinability is another key technological added value of many glass–ceramics for such applications.

Even though the thermodynamic and kinetic aspects of sintering and crystallization are crucial for designing industrial materials, the theoretical understanding of these basic processes still is incomplete. Future development of new rigorous or even computational models will hopefully enable more accurate quantitative predictions of glass and glass–ceramic properties.

As a general comment, we think that the progress of glass/glass–ceramic science and technology in these so advanced fields will benefit from an ever stronger combination between theoretical aspects and computational approaches. Mathematical and numerical modelling is very helpful not only to shed light on the basic aspects of glass structure, but also to better describe and, to some extent, predict the functional properties of glasses. Conventional empirical, trial-and-error approaches for developing new glasses and characterizing their properties are expensive and time-consuming. In order to overcome these limitations, modelling has been emerging as a significant component of research in glass science and technology. Just to mention a famous and well-known example, the software SciGlass is able to implement a series of empirical and semi-empirical models based on large datasets to estimate the main physical, thermal and mechanical properties of a variety of glasses after receiving the oxide composition (wt% or mol%) as the input. These predictions are very useful to perform a preliminary selection of glass candidates for a given application and refine the composition in the attempt to achieve the desired target properties [98].

At present, a variety of computational techniques are available to researchers, such as those based on machine learning or molecular dynamics. A number of models are available as well, from purely empirical models relying on the mathematical interpolation of experimental data to ab-initio methods needing a detailed description of the material's electronic structure [99].

While many of these modelling approaches are potentially applicable to any class of materials, some methods (e.g. the topological constraint theory that describes the connectivity of disordered glassy networks) have been specifically developed for modelling amorphous or partially-amorphous systems such as glasses and glass–ceramics. The appropriate choice of modelling technique depends on the nature of the system under investigation, the desired properties to estimate, the quality and availability of existing data, and the level of physical understanding behind the structure–property relationships in that system. A combination of multiple modelling approaches can often be beneficial. Indeed, all models need to be validated by experiments; therefore, cross-disciplinary collaboration among glass researchers, mathematicians and data/computer scientists will be more than ever necessary.

References

1. Doremus, R.H.: Glass Science. Wiley, New York (1994)
2. Gutzow, I., Schmelzer, J.: The Vitreous State. Springer-Verlag, Berlin (1995)
3. Paul, A.: Chemistry of Glasses, 2nd edn. Chapman and Hall, London (1990)
4. Rawson, H.: Inorganic Glass Forming Systems. Pergamon Press, New York (1967)
5. Shelby, J.E.: Introduction to Glass Science and Technology. The Royal Society of Chemistry, Cambridge (1997)
6. Varshneya, A.K.: Fundamentals of Inorganic Glasses. Academic Press, London (1994)
7. Vogel, W.: Structure and Crystallization of Glass. Pergamon press, Lipzig (1971)
8. Zarzycki, J.: Glasses and the Vitreous State. Cambridge University, Cambridge (1991)
9. Phillips, C.J.: Glass, the Miracle Maker: Its History, Technology, Manufacture and Applications, 2nd edn. Pitman, New York (1948)

10. Höland, W., Beall, G.: Glass-Ceramic Technology. The American Ceramic Society, Westerville, Ohio (2002)
11. McMillan, P.W.: Glass-Ceramics. Academic Press, London (1979)
12. Le Bourhis, E.: Glass: Mechanics and Technology. Wiley-VCH Verlag, Weinheim (2008)
13. Scholze, H.: Glass: Nature. Structure and Properties. Springer, Berlin (1991)
14. Harper, C.A.: Handbook of Ceramics, Glasses and Diamonds. McGraw-Hill, New York (2001)
15. C162—Compilation of ASTM Standard Definitions. The American Society for Testing Materials, Philadelphia (1945)
16. Varshneya, A.K., Mauro, J.C.: Comment on Misconceived ASTM Definition of Glass by A.C. Wright. Glass Technol. Eur. J. Glass Sci. Technol. Part A **51**(1), 28–30 (2010)
17. Zanotto, E.D., Mauro, J.C.: The glassy state of matter: Its definition and ultimate fate. J. Non-Cryst Solids **471**, 490–495 (2017)
18. Schmelzer, J.W.P., Tropin, T.V.: Glass transition, crystallization of glass-forming melts, and entropy. Entropy **20**, 103 (2018)
19. Ojovan, M.I.: Ordering and structural changes at the glass–liquid transition. J. Non-Cryst Solids **382**, 79–86 (2013)
20. Goldschmidt, V.M.: Geochemische Vertailungsgesetze der Elemente. Skrifter Norske Videnskaps-Akademie, Oslo (1926)
21. Zachariasen, W.H.: The atomic arrangement in glass. J. Am. Chem. Soc. **54**(10), 3841–3851 (1932)
22. Warren, B.E.: X-ray determination of the structure of glass. J. Am. Ceram Soc. **17**(1–12), 249–254 (1934)
23. Wright, A.C.: Neutron and X-ray amorphography. J. Non-Cryst. Solids **106**(1–3), 1–16 (1988)
24. Jiang, Z.-H., Zhang, Q.-Y.: The Structure of Glass: A Phase Equilibrium Diagram Approach. Prog. Mater. Sci. **61**, 144–215 (2014)
25. Cabral, A.A., Cardoso, A.A.D., Zanotto, E.D.: Glass-forming ability versus stability of silicate glasses I. Experimental. J. Non-Cryst. Solids **320**, 1–8 (2003)
26. Hrubÿ, A.: Evaluation of glass-forming tendency by means of DTA. Czech J. Phys. **22**, 1187–1193 (1972)
27. Turnbull, D., Cohen, M.: Crystallization Kinetics and Glass Formation. Butterworths, London (1960)
28. Avramov, I., Zanotto, E.D., Prado, M.O.: Glass-forming ability versus stability of silicate glasses II. Theoretical demonstration. J. Non-Cryst. Solids **320**, 9–20 (2003)
29. Uhlmann, D.R.: A kinetic treatment of glass formation. J. Non-Cryst. Solids **7**(4), 337–348 (1972)
30. Weinberg, M.C., Uhlmann, D.R., Zanotto, E.D.: Nose method' of calculating critical cooling rates for glass formation. J. Am. Ceram. Soc. **72**, 2054–2058 (1989)
31. Weinberg, M.C., Zanotto, E.D.: Calculation of the volume fraction crystallised in nonisothermal transformations. Phys. Chem. Glasses **30**(3), 110–115 (1989)
32. Uhlmann, D.R., Zelinski, B.J.J., Zanotto, E.D., Weinberg, M.C.: Sensitivity of critical cooling rate to model and material parameters. Paper presented at: XV International Congress on Glass, Leningrad, USSR (1989)
33. Weinberg, M.C.: Glass-forming ability and glass stability in simple systems. J. Non-Cryst. Solids **167**(1–2), 81–88 (1994)
34. Cabral, A.A., Jr., Fredericci, C., Zanotto, E.D.: A test of the Hrubÿ parameter to estimate glass-forming ability. J. Non-Cryst. Solids **219**, 182–186 (1997)
35. Panda, P.C., Raj, R.: Sintering and crystallization of glass at constant heating rate. J. Am. Ceram. Soc. **69**, 1564–1566 (1989)
36. Ferraris, M., Verné, E.: Viscous phase sintering of particle-reinforced glass matrix composites. J. Eur. Ceram. Soc. **16**, 421–427 (1996)
37. Lara, C., Pascual, M.J., Duran, A.: Glass-forming ability, sinterability and thermal properties in the system RO–BaO–SiO_2 (R=Mg, Zn). J. Non-Cryst. Solids **348**, 149–155 (2004)
38. Baino, F., Ferraris, M., Bretcanu, O., Verné, E., Vitale-Brovarone, C.: Optimization of composition, structure and mechanical strength of bioactive 3-D glass-ceramic scaffolds for bone substitution. J. Biomater. Appl. **27**, 872–890 (2013)

39. Hench, L.L., West, J.K.: The sol-gel process. Chem. Rev. **90**, 33–72 (1990)
40. Baino, F., Fiume, E., Miola, M., Verné, E.: Bioactive sol-gel glasses: Processing, properties, and applications. Int. J. Appl. Ceram. Technol. **15**, 841–860 (2018)
41. Rao, K.J.: Structural Chemistry of Glasses. Elsevier Science & Technology Books, Amsterdam (2002)
42. Jones, G.O.: Glass. Wiley, Methuen (1956)
43. Kurkjian, C.R., Prindle, W.R.: Perspectives on the history of glass composition. J. Am. Ceram. Soc. **81**(4), 795–813 (1998)
44. Yamane, M., Asahara. Y.: Glasses for Photonics. Cambridge University Press, Cambridge (2004)
45. Wright, A.C., Bachra, B., Brunier, T.M., Sinclair, R.N., Gladden, L.F., Portsmouth, R.L.: A neutron diffraction and MAS-NMR study of the structure of fast neutron irradiated vitreous silica. J. Non-Cryst. Solids **150**(1–3), 69–75 (1992)
46. Kirk, R.E., Othmer, D.F., Grayson, M., Eckroth, D.: Encyclopedia of Chemical Technology, 5th edn. Wiley, New Jersey (2007)
47. Bergeron, C.G., Risbud, S.H.: Introduction to Phase Equilibria in Ceramics. The American Ceramic Society, Westerville, Ohio (1984)
48. Morey, G.W., Bowen, N.L.: Ternary system sodium metasilicate-calcium metasilicate-silica. J. Glass Technol. Soc. **9**(35), 226–264 (1925)
49. Morey, G.W.: The devitrification of soda-lime-silica glasses. J. Am. Ceram. Soc. **13**, 683–713 (1930)
50. Segnit, E.R.: Further data on the system Na_2O–CaO–SiO_2. Am. J. Sci. **251**(8), 586–601 (1953)
51. Toropov, N.A. et al.: Phase Diagrams of Silicate Systems: Handbook; third Issue; Ternary Systems, Nauka, Leningrad (1972); Translation, Army Foreign Science and Technology Center Charlottesville, Virginia (10 June 1974)
52. Artamonova, M.V., Aslanova, M.S., Buzhinsky, I.M.: Chemical Technology of Glasses and Glass-Ceramics. In: Pavlushkin, N.M. Stroiizdat, Moscow (1983)
53. Baino, F., Quaglia, A.: Evidences of Glass-Ceramic White Opaque Tesserae from Roman Age: A Thermo-Analytical Approach. Mater. Lett. **74**, 194–196 (2012)
54. Pavlushkin, N.M.: Principles of Glass-Ceramic Technology. Stroiizdat, Moscow (1979).(in Russian)
55. Réaumur, R.A.M.: Art de faire une nouvelle espèce de porcelaine, par des moyens extrêmement simples et faciles, ou de transformer le Verre en Porcelaine (Art of a new porcelain species, extremely simple and easy ways, or how to transform Glass into Porcelain). Hist. Acad. R Sci. Paris **1**, 370–388 (1739)
56. Woodard, K.L.: Profiles in Ceramics. S. Donald Stookey. Am. Ceram. Soc. Bull. **3**, 34–39 (2000)
57. Stookey, S.D.: Chemical Machining of Photosensitive Glass. Ind. Eng. Chem. **45**(1), 115–118 (1953)
58. Stookey, S.D.: Catalyzed crystallization of glass in theory and practice. Ind. Eng. Chem. **51**(7), 805–808 (1959)
59. Zanotto, E.D.: A bright future for glass-ceramics. Am. Ceram. Soc. Bull. **89**(8), 19–27 (2010)
60. Tammann, G.: Glasses as super-cooled liquids. J. Soc. Glass Technol. **9**, 166–185 (1925)
61. Strnad, Z.: Glass-Ceramic Materials. Elsevier, Amsterdam (1986)
62. Lewis, M.H., Metcalf-Johansen, J., Bell, P.S.: Crystallization mechanism in glass-ceramics. J. Am. Ceram. Soc. **62**(5–6), 278–288 (1978)
63. Pinkney, L.R.: Glass ceramics. In: Buschow, K.H.J., Cahn, R., Flemings, M.C., Ilschner, B., Kramer, E.J., Mahajan, S., et al. (eds.) Encyclopedia of Materials: Science and Technology, Pergamon, Oxford (2001)
64. Borrelli, N.F.: Photosensative Glass and Glass Ceramics. Taylor & Francis Group, Corning Inc., Corning, New York, USA, CRC Press (2017)
65. Karamanov, A., Pisciella, P., Pelino, M.: The effect of Cr_2O_3 as a nucleating agent in iron-rich glass-ceramics. J. Eur. Ceram. Soc. **19**, 2641–2645 (1999)
66. Besborodov, M.A.: Glass-Ceramic Materials. Nauka i Technika, Minsk (1982) (in Russian)

67. Rawlings, R.D.: Production and properties of silceram glass-ceramics. In: Glass-Ceramic Materials, Fundamentals and Applications Series of Monographs on Material Science, Engineering and Technology. Mucchi Editore, Modena, pp. 115–133 (1997)
68. Zhunina, L., Kuzmenkov, M., Yaglov, V.: Pyroxene Sitalls. University Publishers, Minsk (1974).(in Russian)
69. Alizadeh, P., Marghussian, V.K.: The effect of compositional changes on the crystallization behavior mechanical properties of diopside-wollastonite glass-ceramics in the SiO_2–CaO–MgO–Na_2O system. J. Eur. Ceram. Soc. **20**, 765–773 (2000)
70. Rezvani, M., Eftekhari-Yekta, B., Solati-Hashjin, M., Marghussian, V.K.: Effect of Cr_2O_3, Fe_2O_3 and TiO_2 nucleants on the crystallization behaviour of SiO_2–Al_2O_3–CaO–MgO (R_2O) glass-ceramics. Ceram. Int. **31**, 75–80 (2005)
71. Karamanov, A., Arrizza, L., Matekovits, I., Pelino, M.: Properties of sintered glass-ceramics in the diopside–albite system. Ceram. Int. **30**, 2129–2135 (2004)
72. Vasilopoulos, K.C., Tulyaganov, D.U., Agathopoulos, S., Karakassides, M.A., Ferreira, J.M.F., Tsipas, D.: Bulk nucleated fine grained mono-mineral glass-ceramics from low-silica fly-ash. Ceram. Int. **35**(2), 555–558 (2009)
73. Vasilopoulos, K.C., Tulyaganov, D.U., Agathopoulos, S., Ribeiro, Karakassides, M.A., Ferreira, J.M.F., Tsipas, D.: Vitrification of low-silica fly-ash. Suitability of the resulting glass-ceramics for architectural or electrical insulator applications. Adv. Appl. Ceram. **108**(1), 27–32 (2009)
74. Vogel, W.: Chemistry of Glass. The American Ceramic Society, Westerville, Ohio (1985)
75. Tulyaganov, D.U., Agathopoulos, S., Fernandes, H.R., Ventura, J.M., Ferreira, J.M.F.: Crystallization of glasses in the system tetrasilicic mica-fluorapatite-diopside. J. Eur. Ceram. Soc. **24**(13), 3521–3528 (2004)
76. Hamzawy, E.M.A.: Crystallization behaviour of fluorphlogopite glass-ceramics. Ceram. Silikaty **45**, 89–96 (2001)
77. Taruta, S., Mukoyama, K., Suzuki, S.S., Kitajima, K., Takusagawa, N.: Crystallization process and some properties of calcium mica-apatite glass-ceramics. J. Non-Cryst. Solids **296**, 201–211 (2001)
78. Höland, W., Vogel, W.: Machinable and phosphate glass-ceramics. In: An Introduction to Bioceramics. In: Hench, L.L., Wilson J. (eds.) World Scientific, Singapore, pp. 125–137 (1993)
79. Chen, X., Hench, L., Greenspan, D., Zhong, J., Zhang, X.: Investigation on phase separation, nucleation and crystallization in bioactive glass-ceramics containing fluorphlogopite and fluorapatite. Ceram. Int. **24**, 401–410 (1998)
80. Deubener, J., et al.: Updated definition of glass-ceramics. J. Non-Cryst. Solids **501**, 3–10 (2018)
81. James, P.F.: Glasses and Glass-Ceramics. Chapman and Hall, London (1989)
82. Rawlings, R.D., Wu, J.P., Boccaccini, A.R.: Glass-ceramics: Their production from wastes—A Review. J. Mater. Sci. **41**, 733–761 (2006)
83. Carter, C.B., Norton, M.G.: Ceramic Materials—Science and Engineering. Springer, New York (2007)
84. Siligardi, C., D'Arrigo, M.C., Leonelli, C.: Sintering behavior of glass-ceramic frits. Am. Ceram. Soc. Bull. **79**(9), 88–92 (2000)
85. Boccaccini, A.R., Stumpfe, W., Taplin, D.M.R., Ponton, C.B.: Densification and crystallization of glass powder compacts during constant heating rate. Mater. Sci. Eng. A **219**, 26–31 (1996)
86. Tulyaganov, D.U., Reddy, A., Kharton, V.V., Ferreira, J.M.F.: Aluminosilicate-based sealants for SOFCs and other electrochemical applications—A brief review. J. Power. Sources **242**(15), 486–502 (2013)
87. Reddy, A.A., Tulyaganov, D.U., Pascual, M.J., Kharton, V.V., Tsipis, E.V., Kolotygin, V.A., Ferreira, J.M.F.: Diopside–Ba disilicate glass–ceramic sealants for SOFCs: Stabilizing thermal parameters and improving adhesion to interconnects by Sr for Ca substitution. Int. J. Hydrogen Energy **38**, 3073–3086 (2013)
88. Reddy, A.A., Tulyaganov, D.U., Kapoor, S., Goel, A., Pascual, M.J., Kharton, V.V., Ferreira, J.M.F.: Study of melilite based glasses and glass–ceramics nucleated by Bi_2O_3 for functional application. RSC Adv. **2**, 10955–10967 (2012)

89. Sabato, A.G., Salvo, M., De Miranda, A., Smeacetto, F.: Crystallization behaviour of glass-ceramic sealant for solid oxide fuel cells. Mater. Lett. **141**, 284–287 (2015)
90. Baino, F., Fiume, E.: Quantifying the effect of particle size on the crystallization of 45S5 bioactive glass. Mater. Lett. **224**, 54–58 (2018)
91. Avrami, M.: Kinetics of phase change I General theory. J. Chem. Phys. **7**, 1103–1112 (1939)
92. Matusita, K., Komatsu, T., Yokota, R.: Kinetics of non-isothermal crystallization process and activation energy for crystal growth in amorphous materials. J. Mater. Sci. **19**, 291–296 (1984)
93. Malek, J.: Kinetic analysis of crystallization processes in amorphous materials. Thermochim. Acta **355**, 239–253 (2000)
94. Malek, J.: Testing method for the Johnson–Mehl–Avrami equation in kinetic analysis of crystallization processes. J. Am. Ceram. Soc. **83**, 2103–2105 (2000)
95. Krause, D., Bach, H. (eds.): Low Thermal Expansion Glass Ceramics. Springer, Berlin (2005)
96. Vladislavova, L., Thieme, C., Zscheckel, T., Patzig, C., Rüssel, C.: Heterogeneous nucleation of Ba_{1-x} Sr_x Zn_2 Si_2 O_7 from a BaO/SrO/ZnO/SiO_2 glass using platinum as nucleation agent. J. Eur. Ceram. Soc. **37**, 4801–4808 (2017)
97. Montazerian, M., Prakash Singh, S., Zanotto, E.D.: An analysis of glass–ceramic research and commercialization. Am. Ceram. Soc. Bull. **94**, 30–35 (2015)
98. Baino, F., Marshall, M., Kirk, N., Vitale-Brovarone, C.: Design, selection and characterization of novel glasses and glass-ceramics for use in prosthetic applications. Ceram. Int. **42**, 1482–1491 (2016)
99. Montazerian, M., Zanotto, E.D., Mauro, J.C.: Model-driven design of bioactive glasses: from molecular dynamics through machine learning. Int. Mater. Rev. **65**, 297–321 (2020)

Glasses and Glass–Ceramics for Biomedical Applications

Francesco Baino, Carla Migneco, Elisa Fiume, Marta Miola, Sara Ferraris, Silvia Spriano, Monica Ferraris, and Enrica Verné

Abstract The invention of bioactive glasses has undoubtedly represented an important watershed in the history of biomedicine, innovatively revolutionizing the key concept of biomaterials. Although 50 years have passed since the first bioactive glass (45S5 Bioglass®), these materials still continue to inspire numerous generations of researchers all over the world, attracted by the promise of numerous possible fields of investigations given by the versatility of glass manufacturing and processing strategies. This allows obtaining final clinical products that are incredibly diverse in terms of chemical characteristics, shape and texture and, therefore, adaptable to different therapeutic needs. The possibility to tune textural properties and degradation rates, perform high-temperature sintering processes without or minimally altering the original properties of the glass, as well as the facile introduction of therapeutically active ions within the composition and the easy surface functionalization led, over year, to the development of multiple pruducts to be used in various clinical fields, including the regeneration of both hard and soft tissues, bacterial/viral infection treatments and development of antitumoral strategies. This chapter opens a wide window on the world of bioactive glasses, starting with the description of their peculiar chemical properties, discussed in relation to the most commonly used manufacturing processes to obtain glass monoliths or particles. Then, an overview on the most common applications of BG-based products will be provided, paying particular attention to porous scaffolds for bone tissue engineering, bioactive coatings, antibacterial glasses and surface functionalization. In conclusion, a comprehensive overview on clinical applications updated to the state of the art will be provided.

Keywords Bioactive glass · Bioactivity · Tissue engineering · Scaffold · Antibacterial · Biomaterials

---

F. Baino (✉) · C. Migneco · E. Fiume · M. Miola · S. Ferraris · S. Spriano · M. Ferraris · E. Verné
Department of Applied Science and Technology, Institute of Materials Physics and Engineering, Politecnico Di Torino, Corso Duca degli Abruzzi 24, 10129 Torino, Italy
e-mail: francesco.baino@polito.it

F. Baino et al. (eds.), *Ceramics, Glass and Glass-Ceramics*, PoliTO Springer Series,
https://doi.org/10.1007/978-3-030-85776-9_5

1 Introduction

Given their inorganic nature, mechanical rigidity and physical characteristics relatively close to hard tissues, bioactive glasses and glass–ceramics have traditionally received much attention for use in bone substitution and repair. The adult human skeleton is made of 206 bones, which perform support actions and also exhibit protective functions to internal, delicate organs.

The mechanical properties of osseous tissue are given by the complex internal microstructure of bone, which is made of an organic phase, composed mainly by collagen, and a mineral phase constituted by carbonated apatite, plus other proteins that stimulate cellular functions.

Unlike soft tissues, most of bony fractures can heal without scar formation and the regenerated bone matches perfectly to the pre-existing tissue. In fact, bones exhibit an intrinsic capacity for self-repair and regeneration. Bone regeneration process occurs during normal fracture healing, but can also be associated to physiological load conditions, which produce micro-damages and lead to continuous remodelling, which is known as bone turnover [1].

Moreover, there are special clinical conditions that require enhancement of bone regeneration, including skeletal reconstruction of large bone defects due to congenital abnormalities, infections (osteomyelitis), trauma, tumour resections, age- and sex-related pathologies such as osteoporosis and avascular necrosis, osteopenia and several dental problems associated to periodontitis [1]. Bone defects may also carry important social and psychological implications to patients, with an obvious impact on their overall life quality [2]. Therefore, bone defects due to trauma and pathological bone resorption or loss are a major challenge and must be considered as a global health problem.

There are different clinical approaches aimed at enhancing bone regeneration when the physiological healing process is not sufficient or somehow compromised. They can be divided into invasive methods, such as bone grafting and induction of cement spacer (Masquelet technique), and non-invasive techniques, which reproduce biophysical stimulation by applying low-intensity pulsed ultrasound and pulsed electromagnetic fields [1].

In orthopaedic and maxillofacial applications, bone grafting is a common surgical procedure to improve bone regeneration and involves the use of transplant materials (mainly autologous or allogenic bone graft substitutes), combined or not with growth factors [3]. The "gold standard" option is indeed represented by autologous bone grafts, where the bone tissue is usually taken from anterior or posterior iliac crests of the patient's pelvis. In this case, side effects due to immunoreactions and infections are greatly reduced. On the other hand, the bone harvesting process requires extra-surgery to patients, with possible second-site complications and a substantial cost increase [4]. These limitations can be overcome by allografts, obtained from human cadavers (and stored in certified bone banks) or—seldom—from living donors. Unlike autografts, tissues coming from another source than own patient may involve immunogenicity and rejection reactions, possibility of infection transmission, additional costs and ethical or religious issues.

A versatile alternative to autologous or allogenic grafting is represented by man-made bone substitutes, which typically consist of injectable powders, pastes or rigid three-dimensional porous structures (scaffolds) made of synthetic or natural biomaterials that stimulate the migration, proliferation and differentiation of bone cells while providing mechanical support for bone regeneration. These approaches are already used in clinical practice and show great promise even when regeneration of large bone injuries is required [5].

In these regard, biomedical glasses have been extensively studied with a growing interest by scientists all over the world due to their appealing technological characteristics and bio-functional properties that allow suitability in advanced regenerative medicine.

2 Brief Story of Bioactive Glass Invention and Development

The origin of biomedical glasses dates back to the late 1960s with the discovery of 45S5 composition, later marketed as Bioglass®, by Professor Larry L. Hench who was beginning his studies about glass–ceramics at the University of Florida. His interest about materials able to regenerate human tissue was triggered by a fortuitous conversation with Colonel Klinker, as reported by Hench in the article "The story of Bioglass®" [6]. Colonel Klinker had just came back to USA from Vietnam where he was enlisted with the Army Medical Corps. Hench described to him his recent studies about polymeric and metal implants and the problem derived from their use. Moreover, he mentioned other experiments about gamma rays applied to vanadia-phosphate semiconductors. These studies caught the attention of the Colonel as they had proved that these new materials could survive to high dose of high-energy radiation. After listening to the description of these new materials and applications, Colonel Klinker formulated a question which will deeply inspire Hench and his future discoveries: "If you can make a material that will survive exposure to high energy radiation, can you make a material that will survive exposure to human body?". After coming back home from Vietnam War, a growing number of people in the USA during those years needed to be treated because of amputated limbs or damaged tissue: thus, the availability of materials able to regenerate defects without being rejected was actually crucial.

In his report, Hench described how he and his co-workers based their research upon a simple hypothesis: "The human body rejects metallic and synthetic polymeric materials by forming scar tissue because living tissue are not composed of such materials. Bone contains a hydrated calcium phosphate component, hydroxyapatite (HA) and, therefore, if a material is able to form a HA layer in vivo, it may not be rejected by the body". On the base of this hypothesis, Prof. Hench and his research group had the intuition to study and test different glass compositions based on the SiO_2–Na_2O–CaO–P_2O_5 quaternary system. Specifically, the formulation $45SiO_2$–$24.5Na_2O$–24.5CaO–$6P_2O_5$ (wt.%), referred to as 45S5 and trademarked as Bioglass® by University of Florida, was selected as the most promising one because of its easiness of melting and its high CaO-to-P_2O_5 ratio which improves material surface ability to react into a physiological environment.

Experiments carried on 45S5 Bioglass® showed an excellent biocompatibility as well as osteoconduction and even osteoinduction abilities: this new material could play an active role in the process of bone tissue regeneration, becoming the first example of surface-active or "bioactive" glasses (BGs) [7]. A number of in vitro and in vivo tests were conducted on 45S5 glass to investigate its bioactivity and bone regeneration potential, proving its extraordinary ability to induce the formation of a nano-crystalline HA layer which forms a strong bone-implant interfacial bond and finally leads to complete restoration of bone tissue.

Since then, many other compositions and different BGs have been studied and developed by scientists all over the world, leading to the development of three main classes of BGs categorized according to the main former oxide, i.e. silicate, phosphate or borate BGs.

45S5 Bioglass® is a silica-based composition and it has been demonstrated that silicon plays a fundamental role in bone regeneration process by promoting the activation of several molecular pathways involved in osteogenesis [8]. In silicate glasses, the network is formed by basic units of SiO_4 tetrahedrons and its connectivity can widely vary into 1-, 2-, and 3-dimensional structures. Each oxygen anion is coordinated by two silicon cations (Si–O–Si), thus resulting in relatively open structures which can be easily broken once in contact with biological fluids.

In 1997, Brink proposed the first borosilicate glass for biomedical application [9]. In that composition, the amount of B_2O_3 was carefully tailored to achieve a pronounced bioactivity. Glasses based on B_2O_3 as network former oxide are very reactive and characterized by lower chemical durability, which allows them to create the surface layer of HA more rapidly than the silica-based ones.

P_2O_5 was used as former oxide in biomedical glasses for the first time by Andersson et al. in 1980 [10]. In nature, the phosphate group [PO_4] is present as tetrahedral structural unit, which is intrinsically asymmetric. P–O–P bonds are easily prone to hydration and, therefore, phosphate glasses are highly soluble in biological fluids according to kinetics ranging from hours to weeks, depending on the glass composition [11].

In general, flexibility of bioactive formulations is an appealing characteristic which has permitted the design and production of a huge number of BGs with different reaction rates in vitro and in vivo.

3 Bioactivity Process

In 1987, the European Society for Biomaterials proposed the definition of bioactive material as "*a material which has been designed to induce specific biological activity*" [2]. This definition may be declined to different application fields in medicine. Focusing on bone regeneration, BGs are highly attractive bone substitutes due to their property of chemically bonding to living bone through the formation of a bone-like HA layer at the implant-bone interface [12].

As first suggested by Hench for silicate glasses (Fig. 1), the bioactivity process can be conceived as a special type of glass corrosion and is governed by complex glass-fluid interactions driven by inorganic chemical (stages 1–5) and biochemical (stages 6–12) mechanisms [13].

In the first reaction, exchange of Na^+ ions from the glass with H^+ and H_3O^+ from body fluids rapidly occurs. The second step involves the breaking of Si–O–Si bonds and formation of Si–OH (silanols) at the interface between glass and bone tissue. In the third stage, condensation of silanols takes place yielding to formation of silica gel, followed by a migration of Ca^{2+} and PO_4^{3-} from the biological fluid

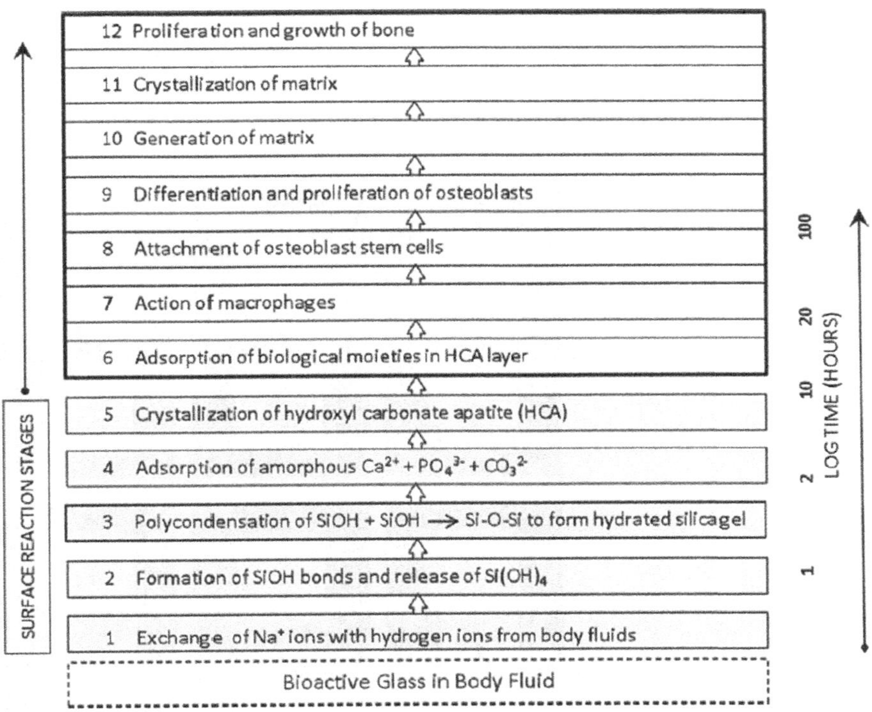

Fig. 1 Sequence of interfacial reactions between bone and a BG. Image reproduced from Hench et al. [14] with permission. Copyright © 2014 Elsevier B.V.

to the surface; as a result, a layer rich of calcium oxide and phosphorus oxide is formed on top of the silica gel layer. Stage number five represents the progressive crystallization of amorphous calcium phosphate to HA (or better hydroxycarbonate apatite) by incorporation of OH^- and CO_3^{2-} anions from the solution.

While stages 1–5 may also occur in vitro in simulated body fluids, stages 6–12 take place in vivo only. Biochemical adsorption of growth factors has been observed on the newly formed nano-crystalline HA layer [14]. Macrophages do not recognize the HA as a foreign material due to its compositional and crystallographic similarity to the mineral phase of bone tissue. Furthermore, bone-like HA stimulates the attachment of stem cells that progressively differentiate into different cells of the bone tissue, allowing the generation of bone matrix. The crystallized matrix represents the final product of the bioactivity process, leading to bone regeneration [14].

This set of reactions is generally accepted for silicate BGs and, under proper adaptations, is valid for borate BGs as well (however, a borate-rich layer forms in stage 3 instead of silica gel). This is not the case of phosphate glasses, where solubility kinetics are typically faster than those of HA formation and re-precipitation.

The bioactivity mechanism can be simplified for some glass–ceramics, on the surface of which HA can form without the presence of silica gel [15]: for example, the apatite and wollastonite crystals in A/W glass–ceramics act as sites for direct nucleation of HA crystals [16] (Fig. 2).

Some studies have convincingly shown that the ionic dissolution products released from BGs stimulate osteogenesis by regulating osteoblast proliferation, differentiation and also gene expression [18].

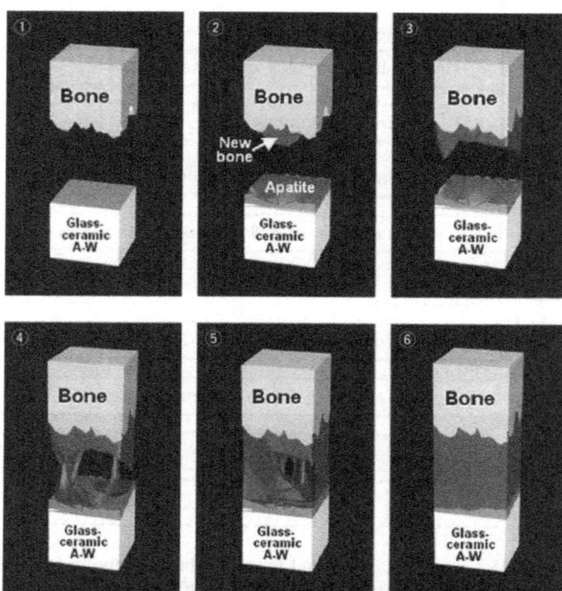

Fig. 2 Bioactivity process in A/W glass–ceramics. Image reproduced with permission from Ref. [17]. Copyright © 1969, Wiley

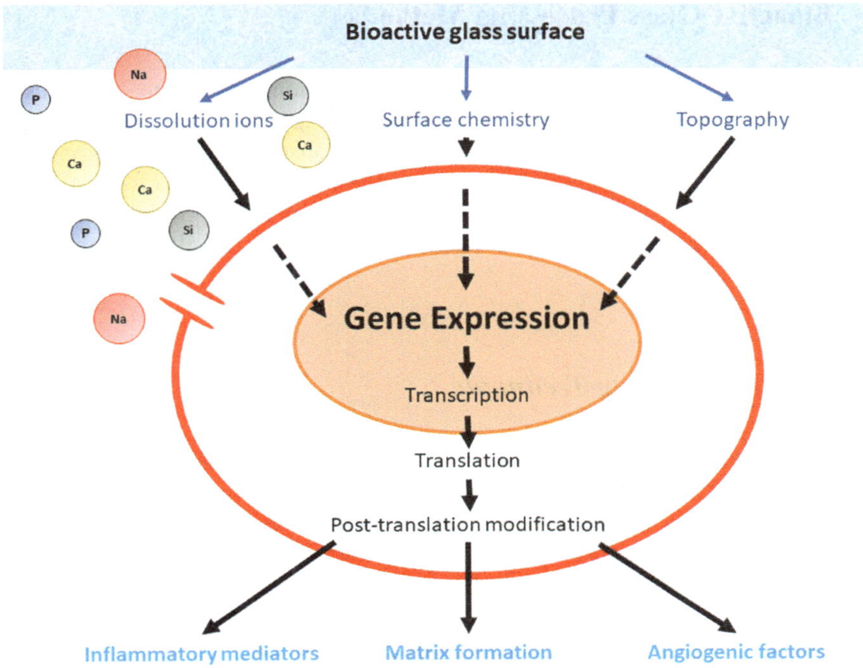

Fig. 3 Main mechanisms proposed to determine bone cell gene expression. Image adapted from Ref. [19] with permission. Copyright © 2006, Springer Science Business Media, LLC

As illustrated in Fig. 3, BGs enhance bone cell gene expression depending on four main factors [19]:

a. Surface chemistry;
b. Surface topography;
c. Rate and type of dissolution ions released;
d. Mechanical properties of glass/bone interfaces.

The bioactivity process can be evaluated through in vitro experiments by soaking the biomaterial into a Simulated Body Fluid (SBF) mimicking the composition of human plasma. The most commonly used SBF follows the Kokubo's formulation [20], which is also recommended in the relevant ISO standard [21]. Immersion studies in SBF allow evaluating the HA-forming kinetics of BGs (stages 1–5); however, in vitro bioactivity tests do not always represent a realistic prediction of the bone-bonding and bone-forming potential of the material in vivo, yielding to false positive or false negative results; further details on this issue and on how to properly adapt the tests in SBF according to the sample geometry (e.g. BG powders, tiles, porous scaffolds) can be found elsewhere [22–24].

4 Bioactive Glass Processing Methods

Glass is a very attractive material for several applications, from optoelectronics to biotechnologies and, for this reason, different processing techniques have been developed. The most common preparation methods of BGs are the melt-quenching and sol–gel techniques [25]. High-temperature post-synthesis treatments are often necessary to consolidate and/or sinter BGs, for example if processing of BG powders is required; in these cases, devitrification may take place, yielding to bioactive glass–ceramic final products [17] (Fig. 4).

4.1 Melt-Quenching Technique

The first BG with 45S5 composition was produced through melt-quench conventional technique [26]. This processing method is the most common way to obtain glasses by fusion of two or more components. Melting procedures begin from the mixing of raw precursors (usually powders of oxides, carbonates or inorganic salts) which should be highly pure to avoid unwanted contamination of the final products. Sometimes, after

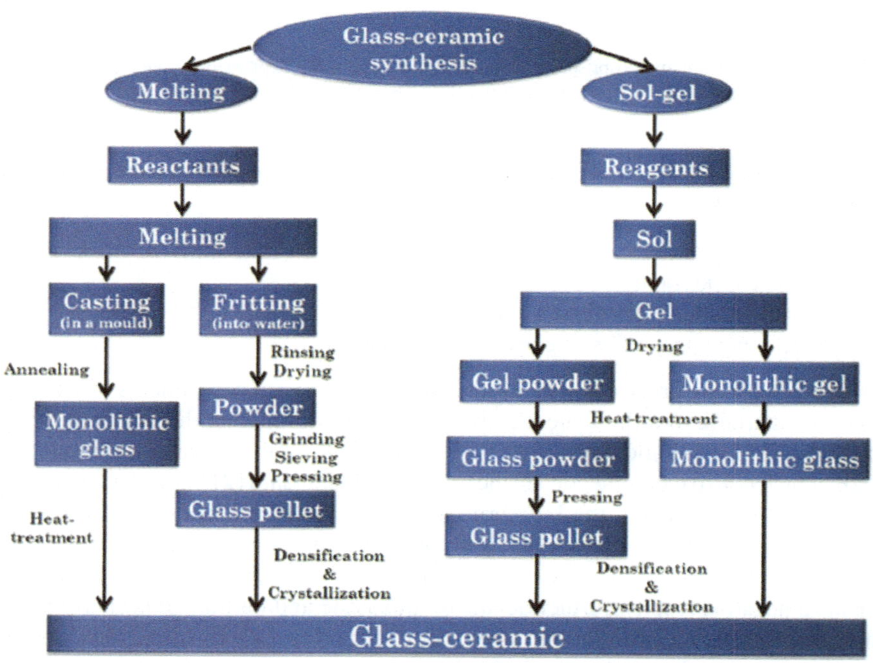

Fig. 4 Overall flowchart of bioactive glass and glass–ceramic production. Image reproduced from Ref. [17] with permission. Copyright © 1969, Wiley

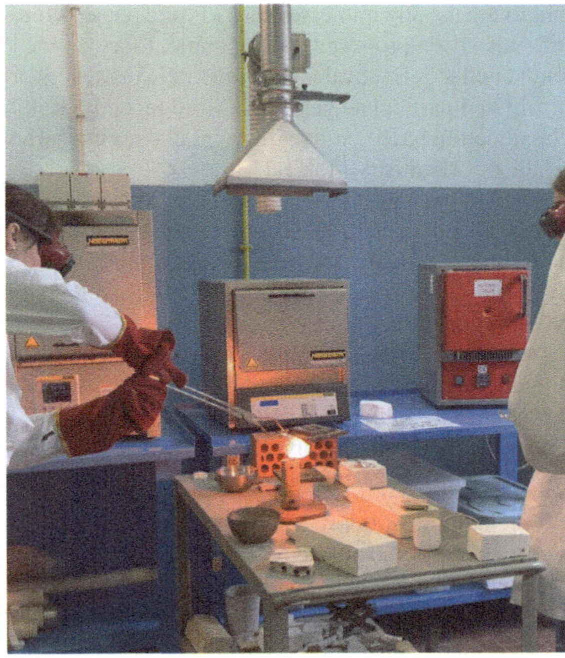

Fig. 5 Casting into a cylindrical mould during melt-quenching process (Glance Group's Lab, DISAT, Politecnico di Torino. Image courtesy of Elisa Fiume)

preliminary mixing of the precursors, the blend is introduced into a ball mill (with or without small amounts of acetone or ethanol) to break agglomerates and improve homogeneity. The resulting mixed powders are dried in air and melted in alumina or platinum crucibles at temperatures from 1200 to 1500 °C for silicate and borate BGs or around 1000–1200 °C for phosphate glasses [27]. Repeated melting procedures can also be applied to improve homogeneity, especially when high amount of glass is produced.

The molten product can be cast in air into graphite or metallic moulds (Fig. 5), thus obtaining a monolithic glass, or can be poured into water to obtain a glass "frit", which is very useful for the production of glass powders being easily pulverisable.

4.2 *Sol–Gel Synthesis*

In 1991, Li et al. first reported the synthesis of silicate BGs by sol–gel process [28]. This technique was first mentioned more than 150 years ago by Graham [29], who observed that the hydrolysis of tetraethyl orthosilicate (TEOS) could lead to the formation of SiO_2-based glasses.

In order to better understand sol–gel procedure, it is necessary to introduce some definitions: the term "sol" refers to a colloidal suspension while the term "gel" generally identifies a more rigid and interconnected network in which pores and silicate chains are usually immersed into a liquid phase. More accurate classifications of gels have also been proposed. For example, Flory divided gels into four classes: ordered and lamellar gels, covalent polymer networks, networks of aggregated polymers and disordered particulate gels [30]. More recently, Kakihana [31] introduced a new gel classification based on five different classes and strictly relied upon the key features of sol–gel chemistry (Table 1).

Depending on the modality of liquid removal, aerogels, xerogels and alcogels can be distinguished. In aerogels, the liquid phase is removed in the form of gas under hypercritical condition; xerogels are monoliths formed after liquid removal by thermal evaporation and, finally, alcogels are defined as gels in which the liquid phase is constituted by alcohols.

Adopting different synthesis parameters (e.g. temperature, pH) and applying different post-processing treatments on the gelled sol permit the formation of several different morphologies in the final glass product [33]. Therefore, sol–gel process is well recognized as a highly versatile approach to produce BGs.

Sol–gel products can be obtained through:

a. gelation of sol-derived colloidal powder
b. hydrolysis and poly-condensation of precursors, such as as alkoxides or nitrates, under hypercritical conditions
c. hydrolysis and poly-condensation of alkoxides in ambient atmosphere.

Three main phases can be defined in sol–gel processing method:

1. Sol preparation
2. Sol gelation
3. Solvent removal.

BGs are usually produced through hydrolysis and poly-condensation process of alkoxide precursors in ambient atmosphere.

There are seven major reaction steps in the sol–gel process for biomedical glass production [34]:

1. Mixing of alkoxide or organometallic reagents at room temperature leads to the formation of the sol through covalent bonding between the elements. In this step, hydrolysis and poly-condensation reactions are concurrent and proceed simultaneously. A structure-directing agent (surfactant) may be optionally introduced in the sol for finely modulating the nanopore size distribution via self-organization of micelles, which will be thermally removed in the step no. 7.
2. Sol casting into a proper mould, if the beaker used for the synthesis is not appropriate or special geometries are required.
3. Gelation: the glass network is formed due to progressive increase of fluid viscosity. Gelation time depends on solvent concentration, oxides/elements involved in the synthesis and water amount; it can be accelerated by the use of a proper acid (e.g. HF, HNO_3) or basic (e.g. NH_4OH) catalyst, depending on the type of synthesis.

Table 1 Classification of gels synthetized via sol–gel procedure

Type of gel	Bonding	Source	Gel schematic
Colloidal	Particles connected by Van der Waals or hydrogen bonding	Metal oxide or hydroxide sols	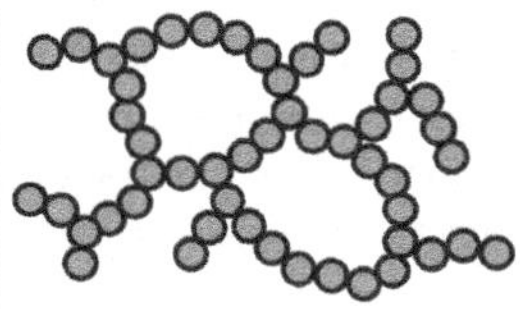
Metal-oxane polymer	Inorganic polymers interconnected via covalent or intermolecular bonding	Hydrolysis and condensation of metal alkoxides, e.g. SiO_2 from tetramethyl orthosilicate	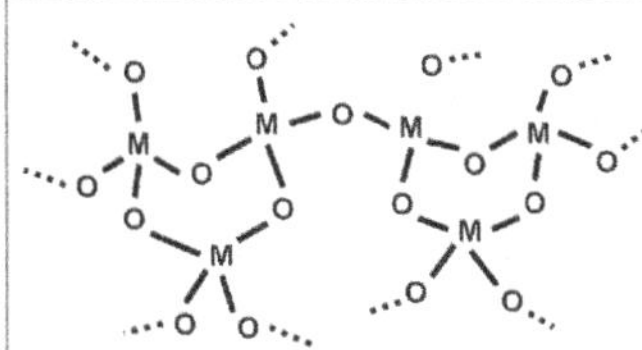
Metal complex	Weakly interconnected metal complexes	Concentrated metal complex solution, e.g. aqueous metal citrate or ethanolic metal urea, often forms resins or glassy solids rather than gels	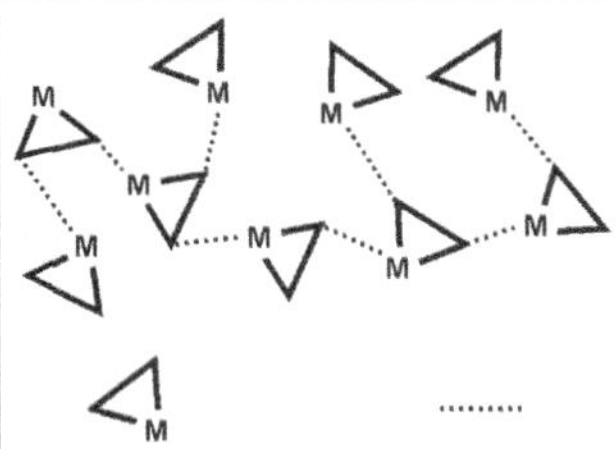
Polymer complex I In situ polymerizable complex (Pechini method)	Organic polymers interconnected by covalent and coordinate bonding	Polyesterification between polyhydroxy alcohol (e.g. ethylene glycol) and carboxylic acid with methal complex (e.g. metal citrate)	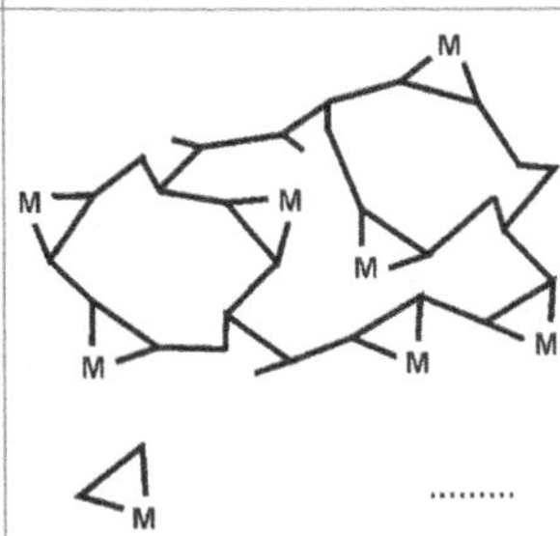
Polymer complex II Coordinating and crosslinking polymers	Organic polymers interconnected by coordinate and intermolecular bonding	Coordinating polymer (e.g.) alginate and metal salt solutions (typically aqueous)	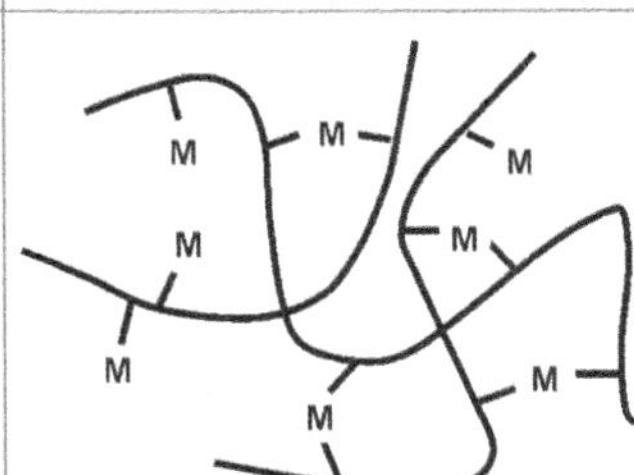

Table adapted from Ref. [32] under a Creative Commons Attribution 3.0 Unported Licence

4. Aging: poly-condensation prevails over hydrolysis reaction causing a decrease in gel porosity and an increase of mechanical properties. This process usually occurs at 25–80 °C for several hours influencing also density, surface area and pore volume of the gel.
5. Drying (liquid phase removal): colloidal gels should be easily dried, but great capillary stress may be originated leading to cracking problems.
6. Dehydration (also known as chemical stabilization): silanol bonds are removed from the network reaching the final chemical stability of the glass.
7. Densification (also called calcination): this process occurs by thermal treatment in furnace at relatively high temperatures (about 500–700 °C, which however are definitely lower than those required for melt-quenching technique).

TEOS is commonly used as precursor for silica while nitrates act as source for modifiers; metal chlorides can also be used to introduce additional cations (e.g. Fe^{2+}/Fe^{3+} or Cu^{2+}). A comprehensive picture of sol–gel processing and the role of precursors/reagents on the glass properties can be found elsewhere [33].

Different studies have shown the effects of H_2O/TEOS or ethanol/TEOS molar ratios and solvent concentrations on TEOS hydrolysis: increase of H_2O/TEOS ratio or decrease of EtOH/TEOS ratio improves solvent polarity and interfacial energy, with the formation of bigger primary particles of silica [35]. Hydrolysis and poly-condensation are the key processes of sol–gel synthesis allowing the formation of the glass network through different bond recombination.

The two chemical reactions are represented below:

a. Hydrolysis is defined as a nucleophilic attack in which –OH group replaces –OR group (Eq. (1)):

$$\equiv \mathrm{Si-OR} + \mathrm{H_2O} \leftrightarrow \equiv \mathrm{Si-OH} + \mathrm{ROH} \quad (1)$$

b. Condensation yields water or alcohol as final products through the formation of silanol bonds (Eq. (2)):

- Condensation with alcohol elimination:

$$\equiv \mathrm{Si-OR} + \mathrm{OH-Si} \equiv \leftrightarrow \equiv \mathrm{Si-OH-Si} \equiv + \mathrm{ROH} \quad (2)$$

- Condensation with water elimination (Eq. (3)):

$$\equiv \mathrm{Si-OH} + \mathrm{OH-Si} \equiv \leftrightarrow \equiv \mathrm{Si-O-Si} \equiv + \mathrm{H_2O} \quad (3)$$

In these reactions, R indicates an alkyl functional group in the form C_xH_{2x+1}.

BGs produced by sol–gel technique have shown an enormous potential in tissue engineering applications compared to the traditional melt-derived materials [36]. Firstly, sol–gel is considered as a technologically appealing method due the lower temperatures required and the higher versatility of final products.

Others advantages of sol–gel BGs over melt-derived BGs include [37]:

1. High specific surface area (above 50 vs. less than 1 m^2/g), associated to higher solubility and higher reactivity (i.e., faster HA-forming kinetics) in biological environment;
2. Nanopore size which can be tailored by modifying the processing parameters;
3. Simpler compositions, that allow avoiding the addition of high amount of alkaline oxides to lower the melting temperature and facilitate the glass processing;
4. Possibility to adjust the composition during the synthesis;
5. Variety of products (e.g. monoliths, particles, hollow spheres etc.) that can be obtained by just modifying some processing parameters;
6. Bioactive properties in a wider compositional range (up to 90 mol.% of silica).

From a chemical perspective, the properties of sol–gel BGs can be further expanded by bonding functional groups on the glass surface due to the high presence of silanol groups [38]. The grafting of functional groups usually occurs to enhance biocompatibility, binding affinity for proteins, adsorption of biomolecules and drug loading capacity.

However, sol–gel BGs have higher brittleness and lower mechanical properties than melt-derived glasses with the same composition due to their inherently porous structure comprising mesopores in the range of 2–50 nm [39]. Interestingly, mesopores are concurrently the strength (improvement of bioactivity, drug loading/release) and the weakness (poor mechanical properties) of such materials. Recently, additive manufacturing strategies have allowed overcoming—at least partially—these drawbacks [40].

5 Crystallization of Bioactive Glasses

The bioactivity mechanism of glasses relies upon the dissolution of the amorphous network when the material is in contact with biological fluids. However, during high-temperature post-processing treatments (e.g. sintering of glass powder compacts), BGs may undergo devitrification [17]. This involves the partial conversion of glass into crystalline domains, yielding the production of a glass–ceramic material. Development of crystalline phases reduces the volume of the glassy fraction and, therefore, has an obvious impact on bioactivity. The nature of the crystalline phases that may nucleate and grow are strongly dependent on glass composition. Heat-treated 45S5 Bioglass® and similar oxide systems typically form sodium-calcium-silicate phases: specifically, 45S5 and S53P4 ($53SiO_2$–$23Na_2O$–$20CaO$–$4P_2O_5$, wt.%) glasses tend to crystallize to combeite-like phases ($Na_2CaSi_2O_6$ [41, 42] and $Na_2Ca_2Si_3O_9$ [43], respectively) above 550–600 °C. Glasses such as 1–98 ($5.9Na_2O$–$7.1K_2O$–$7.6MgO$–$23.9CaO$–$0.9B_2O_3$–$0.9P_2O_5$–$53.8SiO_2$, mol.%) crystallize as wollastonite ($CaSiO_3$) upon heating above 800 °C [27, 44]. It is also possible that some systems form diopside crystals ($CaMgSi_2O_6$) [45–47]. The presence of crystalline phases can significantly slower the formation rate of HA on the glass surface; however, even if 100%

conversion is achieved, fully crystallized 45S5 Bioglass® was still reported to be weakly bioactive [48]. The HA-forming ability was also maintained to some extent in both partially and fully crystallized glasses having similar composition to 45S5 [49]. In general, the HA-forming kinetics are dictated by the amount of crystalline phase. If the crystalline fraction in the 45S5 system is below 60%, HA forms in less than 20 h on the glass surface; otherwise, it takes 25 h or more. For the sake of comparison, HA starts to form on fully amorphous melt-derived 45S5 glass by less than 10 h in SBF.

Both sodium-calcium-silicate and wollastonite crystals have been proved to dissolve in vitro when immersed into SBF, but the dissolution of wollastonite is much slower compared to the former phase [50]. On the contrary, wollastonite is significantly more reactive than diopside in SBF [51].

45S5 glass has a dramatically narrow sintering window, which means that glass powder compacts of this glass cannot be practically sintered without undergoing concurrent devitrification (i.e., sinter-crystallization occurs unavoidably). Some studies carried out over the last years have been addressed to developing BG compositions that can undergo high-temperature thermal treatments without devitrification, in order to obtain, for example, good sinterability while maintaining high bioactive properties. A successful example is represented by the 13–93 glass composition ($53SiO_2$–20CaO–5MgO–$4P_2O_5$–$12K_2O$–$6Na_2O$, wt.%) [52, 53], which has received FDA approval in the USA and CE marking in Europe for clinical use.

Devitrification of BGs is also dependent on dimensional effects: the smaller the glass particle size, the lower the temperature for crystallization onset (in other words, crystallization is "anticipated" at lower temperatures) [54, 55]. This behaviour is generally common to all glasses and its details are described in the Chap. 4 of the present book.

Crystalline phases were also reported to nucleate and grow in a sol–gel multicomponent $47.5SiO_2$–20CaO–10MgO–$2.5P_2O_5$–$10K_2O$–$10Na_2O$ (mol.%) BG after calcination at different temperatures, without any significant impact in vitro bioactivity [56]. These gel-derived glass–ceramic materials, in spite of their partially crystalline nature, exhibited a faster HA-forming ability in SBF as compared to their melt-derived counterparts.

Bioactive glass–ceramics are highly appreciated in dental applications, as the crystalline phases play a role in improving the mechanical properties of parent BGs; for example, the toughening effect of crystals is key to withstand cyclic masticatory loads and extend implant lifetime [57].

The increase of mechanical properties in bioactive glass–ceramics as compared to the parent BGs is also beneficial in porous scaffolds, which are often produced by sinter-crystallization of glass powders.

6 Role of Dopants

Incorporating different ions into the BG composition allows modulating some physico-chemical properties (e.g. solubility) while conferring specific therapeutic action [58].

Incorporation of special ions in the glass network is called doping process and has been widely employed for the production of BG-based multifunctional products for various clinical applications [59].

A doping element, by definition, is an additional incorporation in the main composition at a very low concentration as compared to the major components, ranging from a few ppm to a few percent units [58].

Recent studies have shown that the controlled introduction of dopants can lead to increased efficiency in performing a specific therapeutic action (e.g., antibacterial properties, angiogenesis) [60]. In other cases, doping may have an effect on glass dissolution kinetics, stability against crystallization, thermal and mechanical properties etc. [61–63].

At the beginning, the dopants were selected according to their similarity in valence with the elements already contained in the glass; later, the choice of the dopants was guided by a more "biological" criterion, depending on the essential trace elements required in the human body and their action on cell and tissue metabolism [64]. Many metal ions, acting as enzyme cofactors, can affect signaling pathways and promote tissue formation, and are thus considered highly-interesting doping materials in biomedicine (Table 2) [18].

7 Three-Dimensional Glass-Based Scaffolds for Regenerative Medicine

7.1 *Properties and Requirements*

Before discussing about BG scaffolds, some concepts related to tissue engineering and its latest developments need to be introduced. Tissue Engineering and regenerative medicine are considered the new frontier of biomedicine for repairing and regenerating damaged biological tissues. These multidisciplinary fields of research concern the development of biocompatible tissue substitutes to be implanted in the injured site and able to stimulate the growth of functional tissue [65].

Figure 6 schematically shows the regenerative process, highlighting the three main elements considered in the tissue-engineering approach:

- The cells relevant to the diseased tissue are the first element to consider before developing a tissue substitute. Cells are responsible of tissue synthesis and trigger regeneration mechanisms.

Table 2 Biological role of the main BG elements and dopants in the human body

Element	Biological activity
Si	• Stimulatory effect on formation of bone tissue • Intake of Si increases bone mineral density • HA precipitation • Helps to stimulate collagen I formation and osteoblast differentiation
Ca	• Favours osteoblast proliferation, differentiation and mineralisation • Activates Ca-sensing receptors in osteoblast cells
P	• Matrix Gla protein (MGP) stimulation
Mg	• Helps to form new bone enhancing osteoblast activity • Increases bone-cell adhesion and stability
Zn	• Shows anti-inflammatory effect • Bone formation in vitro by activation of protein synthesis in osteoblasts • Increases ATP activity
Sr	• Beneficial effects on bone formation in vivo • Anti-resorption effect on bone (for osteoporosis)
Cu	• Stimulates proliferation of human endothelial cells and, in general, angiogenesis • Antibacterial properties
Ag	• Antimicrobial properties • Anti-inflammatory properties
Co	• Potent pro-angiogenic effect
Li	• Treatment of both bipolar and unipolar depressive disorder • Enhances immunological activities of monocytes and lymphocytes

- Scaffolds are three-dimensional (3D) porous structures that provide support to cells allowing them to adhere, migrate, proliferate and differentiate.
- Biological, chemical or physico-mechanical signals influence cell pathways during each steps of cell proliferation and differentiation.

The term "scaffold" literally means "supporting framework" and was originally introduced in the field of Civil Engineering. In tissue engineering, scaffolds can be defined as (porous) materials that have been engineered to cause desirable cellular interactions contributing to the formation of new functional tissues for medical purposes.

In bone repair applications, scaffolds should mimic as much as possible the 3D trabecular architecture of healthy cancellous bone (Fig. 7) in order to optimize the integration with the host tissue and provide a suitable template for tissue regeneration. Given the broad variability in the structural and mechanical features of bones due to sex, age, activity and pathologies of patients, a "universal" scaffold does not exist but some design recommendations can be identified, regardless of the biomaterials used. The basic functions that a scaffold should primarily perform are [66, 67]:

(1) Providing the correct anatomic geometry, matching the defect size and shape
(2) Withstanding the mechanical loads typical of the interested site
(3) Stimulating tissue regeneration.

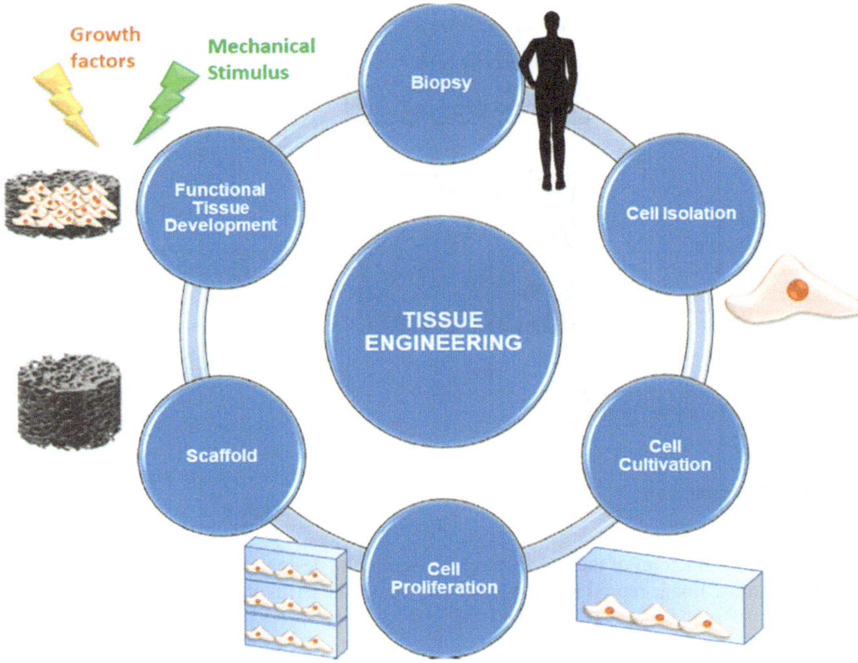

Fig. 6 Schematic illustration of tissue-engineering approach from patient biopsy to tissue substitute implantation. Image reproduced from Ref. [65] nder the terms and conditions of the Creative Commons Attribution (CC BY) license

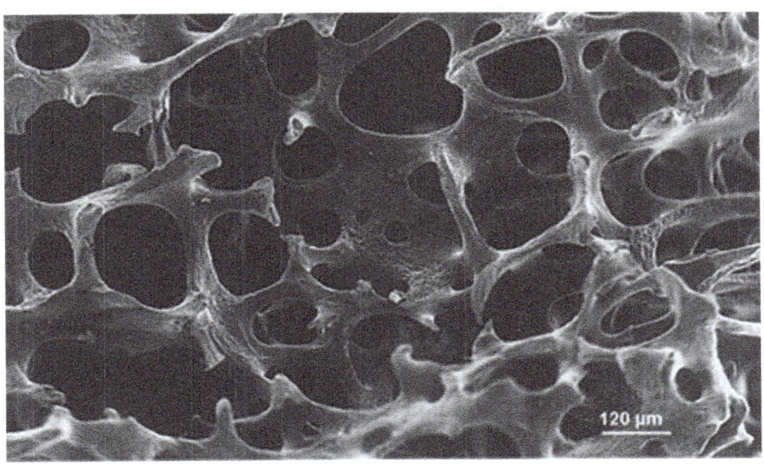

Fig. 7 SEM imaging of bone acquired in secondary electron mode shows the microarchitecture of trabecular bone. Image adapted with permission from Ref. [68]. Copyright © 2011, International Osteoporosis Foundation and National Osteoporosis Foundation

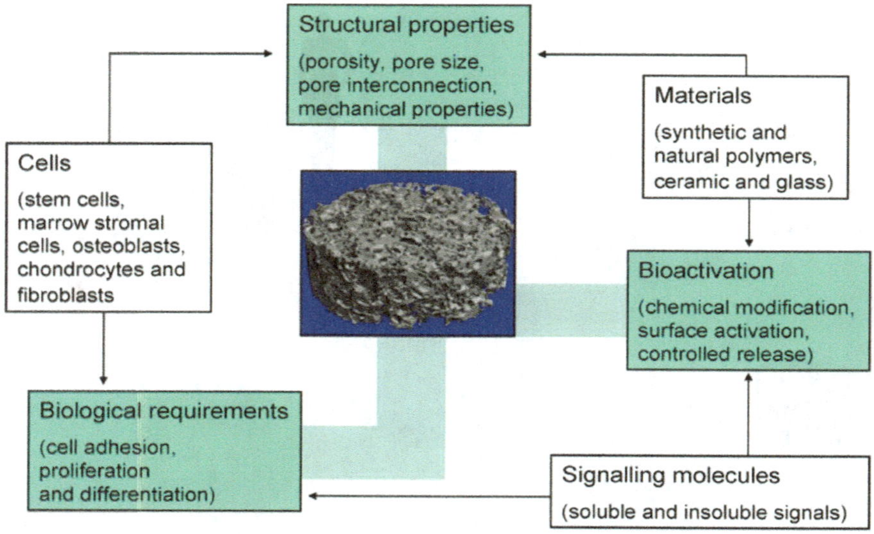

Fig. 8 Schematic diagram of the key factors involved in the design of optimal scaffolds for bone tissue engineering. Image reproduced from Ref. [69] under the terms and conditions of the Creative Commons Attribution License (CC BY-NC-SA 3.0)

According to the final application, scaffolds should match the structural and mechanical properties of host tissue and optimize the micro-environment of the defected site. Focusing on bone repair, the design and development of an "ideal" scaffold should account for different requirements (Fig. 8), including [69–71]:

(1) Biocompatibility and bioactivity (i.e. HA-forming ability)
(2) Capability to bond the host tissue without scar formation
(3) Porous and interconnected structure
(4) Mouldability in different shapes and sizes
(5) Suitable degradation rate
(6) Maintenance of mechanical properties
(7) Easy fabrication with affordable cost
(8) Sterilization without damage or deterioration.

The main properties required for bone tissue-engineering scaffolds are also collected and shortly discussed in Table 3.

BGs, being also osteoinductive (i.e. able to stimulate osteogenesis via ionic dissolution products) [8, 73], are highly promising materials for making scaffolds.

The first BG-based macroporous scaffolds were fabricated in the early 2000s by applying foaming techniques to melt-derived 45S5 glass (H_2O_2-driven foaming) [74] or sol–gel glasses (surfactant-mediated foaming) [75]; the latter approach actually yielded achieving hierarchical systems with macro- (> 100 μm) and meso-scale (2–50 nm) porosity [76].

Table 3 Overview of the key properties for a scaffolds aimed at regenerating bone [72]

Property	Effect/explanation
Ability to deliver cells	The material should not only be biocompatible but also foster cell attachment, differentiation and proliferation
Osteoconductivity	Osteoconductivity not only allows avoiding the formation of fibrous tissue around the implant (encapsulation) but also brings about a strong bone between scaffold and host bone
Biodegradability	The composition of the material, combined with the porous structure of the scaffold, should lead biodegradation/dissolution in vivo at rates appropriate to tissue regeneration
Mechanical properties	The mechanical strength of the scaffold, which is determined by both the properties of the biomaterial and the porous structure, should be sufficient to provide mechanical stability and withstand loads at the implant site prior to synthesis of the new extracellular matrix by cells
Porous structure	The scaffold should have an interconnected porous structure with total porosity >50 vol.% and pores sizes between 100 and 500 μm for allowing cell penetration, tissue ingrowth and vascularization
Fabrication	The material should possess a certain technological versatility, for example, being readily produced into irregular shapes so that the scaffolds can match the defect geometry in the bone of individual patients
Commercialization potential	The synthesis of the basic material and the scaffold fabrication process should be reproducible and reliable. The scaffold should also be sterilisable without losing its properties and be marketable at an affordable cost

Since then, a lot of experiments have been carried out to produce BG-based scaffolds in myriads of shapes and sizes in order to best fit into the damaged bone. With the evolution of manufacturing processes, major issues concerning the intrinsic brittleness of glass and glass–ceramic materials have been overcome—at least partially—and scaffolds with mechanical properties comparable to those of human bone have been produced [72].

The concurrent need for adequate mechanical properties and a highly-porous structure of interconnected macropores, which are key in tissue engineering applications, is a big challenge for BG-derived scaffolds. These two factors are intrinsically correlated as high porosity in glass–ceramic scaffolds results in low mechanical properties. A deep analysis carried out by Gerhard and Boccaccini [69] on many types of BG scaffolds has shown a negative linear relationship between scaffold porosity and compressive strength, characterized by coefficients of determination R^2 between 0.80 and 0.99. This result means that the systematic impact of porosity on the variability of compressive strength is at least 80% (Fig. 9). A power-law relation was observed between elastic or shear modulus and total porosity in highly-porous (> 50 vol.%) 45S5-based glass–ceramic foams, while the Poisson's ratio was a linear function of porosity [77]. The relation between tensile strength and porosity in foam-like

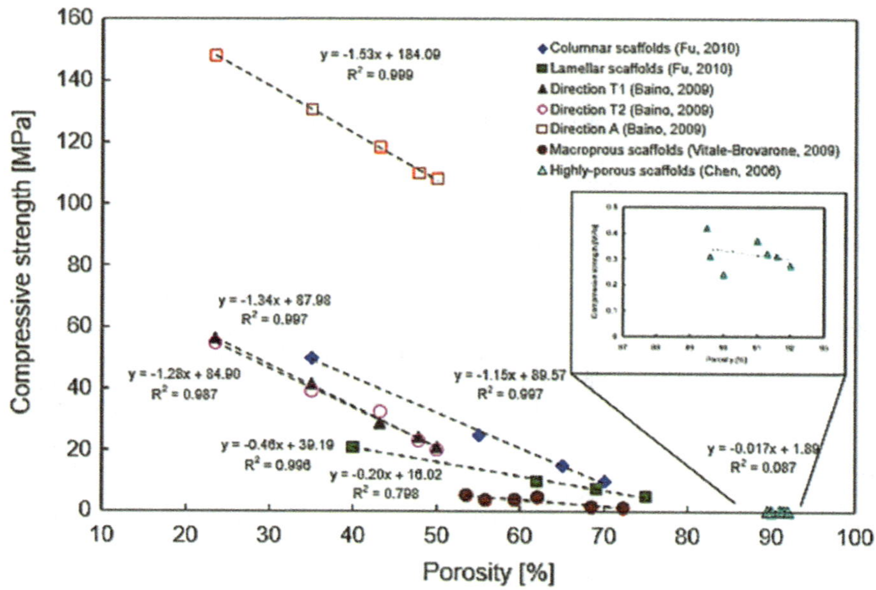

Fig. 9 Compressive strength vs. porosity curve for glass–ceramic scaffolds. The negative slope shows how increase in porosity percentage reduces mechanical compressive strength according to negative linear relationships. It can be noticed also that for very high values of porosity (≈ 85–95%) the relation does not fit a linear curve and the mechanical performances of the scaffold are dramatically poor (< 0.4 MPa), being unsuitable for biomedical applications. Image reproduced from Ref. [69] under the terms and conditions of the Creative Commons Attribution License (CC BY-NC-SA 3.0)

silicate BG scaffolds was reported to obey more complex models based on quantized fracture mechanics [78].

In this scenario, one possible solution to solve the mechanical drawbacks of BG-derived scaffolds relies on the use of multifunctional composite structures, which are typically produced by (i) the integration of a biodegradable polymer matrix with BG particles as filler phase or (ii) a polymeric coating on BG surface [79]. In this way, the polymeric component strongly improves the mechanical properties of BGs (strength and toughness), acting as a "glue" in keeping glass particles together while the scaffold starts to fail [80, 81].

However, some recent studies have shown that composite BGs and polymeric coatings introduce significant side effects in scaffolds features. For example, the polymeric coating may negatively affect scaffold bioactivity performances due to its covering role. Furthermore, the introduction of the coating also influences environment conditions by its premature degradation which releases acidic products that strongly reduce pH. Finally, mechanical strength and properties do not remain constant after the scaffold implantation in vivo, but rapidly decrease because of the interaction between polymeric coating and glass and the influence of reciprocal degradation mechanism [82].

7.2 *Fabrication Methods*

Since the fabrication of the first BG scaffold in 2001 [74], a lot of researchers had focused their studies on the optimization of manufacturing processes in the attempt of obtaining an "ideal" scaffold [83]. In principle, manufacturing processes should be easily repeatable and should give the same outputs, guaranteeing constant characteristics of scaffolds and allowing potential mass production. The processing route should also be economically sustainable and safe for all workers involved in the process.

In the last two decades, different technologies have been developed to produce glass-based scaffolds. Manufacturing methods may be divided into two different major categories, known as conventional methods and additive manufacturing techniques (Table 4).

Conventional methods are characterized by a top-down approach in which the realization of the desired form occurs by the progressive removal of material from a bigger bulk piece (e.g. sacrificial pore-forming agents). On the contrary, additive manufacturing technologies (AMTs) indicate those bottom-up approaches where 3D

Table 4 Overview of the manufacturing techniques used for the production of glass-based scaffold in bone tissue engineering [72]

Manufacturing methods	Technological class	Specific methods
Conventional	Foaming techniques	Gel-casting foaming, sol–gel foaming, H_2O_2 foaming
	Thermal consolidation of particles	Organic phase burning-out: polymeric porogens, starch consolidation, rice husk method
	Porous polymer replication	Coating methods, foam replication
	Freeze-drying	Freeze-casting of suspensions, ice-segregation-induced self-assembly
	Thermally induced phase separation	
	Solvent casting and particulate leaching	
Additive manufacturing	Selective laser sintering	
	Stereolithography	
	Direct ink writing	3D printing, ink-jet printing, robocasting

structures are fabricated by progressively adding materials in the form of thin layers to obtain the desired morphology. A very comprehensive picture of BG fabrication strategies can be found elsewhere [72].

Historically, foaming methods were the first routes used to produce glass-based 3D scaffolds. These techniques take their name from the use of a foaming agent in the manufacturing process. The foaming agent is generally introduced in a slurry to create air bubbles which are responsible of porosity in the final product. The use of these techniques may lead to some side effects, such as high brittleness of BG-derived scaffolds, low pore interconnectivity, and lack of pores in the outer layer due to formation of an external compact "skin". Gel-cast foaming, sol–gel foaming and H_2O_2 foaming belong to this family of manufacturing methods.

Gel casting employs an organic monomer that is polymerized to cause the in-situ gelation of a foamed aqueous slurry containing melt-derived glass particles [84].

Unlike gel-casting, where the glass has been previously prepared, sol-gel foaming involves the formation of the 3D network of macropores simultaneously to the glass synthesis. A surfactant is added to the sol and, upon vigorous stirring, gelation occurs along with incorporation of air bubbles; thermal treatment allow consolidation of solid skeleton/glass formation [75]. The development of this method was due to the initial difficulty of producing melt-derived BGs that could undergo another high-temperature thermal step without crystallizing, with a partial loss of bioactivity. Although most sol–gel BG compositions are relatively simple, being based on binary (SiO_2–CaO) [85] or ternary (SiO_2–CaO–P_2O_5) [86] systems, Ag-doped [87] and Fe-doped [88] scaffolds were also produced by this method to impart antibacterial and magnetic extra-functionalities, respectively.

H_2O_2 foaming process includes the use of a peroxide solution as foaming agent. It was observed that pore interconnectivity, pore size and total porosity of the final 3D scaffold increased with increasing H_2O_2 content, but controlling these increments was a difficult task [89].

Methods based on thermal consolidation of particles include all those processes which require the introduction of sacrificial porous templates or particles, usually polymeric, before the sintering procedure of the green body [72]. These techniques allow properly tailoring the porosity degree and characteristics by controlling process parameters and varying the type of templates (e.g. polyurethane sponge [90], marine sponge [91], stale bread [92]) or particles (e.g. starch [93], polyethylene [94]) used. In principle, both melt-derived and sol–gel glass particles can be used to produce the green. The process could also occur without the introduction of any sacrificial particles or template but just varying the sintering parameters; in this case the final porosity comes from inter-particle voids only [95]. A special variant of these methods involves concurrent foaming and glass particle sintering: for example, if dolomite fine powder is used as a foaming agent in the green compact bodies, the result is qualitatively similar to that obtained by sol–gel foaming but the scaffolds are mechanically stronger [96].

As an alternative to the use of organic particles/foams as porogen agents, freeze-drying methods have been developed in which the formation of ice crystals generates porosity in the final 3D scaffolds [97]. In these methods, the suspension containing glass particles is frozen and then the solvent crystals are removed, leaving a porous structure in the scaffold that will be thermally consolidated during the final processing step.

Another conventional method involves thermally induced phase separation, which relies on the change of solubility between different polymers depending on temperature variations. This process mainly produces polymeric scaffolds, but it can be also extended to the fabrication of polymer/glass porous composites [98]. The main step is addressed to polymer solution cooling which reveals phase separation; then, porous structures can be obtained by selective phase removal.

Unlike conventional methods, AMTs include all the techniques which involve the use of a CAD model or a computed tomography (CT) reconstruction as a template for the final product [99]. There are two main classes of AMTs to produce ceramic and glass materials, i.e. direct fabrication techniques, which produce sintered ceramic or glass parts without needing any further thermal treatments, and indirect fabrication techniques that involve layer-wise building of the scaffold (printing), thermal debinding and sintering.

Selective laser sintering belongs to the class of direct fabrication techniques because just one step is necessary to produce the 3D scaffold. In fact, the CAD model of the object is followed by a computer that controls a laser over a bed of glass powders, which can be locally melted or sintered to produce the desired path [100].

Stereolithography probably represents the most accurate AMT, reaching high resolution values up to 20 μm. In this process a UV-photocurable liquid polymer, a UV-laser and a movable platform are used to build layer-by-layer the 3D object [101]. The major limitation of this technique is the poor availability of UV-curable polymers with appropriate rheological characteristics.

Direct ink writing methods include many different AMTs, such as 3D printing, ink-jet printing and robocasting [102]. In all these techniques, a pattern-generating device (print head or nozzle) builds up a 3D object by following computer instructions from a script file or virtual CAD model. While in 3D printing the ink is formed by the binder and glass particles are in the building bed, in ink-jet printing the ink contains all the components. Both in 3D printing and ink-jet printing, the ink flow through the nozzle is generated by acoustic, electrical or piezoelectric systems (or their combinations). Generally speaking, the expression "3D printing" is often used to indicate any technique belonging to direct ink writing methods. Robocasting is probably the most common AMT and involves the extrusion of a continuous filament by pressurized air [53, 103]. Its main advantage is the possibility to change ink viscosity through chemical and physical processes in order to achieve strong 3D porous structures. Robocasting also allows printing polymer/BG filaments, thus obtaining composite scaffolds [104].

8 Bioactive Glass Coatings

The need for bioactive coatings is of particular interest when inert materials are implanted in the patient's bone. In fact, after being implanted, nearly-inert ceramics (e.g. alumina) or metals are typically encapsulated within fibrous tissue without establishing a chemical bond with host bone. BG coatings have the potential to overcome this limitation as they can improve the stability of underlying implant by tightly bonding it to the host bone. Furthermore, if metallic implants are used, BGs protect the metal from corrosion and avoid the release of toxic metallic cations in vivo [105]. The glass composition, which dictates the bioactive behavior, should be carefully designed if the intended application is for coating: in fact, glasses with high bioactivity are also prove to quickly dissolve in the biological fluids, thereby causing instability of the implant lying underneath. This is probably the major reason why the use of BG coatings is still limited compared to other bioceramics, such as non-resorbable thermal-sprayed HA [106].

Another key point to consider is the mismatch between the thermal expansion coefficient (TEC) of BG and substrate. Ideally, the TEC of BG should be as close as possible to that of the substrate to prevent the glass pulling away from the implant upon thermal processing (e.g. sintering) [107]. However, the TECs of 45S5 Bioglass® (15×10^{-6} °C^{-1}) and of most of silicate BGs are significantly higher than that of titanium alloys (about 9×10^{-6} °C^{-1}), which are commonly used in orthopaedic and dental prosthetics. Therefore, a great challenge of the next few years will be the development of new BGs with more suitable TEC and dissolution rate for use as coating materials.

Conventional techniques for producing BG coatings include manual deposition or dipping followed by glass particle sintering (enameling and glazing) [108–114] and thermal spraying [115–117]. In the last years, new approaches (e.g. multilayer BG coatings to achieve a good compromise between adequate TEC, slow dissolution rate and bioactivity [118]) and fabrication methods (e.g. electrophoretic deposition [119], radio-frequency sputtering [120]) have been experimented to produce well-adherent and durable coatings on a variety of materials and implants, including scaffolds, suture wires, surgical screws and ocular implants. Composite coatings were also produced where the polymeric phase, acting as a glue, ensures a good adhesion of the coating to substrates of complex geometries [121, 122]; in this case, thermal post-processing must be avoided to preserve the organic phase. A comprehensive overview of BG coatings has been provided by Baino and Verné [123]; in this regard, a special update of that work deserves to be reported here about antiviral coatings, the potential of which became so apparent from winter 2019 in the frame of the global battle against Coronavirus SARS-CoV-2.

During the pandemic of COVID-19 caused by the new Coronavirus SARS-CoV-2, the confinement measures slowed down the contagion but did not completely avoid the disease diffusion. The individual protection equipment (e.g. facial masks) as well as the filters for air conditioning systems and medical respiratory devices do not possess an intrinsic antimicrobial/antiviral action and, thus, are susceptible to microbial/viral colonization. An efficient antimicrobial/antiviral technology on air filtering media is crucial for maintaining a safe air environment and protecting people, in particular when lockdown is eased. In this regard, a silver nanocluster/silica composite coating deposited with a patented co-sputtering process [124–126] demonstrated its antibacterial, antifungal and antiviral behavior in different applications, such as biomedical implants [127], natural and technical textiles [128, 129], mobile phones [130], air filter [131] and aerospace structures [132]. Recently, this coating has proved to elicit an antiviral effect towards Coronavirus SARS-CoV-2 [133, 134]. In principle, this coating can be deposited on every kind of substrates (i.e. metallic, ceramic, polymeric and glass surfaces), including filtering devices and textiles. It can hence provide an effective contribution to safety of crowded areas like supermarkets, production sites, schools, hospitals etc., where surfaces are exposed to many contacts with body parts every day. This coating can also increase the working life of filtering masks and filtering media, thereby reducing the waste production related to their disposal. This thin antimicrobial/antiviral silver nanocluster/silica composite coating (less than 200 nm) deposited on a disposable facial FFP3 mask in non-woven fabric is shown in Fig. 10. The coating, made of silver nanoclusters well embedded in a silica glass matrix, exhibits the typical morphology displayed in Fig. 11, as already observed in several previous works [127, 131].

While most studies on coatings underline the need for obtaining pore-free layers, Verné et al. [135] innovatively applied the concept of "trabecular metal", which is already used in clinics for improving fixation of orthopaedic and dental devices, to the field of BG surface coatings. Specifically, they claimed the concept that a single-piece ceramic acetabular cup can be fixed to the patient's bone by means of a porous BG coating with trabecular structure. Fabrication of early prototypes of such an implant was a highly challenging task: specifically, dipping procedures combined with airbrush spraying of BG slurries were used to manufacture the non-porous interlayer on the curved surface of the cup [136], while the outer trabecular coating was produced by properly adapting the sponge replica method to the 3D radial geometry of the ceramic prosthesis [137, 138]. A second-generation prototype

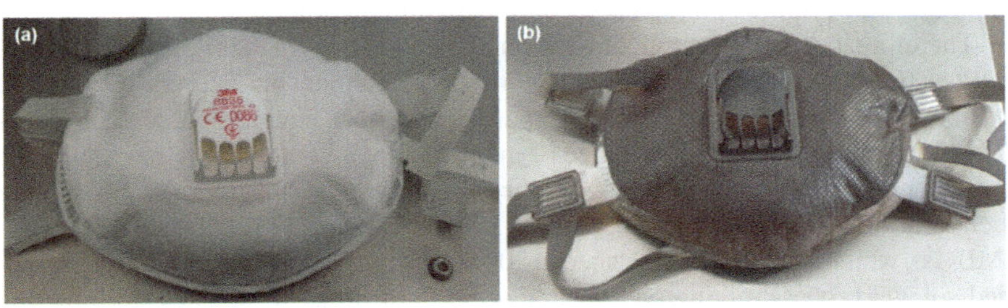

Fig. 10 Disposable facial FFP3 mask in non-woven fabric **a** uncoated (as such) and **b** coated by a co-sputtered antimicrobial/antiviral silver nanocluster/silica composite layer (original photos, courtesy of Monica Ferraris)

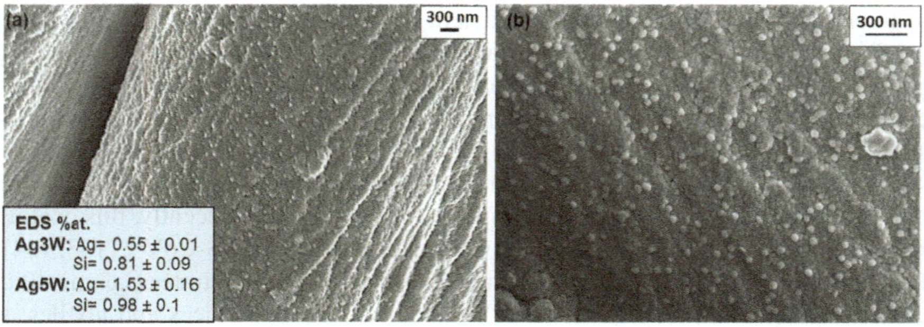

Fig. 11 Typical morphology (FESEM images) of a coated textile at lower (**a**) and higher (**b**) magnification: bright spots are silver nanoclusters embedded in the silica matrix (original SEM micrographs, courtesy of Monica Ferraris)

was produced by laser cladding of the BG particles directly on the outer surface of the cup [139].

9 Bioactive Glasses as Platforms for Controlled Drug Release

Drugs and growth factors, being organic by nature, cannot be incorporated in the glass during the manufacturing process if the material is produced by melting due to thermal degradation of biomolecules. On the contrary, drugs can be mixed into the starting sol if the glass is synthesized by a low-temperature sol–gel process [140]; however, this approach suffers from the obvious impossibility of performing calcination, which would be required for the glass stabilization and the removal of organic by-products. Another approach involves the introduction of therapeutic molecules into the organic phase of polymer/BG composites [141]; however, this approach may require complex procedures for drug incorporation.

Therefore, post-synthesis incorporation of biomolecules in the BG, once the material has been produced and no additional thermal treatment is required, seems to be the most feasible strategy. Silicate mesoporous materials, having an ordered arrangement of nanopores matching the size of many therapeutic molecules, have attracted great attention in the biomedical community due to their capacity of acting as efficient platforms for uptake and controlled release of drugs [142]. Mesoporous BGs (MBGs), which were first synthesized in 2004 [143]. are typically obtained by a modified sol–gel process relying on cooperative self-assembly of micelles (forming the mesoporores) and oligomeric silica species (forming the glass network). The surfactant and other by-products are removed by calcination, leaving behind an ordered arrangement of mesopores with channel diameter from 5 to 20 nm. The critical aspects of MBG synthesis have been recently discussed elsewhere in detail [144].

MBGs have faster HA-forming kinetics than conventional sol–gel BGs due to the high pore volume and surface area, which also allow achieving high loading efficiency of biomolecules into the mesopores as well as slow and controllable drug release kinetics [145]. The formation of a HA layer on the walls of mesopores implies a partial occlusion of the channels, which decreases the burst release effect and the overall release rate, thereby allowing a prolonged therapeutic effect to be obtained [146].

The uptake ability (and subsequent release) of biomolecules is affected by the composition, specific synthesis method and final form (e.g. spheres, fibres, coating ...) of MBGs. It was observed that the increase of CaO content in the MBG formulation led to the enhancement of loading efficiency and decrease of drug release rate and burst effect [147]. The explanation was that the drug molecule (tetracycline in their study) was chelated with calcium on the pore wall, which made it difficult to be released. Xia and Chang [148] observed a similar trend after comparing the gentamicin loading/release kinetics of different MBG compositions.

Mesopore size and volume are strongly dependent on the type of surfactant used during the MBG synthesis. It was observed that Pluronic P123-templated MBG had higher pore volume and specific surface area compared to Pluronic F127-derived MBG, hence the former material exhibited a significantly higher drug (metoclopramide)-loading efficiency (47.3%) compared to the latter (16.6%) [149]. A similar trend was also reported by Arcos et al. [150] in the case of triclosan-loaded P123- or F127-templated MBGs (loading efficiency: 9.7 vs. 9.1%); drug uptake could be further improved to 10.7% by using CTA-Br as a structure directing agent, leading to smaller pores that fit better with the size of the drug molecule.

The form in which the material is produced is a third factor influencing the drug release kinetics of MBG-based systems. In general, a sufficient drug uptake/release ability may always be obtained as well as an adequate therapeutic action, provided that the biomolecules may have access to the mesopores during the loading phase [151].

MBGs have been traditionally used as bifunctional biomaterials, combining bioactivity and drug release function; recently, they have also been proposed as trifunctional platforms for the additional controlled release of therapeutic ions able to promote, for example, pro-angiogenic or antibacterial effects [60, 152].

10 Antibacterial Bioactive Glasses

The introduction of meticulous hygienic protocols and the systemic administration of antibiotics have remarkably reduced the risk of infection development. Nevertheless, the bacterial contamination of implants still remains a serious complication in surgery, causing often re-operation, damage to patients and prolonged recovery.

Synthetic materials possessing antibacterial properties have been intensively investigated over the last decades; among them, bioactive glasses have become increasingly attractive due to their specific properties, such as their compositional versatility, surface reactivity and tailorable degradability.

Bioactive glasses with different composition (silicate-, borate- or phosphate-based systems) and different architectures (e.g. macro- or mesoporous glasses) have been proposed and investigated for their antimicrobial properties for both hard and soft tissue applications, following different strategies schematized in Fig. 12.

The most investigated approach to impart antimicrobial properties to a bioactive glass is the introduction in its composition of elements with antimicrobial properties, such as Ag, Cu, Zn or Ga. Doping elements can be introduced as reactants during the glass synthesis by melt-quenching process or via sol–gel technique [153–156]; otherwise, they can be incorporated [157–159].

The most commonly-explored antibacterial elements are silver, copper and zinc [160–163]; however, other elements, such as Ga, F, Sr, Ce, Bi and recently Te have been investigated [164–167]. The extensive use of elements with antibacterial effect is due to their broad range of activity toward Gram-positive and Gram-negative strains and their poor aptitude to induce bacterial resistance. In particular, silver-doped glasses have been synthesized both by melt-quenching method and sol–gel technique, as well as by ion-exchange process. Several efforts have been focused on

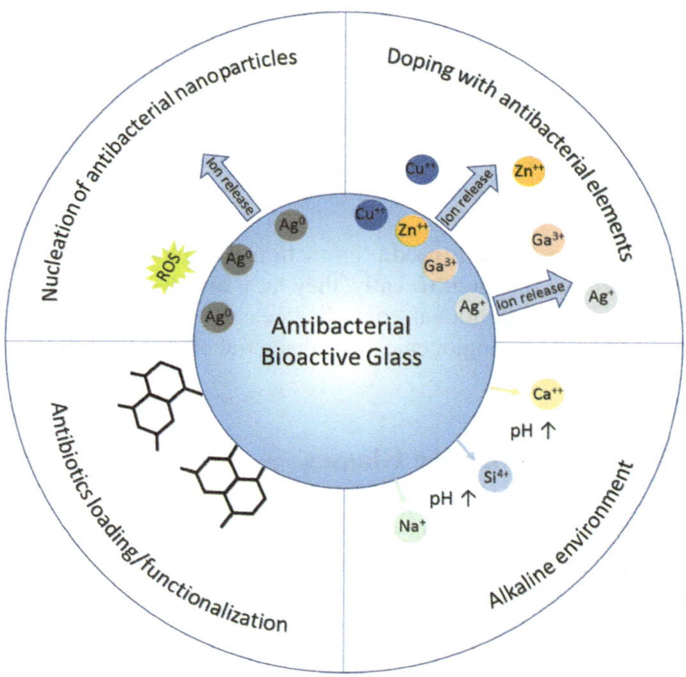

Fig. 12 Different strategies adopted to develop antibacterial bioactive glasses

careful tailoring of the silver amount in order to avoid toxic effects while maintaining unaltered the main properties of glass, especially the bioactivity [59, 168]. Due to the difficult control of the silver content and release, which were sometimes observed using the melt-quenching and sol–gel processes, the ion exchange technique looks very promising. This method allows preserving the structure of the pristine material (glass or glass–ceramic), maintaining the glass bioactivity and reaching a gradual and controlled release of silver ions [157, 159, 169, 170]. In general, the studies dealing with Ag-doped glasses revealed a good bacteriostatic or bactericidal behavior of the Ag-doped material towards both Gram-positive and Gram-negative bacteria (also biofilm producing) depending on the introduced silver amount, thus suggesting their potential application as base materials for a variety of bone substitution devices (e.g. bulk, coatings, macroporous scaffolds) [171, 172]. The high versatility of bioactive glasses doped with Ag ions allowed their investigation also as dispersed phase into polymeric bone cements [173–176].

However, the widespread use of silver-containing devices and the recent discovery of bacterial resistance to silver [177] is pushing researchers to investigate the antibacterial properties of other glass compositions. Therefore, copper or zinc-containing bioactive glasses have also been deeply investigated. In this case, the introduced amount of Cu or Zn must be carefully tailored in order to not compromise the glass bioactivity [59, 178]. Cu-containing bioactive glasses have been mainly obtained by sol–gel process (silica-based glasses) by introducing up to 10 mol% of copper [179, 180]. The traditional melt-quenching process was also adopted for silica-based, phosphate and borate bioactive glasses [156, 162, 181] evidencing that the release of copper ions significantly reduced the bacterial adhesion and proliferation. Recently, the ion-exchange process in aqueous solution has also been proposed to introduce copper in the surface of bioactive glass powders [158], demonstrating the ability of Cu-doped bioactive glasses to limit the *S. aureus* adhesion and proliferation, without affecting the bioactive properties. Furthermore, as for Ag-doped ones, Cu-doped bioactive glasses have been successfully investigated as dispersed phase in composite bone cements [182].

Zn-doped bioactive glasses with antibacterial properties have been investigated [163, 183] showing also in this case a dose-dependent antimicrobial effect. However, the performed studies evidenced an important influence of the ZnO amount in the leaching properties and reactivity of the glasses [184]. Finally, a comparison between the antimicrobial performances of Ag, Cu or Zn-doped glasses demonstrated that silver-containing bioactive glasses have better antibacterial effect than Cu- or Zn-doped glasses [156, 185].

Aiming to better understand the "killing" mechanism of antibacterial elements and to assess the differences among ion release and "contact-killing" mechanism, bioactive glasses containing nanoparticles with antibacterial properties have been recently developed. The formation of antimicrobial nanoparticles can be obtained during the glass synthesis [186, 187] or their nucleation can be promoted directly on glasses surfaces after the material synthesis [188, 189]. Even if the antibacterial action of nanoparticles is not yet fully understood and the efficacy of nanoparticles versus ions is still debated [190], the performed studies revealed a significant in vitro antibacterial activity against the most common bacterial strains.

Another approach to confer antibacterial properties to bioactive glasses is to functionalize or load them with antibiotics, as reported also in Sect. 11. The glass surface can be activated to exhibit many hydroxyl groups, useful for the grafting of several drugs; moreover, the reactivity of the glass in aqueous solution, as well as in SBF, can be exploited to incorporate antibiotics. For example, Miola et al. explored the possibility to graft an antibiotic (carbenicillin) on a bioactive glass surface by taking advantages of the bioactivity process, which occurs by dipping the glass in SBF [191]. The authors confirmed that carbenicillin can be easily incorporated and released with different kinetics on a self-formed silanols/silica gel layer by overworking the bioactivity process.

Mesoporous bioactive glasses, as reported in Sect. 9, have also been explored as smart drug delivery systems. They are characterized by highly ordered mesoporous channels useful to confine antibiotics and release them in a controlled manner. Different antibiotics, such as gentamicin, tetracycline hydrochloride, vancomycin, ceftriaxone and sulbactam sodium have been incorporated into mesoporous glasses. It has been demonstrated that the drug incorporation and release, and as a consequence the antimicrobial activity, are influenced by the use of different surfactants, which in turn impact on the pore dimension and the surface area [192–196]. Although the effectiveness of antibiotic-loaded bioactive glasses has been demonstrated both in vitro and in vivo, the prevalence of antibiotic-resistant bacterial strains still remains a major problem.

The last strategy exploited to limit bacterial contamination relies on the release of ionic compounds once bioactive glasses are immersed in aqueous-based solutions. It has been shown that bioactive glasses are able to increase the local pH due to the ion-exchange mechanism that occurs with protons in simulated (in vitro) or body fluids (in vivo). Specifically, it was proved that the alkaline microenvironment caused by the ion leaching limits the bacteria growth by altering their morphology and changing the expression pattern of numerous genes and proteins. Moreover, the release of ions that characterize a bioactive glass, such as silicon, sodium, calcium and phosphate ions, increases the salt concentration and enhances osmotic pressure, thus affecting bacterial proliferation [197, 198] section. The antimicrobial effect of bioactive glasses due to the release of ionic compounds has been reported by several authors in vitro [199, 200], also towards multi-drug resistant bacterial strains able to form biofilm [201]. However, the in vivo efficacy of the proposed mechanisms still has to be confirmed, since different works demonstrated that some bioactive glass compositions did not show antimicrobial efficacy in vivo due to the buffering of the biological environment [202, 203].

In conclusion, the results obtained by the different strategies highlight the promising features of bioactive glasses in reducing bacterial contamination. However, further investigations are needed to deepen the knowledge of the mechanisms involved in the different approaches, enhancing the efficacy of bioactive glasses in eradicating the biofilm and limit the infection development.

11 Functionalization of Bioactive Glasses

Bioactive glasses easily expose hydroxyl (–OH) groups upon contact with water-based media [14, 204, 205]. This is the first step of their bioactivity mechanism, but it can be also exploited for the effective grafting of specific moieties to their surface [204, 206]. Surface functionalization of bioactive glasses, which at the beginning was less explored than the one of metals and polymers, is gaining increasing interest in the scientific community, as demonstrated by some recent reviews [207, 208]. Surface functionalization represents an interesting and versatile strategy to combine well known properties of bioactive glasses (e.g., bioactivity as the ability to induce apatite precipitation in contact with physiological fluids and bone-bonding ability, as well as specific ion release) with specific properties of the grafted moieties designed for selected applications. In fact, it has been widely reported that surface grafting of active molecules does not inhibit the glass bioactivity [209–211].

Functionalization can be achieved through addition of surface functional reactive groups with a specific biological response or acting as linkers for a further step of grafting. Active biomolecules can be grafted directly to the glass surface or by means of proper spacers, as schematized in Fig. 13.

As far as reactive groups are concerned, the most widely explored ones are amino groups (–NH_2) introduced on the glass surface by means of silanization, mainly with 3-aminopropyltriethoxysilane (APTES) [204, 212].

Silanization can be performed to exploit the peculiar properties of the amino group itself, such as regulation of bioactivity [209, 213], enhanced protein absorption [214, 215], and cytocompatibility [216], or for covalent grafting of active molecules [206, 210, 211, 217]. Most of the literature is focused on the silanization of silica-based bioactive glasses; however, APTES grafting on borate and phosphate glasses has also been reported [218, 219]. It was found that, even upon optimization of surface modification, the level of APTES grafting in the last case was lower than in typical

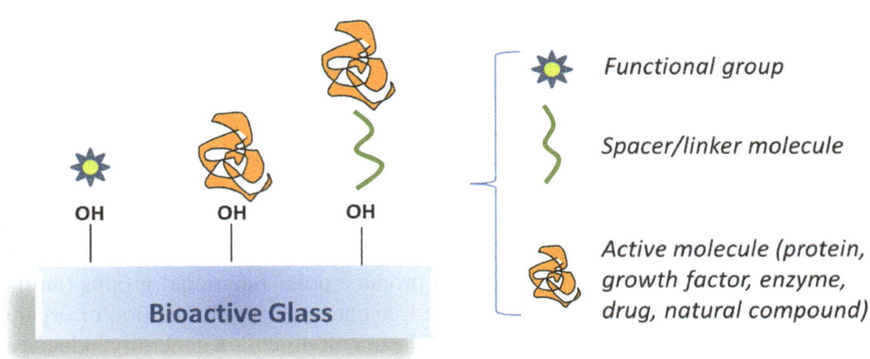

Fig. 13 Scheme of the surface functionalization strategies on bioactive glasses

silica-based glasses, most likely due to the lower content of –OH group on the surface and congruent dissolution of phosphate glasses.

Silanization induces a switch of zeta potential toward positive values because of the presence of the amino group and a strong beneficial effect on the fibronectin adsorption capacity, which is the highest for the NH_2 groups than for all functional groups (such as CH_3, COOH, OH) [220]. The hydrophilic-hydrophobic balance of the silanized surface, with respect to a simple hydrophilic glass surface, can be beneficial for protein adsorption.

Concerning protein adsorption, it must be underlined that 45S5 Bioglass® and bioactive silicate glasses adsorb a larger amount of bovine serum albumin () than bioinert glasses in a wide range of pH. The BSA adsorption on 45S5 Bioglass® is pH-dependent, with a larger amount of adsorbed BSA at lower pH such as in fractured or injured bone tissues (pH = 5). This could be beneficial to reduce the inflammatory response, but several attempts are reported to promote it further or to selectively enhance adsorption of adhesive proteins such as fibronectin. In contrast to bioinert glasses, the surface properties of BGs are time-dependent and change upon contact with the biological fluids due to ion exchange and bioactivity mechanism. After preconditioning in SBF, BSA adsorption is significantly enhanced. However, the trend of pH-dependent adsorption is attenuated without a smart effect in an inflammatory chemical environment [221]. It has been found that increasing the negative zeta potential of 45S5 Bioglass® could cause a significant decrease in the amount of adsorbed serum proteins while surface crystallization of 45S5 Bioglass® could inhibit protein adsorption.

Peptides can be grafted to bioactive glasses using dopamine as a coupling agent, to get antifouling surfaces [222]. On stable glasses, this strategy is often coupled to an antifouling micro- and/or nano-topography (through lithography or laser surface structuring), but in the case of bioactive glasses this is usually ineffective because of the fast growth of a hydroxyapatite layer on the surface. The use of single-molecule force spectroscopy (SMFS) based on atomic force microscopy can be of great interest to directly measure the adhesion force between a silica-binding peptide and glass surface at single molecule level [223]. The peptide-surface interactions can be due to Van der Waals force, hydrogen bond, electrostatic attraction, hydrophobic effect, etc. The measurement of the adhesion force can be of interest to discriminate among these different types of bonding.

Considering that blood compatibility of glasses should usually be further improved, silanization and grafting with zwitterionic polymers can be useful to significantly inhibit platelet adhesion and whole blood cell attachment, when it is required [224] this is not usually the case of bioactive glasses, but it could be further explored.

Plasma modification can be applied to produce polar functional groups (amine, carbonyl or carboxylate) changing surface free energy, with elimination of organic impurities and formation of cross-linking. Plasma modification generally induces an increment of wettability, but hydrophobic glass surfaces can be achieved by using helium [225].

The first attempts of surface functionalization of bioactive glasses were focused on the grafting of proteins, enzymes, and growth factors, such as Bovine Serum

Albumin [226, 227], Bone Morphogenetic Proteins [206, 217], collagen [228] or alkaline phosphatase [210, 211] to cite some examples. These compounds are mainly intended for the achievement of a fast bone integration and regeneration.

More recently, the interest moved through natural compounds of vegetal origin [208], such as simple model polyphenols (gallic acid) [229–231], polyphenols from grapes and tea [232–234], sage [235], algae [236], bud extracts [237] or curcumin [238] to cite some examples. Since this kind of molecules show a variety of additional properties, such as antioxidant, antibacterial, anti-inflammatory, anticancer and even osteo-stimulatory properties, great effort has been done to optimize the grafting conditions in order to assure their effective immobilization while preserving their therapeutic activity. Compared to the above-described strategies based on proteins and growth factors, the use of natural compounds is more focused on the modulation of the biological response (considering also inflammation control) rather than to the achievement of fast bone growth. For example, it was assessed that grafting of gallic acid on the surface of an iron oxide-containing bioactive glass–ceramic may influence the bioactivity of the pristine material, while an evident effect was observed on the redox activity, as the ability of the glass–ceramic to catalyze $HO^{\bullet}$ radical release in the presence of H_2O_2 was significantly increased by gallic acid grafting. Moreover, grafting gallic acid acts as a pro-oxidant, probably reducing Fe^{3+} to Fe^{2+}. Furthermore, it has been reported that green tea extracts grafted on the surface of a bioactive glass show a selective cytotoxic action against bone cancer cells and that this action can be related to the production of reactive species in the cancerous cells causing selective DNA damage, while evidencing an anti-inflammatory action on healthy osteoblast cells.

In addition to their peculiar features, these molecules, due to their chemical reactivity as reducing agents, can be effectively used to obtain an in situ green formation of antibacterial metal nanoparticles (e.g. Ag nanoparticles) directly on the glass surfaces [188, 189, 236]. This step can be considered to increase the antibacterial activity of the surface and to combine multifunctional actions on the surface.

Finally, surface functionalization with drugs can be mentioned. As far as mesoporous bioactive glasses are concerned, their use as substrates for drug release has already been discussed in Sect. 9 of the present chapter. In addition, the direct grafting of antibiotics [191] (Sect. 10) and chemotherapeutics [239–241] to the surface of non-porous bioactive glasses should be cited, as well as the glass covalent functionalization with model molecules (cysteamine and 5-aminofluorescein) by means of a covalent reaction with pH-sensitive organic molecules, in order to induce a triggered stimuli-responsive drug release [242]. The rationale and the experimental strategies are analogous to the grafting of the previously discussed active molecules, with the final aim to obtain localized antibacterial or anticancer properties, and—generally speaking—the specific delivery of drugs for treating bone diseases. Compared to a systemic administration of the same active principles, the local delivery through the implant surface can reduce the dose necessary to reach a therapeutic effect and, consequently, the potential toxicity.

Surface functionalization of bioactive glasses, as well as of other biomaterials, has to face the regulatory procedures (e.g. certification and sterilization) in order to be suitable to be applied onto real biomedical implants. As far as certification is concerned, the presence of active principles, which can be released upon contact with physiological fluids, can change the device class with a consequent increase in the certification times and costs. This point should be considered in the development of innovative surfaces and, in many cases, it could become a barrier for the lab to market transfer. Moreover, the presence of organic molecules, which are often sensitive to heat, radiation and chemicals, can make difficult the sterilization of functionalized surfaces by using conventional techniques. The possibility to sterilize the functionalized materials by a standard procedure is actually poorly explored, but it should be evaluated for the development of biomaterials intended for implantation. Despite some uncertainness, early evidence of the possibility to apply the conventional sterilization techniques to functionalized bioactive glasses has been reported [243].

12 Clinical Applications of Bioactive Glasses

Recent estimates have shown that, since FDA approval in 1985, Hench's 45S5 Bioglass® was implanted in more than 1.5 million patients worldwide to mainly repair bone and dental defects [244]

The first 45S5 Bioglass® implant approved for clinical use in the USA had the purpose of substituting the small bones of the middle ear in order to treat conductive hearing losses [245]. This glass, being able to bond both to bone and to the collagen fibres of tympanic membrane, allowed sound conduction from the eardrum to the internal structures of the ear. After FDA approval in 1985, this device was marketed under the name of "Bioglass® Ossicular Reconstruction Prosthesis" or "Middle Ear Prosthesis" MEP®. Although exhibiting good performance in the short and mid-term, this 45S5 glass implant underwent dissolution and fragmentation in the long-term [246]. Therefore, this device was taken off the US market in 2000.

45S5 Bioglass® was also used to fix cochlear implants to the temporal bone of profoundly deaf patients suffering from irreversible damage to the cochlea. This device was commercialized as Bioglass®-EPI (Extracochlear Percutaneous Implant) in the late 1980s but was then taken off the market due to the same drawbacks already reported for MEP® prosthesis [247].

In 1988, 45S5-based Endosseous Ridge Maintenance Implant (ERMI®) was launched on the market and, still today, it is applied in periodontal surgery. This implant consists of a glass cone to be inserted into fresh tooth extraction area, thereby replacing the tooth root and giving an adequate support to dentures [248].

In 1993, 45S5 Bioglass® particulate (90–710 μm) was approved by the FDA for repairing jaw bone defects associated to periodontal diseases; this product is known under the tradename of PerioGlas® [249]. A similar type of 45S5 particulate, commercialized as NovaBone® (NovaBone Products LLC), is used to repair bone defects in maxillofacial or orthopaedic non-load-bearing sites [249]. Furthermore, 45S5 Bioglass® is also commercialized as porous glass–ceramic sintered blocks.

In fact, the sinterability window of 45S5 Bioglass® is so narrow that it cannot be sintered without undergoing devitrification [250].

Recently, 45S5 Bioglass® has also been used to make oral hygiene products. In 2004 NovaMin®, a 45S5 Bioglass® fine particulate (average size 18 μm) was added to a toothpaste in order to treat dental hypersensitivity, which nowadays affects about one-third of world population [251]. The aim of NovaMin® is to occlude dentinal tubules and remineralize the tooth surface, thus eliminating the cause of the disease [252]. This device was also employed for tooth whitening treatments [253].

In addition to 45S5 Bioglass®, other FDA-approved or CE-marked BGs are available on the market, too. Most of commercial BGs are characterized by a SiO_2-based composition, containing some additional modifiers which increase the bioactivity or confer special characteristics/therapeutic effects to the BG. For example, 13–93 BG ($53SiO_2$–$6Na_2O$–$12K_2O$–5MgO–20CaO–$4P_2O_5$ wt.%) has a larger sinterability window than 45S5 composition, which allows obtaining fully amorphous products from it (e.g. porous scaffolds) with excellent bioactivity and mechanical properties even comparable to cortical bone [53].

A wide range of commercial glass–ceramics are in current use in dentistry for both dental root/alveolar bone surgery (if bioactive) and restorative/aesthetic applications (if inert) [57].

Interest in BGs further increased over the last 15 years since it was found that ionic dissolution products released from glasses can stimulate angiogenesis, which plays a pivotal role in major healing processes in the body [254]. Currently, there are only two BG-based commercial products with a specific pro-angiogenic function. One is addressed to skin regeneration and is trade-named as DermaFuse™/Mirragen™. It consists of borate BG nanofibers (13-93B3 glass, $53B_2O_3$–$6Na_2O$–$12K_2O$–5MgO–20CaO–$4P_2O_5$ wt.%) that aim at accelerating wound healing by imitating the microstructure of a fibrin clot [255]. This product is also used under the tradename of "RediHeal" in veterinarian medicine. Interestingly, this BG is effective also in the treatment of choric diabetic ulcers on the skin which are irresponsive to pharmacological therapy. The second device is a composite orbital implant that was approved by FDA in 2002: it comprises a porous polyethylene sphere coated with NovaBone® particles having a stimulatory effect on fibrovascularization (Medpor®-Plus™) [256].

In addition to DermaFuse™/Mirragen™, other BG-based commercial products for wound healing applications include resorbable Ag-doped phosphate glasses combined with a polymeric adhesive for wound care film dressing (Antimicrobial Arglaes®) or with alginate for topical powders (Arglaes® powder). In both cases, they control infection by constantly releasing silver, which has a potent antibacterial effect.

Bioactive and magnetic glass–ceramics have been emerging in the research scenario since several years, and are gaining increasing interest for the treatment of bone cancer by magnetic induction of hyperthermia [257–259]. Specifically, magnetite-containing bioactive glass–ceramics have been developed both as bulk materials and as dispersed phase into acrylic bone cements, revealing bioactive properties, negligible iron release, cytocompatibility and pro-osteogenic activity, with a synergistic effect between the bioactivity of the materials and cell mineralization in the formation of apatite crystals on their surface. Furthermore, these glass–ceramics revealed the capacity of generating heat under exposure to alternating magnetic fields and, in turn, of inducing in vitro cellular heating, thus causing tumour cell death by apoptosis while preserving the viability of normal cells [260–268].

Although not being "bioactive" according to the Hench's definition, radioactive glasses for cancer treatment deserve to be mentioned as well [269]. Insoluble Y_2O_3–Al_2O_3–SiO_2, glass microspheres (diameter around 25 μm, trade-named as TheraSphere® or TheraGlass®) were approved by the FDA in 1999 for radioembolization of hepatocellular carcinoma and metastatic liver cancer. Before arterial infusion, the glass beads are exposed to neutrons to create ^{90}Y, a radioisotope that has a short half-life (64 h) and decays to stable ^{90}Zr through emitting β-rays. As a result, a localized dosage of up to 15,000 rad can be locally delivered to kill cancer cells, with high benefits to the patient (3000 rad is the maximum dosage allowed under external radiotherapy).

Besides the uses described above, there is a number of emerging applications in contact with non-calcified tissues and soft organs which have not reached yet clearance for clinical use, including cardiac tissue regeneration, artificial lungs, epithelial restitution in gastric mucosa and intestine, peripheral nerve regeneration and corneal repair [270–272]; the overall timeline of BG applications is reported in Table 5. These novel experimental studies are contributing to further expand the use of BGs in medicine and stimulate unconventional research, which could have a great impact on human life and health in the near future.

13 A Forecast for the Future

The medical industry is constantly searching for new, better, and more cost-effective solutions; thus, advancements in the medical industry are progressing faster than just a few years ago due to the introduction of advanced materials.

In this scenario, BGs integrate with the human body in diverse ways to support human health. As aging populations and evolving healthcare approaches shift the medical landscape, increasing opportunities for both established and innovative technologies about biomedical glasses can be forecast. According to a report published by the American Ceramic Society in December 2020 [273], the global market for implantable biomaterials was estimated to be around $110 billion in 2019. In terms of the future of healthcare, regenerative medicine is a big business. The global market for tissue engineering and regeneration was valued at $25 billion in 2018 and is forecast to reach $109.9 billion by 2023—very close to the global estimate for the whole biomaterials market in 2019!—representing an impressive growth rate. While bone is indeed a significant focus of the BG and bioceramic market, the attention is moving to soft tissues as well, including strategies to repair injured cardiac, muscle, neural, and skin tissues. There is potential for many different types of materials in this broad field which can be combined with BGs. In fact, in the field of regenerative medicine and tissue engineering, there is no one material that is going to tackle all the

Table 5 Chronological overview of the key commercial/clinically-approved applications of BGs in medicine

Year	Achievement/application
1969	Invention of the 45S5 glass composition (45S5 Bioglass®)
1977	Replacing of middle ear small bones using Ceravital® glass–ceramics
1978	Ocular implant (biocompatibility with corneal tissue)
1985	Approval by FDA of the first 45S5 Bioglass® implant
1987	Treatment of liver cancer (radioactive glasses)
1988	Clinical use of the 45S5 Bioglass®-based Endosseous Ridge Maintenance Implant (ERMI) in human patients
1993	FDA approval of PerioGlas (45S5 Bioglass® particulate used for bone and dental repair)
1998	Peripheral nerve repair
1999	FDA approval of radioactive glasses (TheraSphere®) for cancer treatment
2000	Wound healing
2002	FDA approval of Medpor®-Plus™ (polyethylene/ 45S5 Bioglass® composite porous orbital implants)
2003	Antibacterial (Zn-containing) bone/dental cements
2004	Lung tissue engineering
2004	Use of mesoporous bioactive glass (MBG) as a drug delivery system
2005	Skeletal muscle and ligament repair
2005	Treatment of gastrointestinal ulcers
2010	Cardiac tissue engineering
2011	Commercialization of a cotton-candy borate bioactive glass for wound healing in veterinarian medicine
2012	Embolization of uterine fibroids
2012	Spinal cord repair
2018	Use of radioactive glasses (TeraSphere®) in patients with metastatic colorectal carcinoma of the liver

challenges. Many of the glass-based strategies to heal tissues actually combine BGs with organic materials, for example in polymer-matrix composites or hydrogels.

An increasing collaboration between materials scientists, biologists and clinicians is key to allow bioactive glass research to really progress. In this regard, understanding the genetic mechanisms and pathways activated by ionic stimuli released from BGs offers the possibility of developing patient-specific therapies, which is a huge challenge for the aging population. Furthermore, therapeutic ions released from BGs open new horizons in the field of implantable or non-implantable antibacterial and antiviral surfaces.

Given the complexity of biological systems, the future of regenerative medicine seems to be addressed to develop biomaterials and technologies able to treat multiple different tissues simultaneously. Although an isolated tissue-specific approach often guides biomaterials developments, components of the human body are known to operate together on several different scales. Therefore, in the attempt to somehow replicate Nature, researchers from academia and industry should focus on manufacturing different material types together to match the really different material types in the body. In this regard, additive manufacturing technologies combined with biofabrication, involving manipulation and printing of BGs along with other biomaterials, biomolecules and even cells, will be an exceptional resource.

References

1. Arvidson, K., et al.: Bone regeneration and stem cells. J. Cell. Mol. Chem. **15**, 718–746 (2010)
2. Williams, D.F.: On the mechanism of biocompatibility. Biomaterials **29**, 2941–2953 (2008)
3. Campana, V., et al.: Bone substitutes in orthopaedic surgery: from basic science to clinical practice. J. Mater. Sci. Mater. Med. **25**, 2445–2461 (2014)
4. Dimitriou, R., Mataliotakis, G.I., Angoules, A.G., Kanakaris, N.K., Giannoudis, P.V.: Complications following autologous bone graft harvesting from the iliac crest and using the RIA: a systematic review. Injury **42**, S3–S15 (2011)
5. Henkel, J., et al.: Bone regeneration based on tissue engineering conceptions—a 21st century perspective. Bone Res. **3**, 216–248 (2013)
6. Hench, L.L.: The story of Bioglass®. J. Mater. Sci. Mater. Med. **17**, 967–978 (2006)
7. Wilson, J., Pigott, G., Schoen, F., Hench, L.L.: Toxicology and biocompatibility of bioglasses. J. Biomed. Mater. Res. A **15**, 805–817 (1981)
8. Xynos, I., et al.: Bioglass® 45S5 stimulates osteoblast turnover and enhances bone formation in vitro: implications and applications for bone tissue engineering. Calcif. Tissue Int. **67**, 321–329 (2000)
9. Brink, M., Turunen, T., Happponen, R., Yli-Urpo, A.: Compositional dependence of bioactivity of glasses in the system Na2O-K2O-MgO-CaO-B2O3-P2O5-SiO2. J. Biomed. Mater. Res. **37**, 114–121 (1997)
10. Andersson, O.H., Liu, G., Karlsson, K.H., Juhanoja, J.: In vivo behavior of glasses in the SiO2–Na2O–CaO–P2O5–Al2O3–B2O3 system. J. Mater. Sci. Mater. Med. **1**, 219–227 (1990)
11. Knowles, J.C.: Phosphate based glasses for biomedical applications. J. Mater. Chem. **13**, 2395–2401 (2003)
12. Hench, L.L.: Bioactive ceramics. Ann. N. Y. Acad. Sci. **523**, 54–71 (1988)
13. Pantano, C.G., Clark, A.E., Hench, L.L.: Multilayer corrosion films on bioglass surfaces. J. Am. Ceram. Soc. **57**, 412–413 (1974)
14. Hench, L.L., Roki, N., Fenn, M.B.: Bioactive glasses: importance of structure and properties in bone regeneration. J. Mol. Struct. **1073**, 24–30 (2014)
15. Salinas, A.J., Vallet-Regí, M.: Bioactive ceramics: from bone grafts to tissue engineering. RSC Adv. **3**, 11116–11131 (2013)
16. Kokubo, T., Ito, S., Sakka, S., Yamamuro, T.: Formation of a high-strength bioactive glass-ceramic in the system MgO-CaO-SiO2-P2O5. J. Mater. Sci. **21**, 536–540 (1986)
17. Montazerian, M., Zanotto, E.D.: History and trends of bioactive glass-ceramics. J. Biomed. Mater. Res. A **104**, 1231–1249 (2016)
18. Hench, L.L.: Genetic design of bioactive glasses. J. Eur. Ceram. Soc. **29**, 1257–1265 (2008)
19. Jell, G., Stevens, M.M.: Gene activation by bioactive glasses. J. Mater. Sci. Mater. Med. **17**, 997–1002 (2006)
20. Kokubo, T., Takadama, H.: How useful is SBF in predicting in vivo bone bioactivity? Biomaterials **27**, 2907–2915 (2006)
21. Mizuno, M.: Implants for Surgery: In Vitro Evaluation for Apatite Forming Ability of Implant Materials (ISO ISO/DIS 23317:2014), last reviewed and confirmed in 2020

22. Bohner, M., Lemaitre, J.: Can bioactivity be tested in vitro with SBF solution? Biomaterials **30**, 2175–2179 (2009)
23. Macon, A.L.B., et al.: A unified in vitro evaluation for apatite-forming ability of bioactive glasses and their variants. J. Mater. Sci. Mater. Med. **26**, 115 (2015)
24. Mozafari, M., Banijamali, S., Baino, F., Kargozar, S., Hill, R.G.: Calcium carbonate: adored and ignored in bioactivity assessment. Acta Biomater. **91**, 35–47 (2019)
25. Kaur, G., Pickrell, G., Sriranganathan, N., Kumar, V., Homa, D.: Review and the state of the art: sol–gel and melt quenched bioactive glasses for tissue engineering. J. Biomed. Mater. Res. **104**, 1248–1275 (2015)
26. Hench, L.L., Splinter, R.J., Allen, W.C., Greenlee, T.K.: Bonding mechanisms at the interface of ceramic prosthetic materials. J. Biomed. Mater. Res. **5**, 117–141 (1971)
27. Vedel, E., Arstila, H., Ylanen, H., Hupa, L., Hupa, M.: Predicting physical and chemical properties of bioactive glasses from chemical composition. Part I. Viscosity characteristics. Glass Technol. Eur. J. Glass Sci. Technol. **49**, 251–259 (2008)
28. Li, R., Clarck, A.E., Hench, L.L.: An investigation of bioactive glass powders by sol-gel processing. J. Appl. Biomater. **2**, 231–239 (1991)
29. Graham, T.: On the properties of silicic acid and other analogous colloidal substances. J. Chem. Soc. **17**, 318–327 (1864)
30. Flory, P.J.: Principle of Polymer Chemistry. Cornell University Press, Ithaca and London (1953)
31. Kakihana, M.: Invited review "sol–gel" preparation of high temperature superconducting oxides. J. Sol-Gel. Sci. Technol. **6**, 7–55 (1996)
32. Danks, A.E., Hall, S.R., Schnepp, Z.: The evolution of 'sol–gel' chemistry as a technique for materials synthesis. Mater. Horiz. **3**, 91 (2016)
33. Baino, F., Fiume, E., Miola, M., Verné, E.: Bioactive sol-gel glasses: processing, properties, and applications. Int. J. Appl. Ceram. Technol. **15**, 841–860 (2018)
34. Hench, L.L.: The sol-gel process. Chem. Rev. **90**, 33–72 (1990)
35. Kesmez, Ö., et al.: Effect of acid, water and alcohol ratios on sol-gel preparation of antireflective amorphous SiO2 coatings. J. Non-Cryst. Solids **357**, 3130–3135 (2011)
36. Arcos, D., Vallet-Regí, M.: Sol-gel silica-based biomaterials and bone tissue regeneration. Acta Biomater. **6**, 2874–2888 (2010)
37. Owens, G.J., et al.: Sol-gel based materials for biomedical applications. Prog. Mater Sci. **77**, 1–79 (2016)
38. Doadrio, J.C., et al.: Functionalization of mesoporous materials with long alkyl chains as a strategy for controlling drug delivery pattern. J. Mater. Chem. **16**, 462–466 (2006)
39. Wu, C., Zhang, Y., Zhou, Y., Fan, W., Xiao, Y.: A comparative study of mesoporous-glass/silk and nonmesoporous-glass/silk scaffolds: physio-chemistry and in vivo osteogenesis. Acta Biomater. **7**, 2229–2236 (2011)
40. Baino, F., Fiume, E.: 3D printing of hierarchical scaffolds based on mesoporous bioactive glasses (MBGs)-fundamentals and applications. Materials (Basel) **13**, 1688 (2020)
41. Lefebvre, L., Gremillardm, L., Chevalier, J., Zenati, R., Bernache-Assolant, D.: Sintering behaviour of 45S5 bioactive glass. Acta Biomater. **4**, 1894–1903 (2008)
42. Bellucci, D., et al.: Macroporous bioglass derived scaffolds for bone tissue regeneration. Ceram. Int. **37**, 1575–1585 (2011)
43. Arstilla, H., et al.: The sintering range of porous bioactive glasses. Glass Technol. **46**, 138–141 (2005)
44. Zhang, D., Vedel, E., Hupa, L., Aro, H.T.: Predicting physical and chemical properties of bioactive glasses from chemical composition. Part III. In vitro reactivity of glasses. Glass Technol. Eur. J. Glass Sci. Technol. **50**, 1–8 (2009)
45. Tulyaganov, D.U., et al.: Synthesis of glass-ceramics in the CaO–MgO–SiO2 system with B2O3, P2O5, Na2O and CaF2 additives. J. Eur. Ceram. Soc. **26**, 1463–1471 (2006)
46. Agathopoulos, S., Tulyaganov, D.U., Ventura, J.M.G., Kannan, S., Saranti, A., Karakassides, M.A.: Structural analysis and devitrification of glasses based on the CaO–MgO–SiO2 system with B2O3, P2O5, Na2O and CaF2 additives. J. Non-Cryst. Solids **352**, 322–328 (2006)

47. Kansal, I., Tulyaganov, D.U., Goel, A., Pascual, M.J., Ferreira, J.M.F.: Structural analysis and thermal behaviour of diopside-fluorapatite-wollastonite-based glasses and glass-ceramics. Acta Biomater. **6**, 4380–4388 (2010)
48. Filho, O.P., Latorre, G.P., Hench, L.L.: Effect of crystallization on apatite-layer formation of bioactive glass 45S5. J. Biomed. Mater. Res. **30**, 509–514 (1996)
49. Peitl, O., Dutra, E., Hench, L.L.: Highly bioactive P 2 O 5 ± Na 2 O ± CaO ± SiO 2 glass-ceramics. J. Non-Cryst. Solids **292**, 115–126 (2001)
50. Arstila, H., Hupa, L., Karlson, K., Hupa, M.: In vitro bioactivity of partially crystallised glasses. Glass Technol. Eur. J. Glass Sci. Technol. A **48**, 196–199 (2007)
51. De Castro Juraski, A., Dorion Rodas, A.C., Elsayed, H., Bernardo, E., Oliveira Soares, V., Daguano, J.: The in vitro bioactivity, degradation, and cytotoxicity of polymer-derived wollastonite-diopside glass-ceramics. Materials **10**, 425 (2017)
52. Fu, Q., Rahaman, M.N., Bal, B.S., Brown, R.F., Day, D.E.: Mechanical and in vitro performance of 13–93 bioactive glass scaffolds prepared by a polymer foam replication technique. Acta Biomater. **4**, 1854–1864 (2008)
53. Liu, X., Rahaman, M.N., Hilmas, G.E., Bal, B.S.: Mechanical properties of bioactive glass (13–93) scaffolds fabricated by robotic deposition for structural bone repair. Acta Biomater. **9**, 7025–7034 (2013)
54. Massera, J., Fagerlund, S., Hupa, L., Hupa, M.: Crystallization mechanism of the bioactive glasses, 45S5 and S53P4. J. Am. Ceram. Soc. **95**, 603–613 (2012)
55. Baino, F., Fiume, E.: Quantifying the effect of particle size on the crystallization of 45S5 bioactive glass. Mater. Lett. **224**, 54–58 (2018)
56. Fiume, E., Migneco, C., Verné, E., Baino, F.: Comparison between bioactive sol-gel and melt-derived glasses/glass-ceramics based on the multicomponent SiO2–P2O5–CaO–MgO–Na2O–K2O system. Materials (Basel) **13**, 540 (2020)
57. Montazerian, M., Zanotto, E.D.: Bioactive and inert dental glass-ceramics. J. Biomed. Mater. Res. Part A **105**, 619–639 (2017)
58. Nedelec, J.-M., et al.: Materials doping through sol–gel chemistry: a little something can make a big difference. J. Sol-Gel. Sci. Technol. **46**, 259–271 (2008)
59. Rabiee, S.M., Nazparvar, N., Azizian, M., Vashaee, D., Tayebi, L.: Effect of ion substitution on properties of bioactive glasses: a review. Ceram. Int. **41**, 7241–7251 (2015)
60. Wu, C., Chang, J.: Multifunctional mesoporous bioactive glasses for effective delivery of therapeutic ions and drug/growth factors. J. Control. Release **193**, 282–295 (2014)
61. Krajewski, A., Ravaglioli, A., Fabbri, B., Azzoni, C.B.: Doping influence on the interaction between a bioactive glass and a simulated physiological solution: Chemical and EPR tests. J. Mater. Sci. **22**, 1228–1234 (1987)
62. Abou Neel, E.A., Knowles, J.C.: Physical and biocompatibility studies of novel titanium dioxide doped phosphate-based glasses for bone tissue engineering applications. J. Mater. Sci. Mater. Med. **19**, 377–386 (2008)
63. Yin, H., et al.: Fabrication and characterization of strontium-doped borate-based bioactive glass scaffolds for bone tissue engineering. J. Alloy. Compd. **743**, 564–569 (2018)
64. Beattie, J.H., Avenell, A.: Trace element nutrition and bone metabolism. Nutr. Res. Rev. **5**, 167–188 (1992)
65. Fiume, E., Barberi, J., Verné, E., Baino, F.: Bioactive glasses: from parent 45S5 composition to scaffold-assisted tissue-healing therapies. J. Funct. Biomater. **9**, 24 (2018)
66. Hutmacher, D.W.: Scaffolds in tissue engineering bone and cartilage. Biomaterials **21**, 2529–2543 (2000)
67. Karageorgiou, V., Kaplan, D.: Porosity of 3D biomaterial scaffolds and osteogenesis. Biomaterials **26**, 5474–5491 (2005)
68. Chappard, D., Baslé, M.F., Legrand, E., Audran, M.: New laboratory tools in the assessment of bone quality. Osteoporos. Int. **22**, 2225–2240 (2011)
69. Gerhardt, L.-C., Boccaccini, A.R.: Bioactive glass and glass-ceramic scaffolds for bone tissue engineering. Materials (Basel) **3**, 3867–3910 (2010)

70. Fu, Q., Saiz, E., Rahaman, M.N., Tomsia, A.P.: Bioactive glass scaffolds for bone tissue engineering: state of the art and future perspectives. Mater. Sci. Eng., C **31**, 1245–1256 (2011)
71. Baino, F., Vitale-Brovarone, C.: Three-dimensional glass-derived scaffolds for bone tissue engineering: current trends and forecasts for the future. J. Biomed. Mater. Res. Part A **97**, 514–535 (2011)
72. Baino, F., et al.: Processing methods for making porous bioactive glass-based scaffolds—a state-of-the-art review. Int. J. Appl. Ceram. Technol. **16**, 1762–1796 (2019)
73. Xynos, I.D., Edgar, A.J., Buttery, L.D.K., Hench, L.L., Polak, J.M.: Gene-expression profiling of human osteoblasts following treatment with the ionic products of Bioglass® 45S5 dissolution. J. Biomed. Mater. Res. **55**, 151–157 (2001)
74. Yuan, H., de Bruijn, J.D., Zhang, X., Van Blitterswijk, C.A., De Groot, K.: Bone induction by porous glass ceramic made from Bioglass® (45S5). J. Biomed. Mater. Res. **58**, 270–276 (2002)
75. Sepulveda, P., Jones, J.R., Hench, L.L.: Bioactive sol–gel foams for tissue repair. J. Biomed. Mater. Res. **59**, 340–348 (2002)
76. Jones, J.R., Lee, P.D., Hench, L.L.: Hierarchical porous materials for tissue engineering. Phil. Trans. R. Soc. A **364**, 263–281 (2006)
77. Baino, F., Fiume, E.: Elastic mechanical properties of 45S5-based bioactive glass-ceramic scaffolds. Materials (Basel) 12 (2019)
78. Baino, F., Pons, E.: Modelling the relationship between tensile strength and porosity in bioceramic scaffolds. Int. J. Appl. Ceram. Technol. **16**, 1823–1829 (2019)
79. Kargozar, S., et al.: Synergistic combination of bioactive glasses and polymers for enhanced bone tissue regeneration. Mater Today Proc. **5**, 15532–15539 (2018)
80. Bretcanu, O., et al.: Biodegradable polymer coated 45S5 bioglass derived glass-ceramic scaffolds for bone tissue engineering. Glass Technol. Eur. J. Glass Sci. Technol. A **48**, 227–234 (2007)
81. Rehorek, L., et al.: Response of 45S5 bioglass_foams to tensile loading. Ceram. Int. **39**, 8015–8020 (2013)
82. Jones, J.R.: Review of bioactive glass: from hench to hybrids. Acta Biomater. **9**, 4457–4486 (2013)
83. Baino, F., Novajra, G., Vitale-Brovarone, C.: Bioceramics and scaffolds: a winning combination for tissue engineering. Front. Bioeng. Biotechnol. **3**, 1–17 (2015)
84. Wu, Z.Y., et al.: Melt-derived bioactive glass scaffolds produced by a gel-cast foaming technique. Acta Biomater. **7**, 1807–1816 (2011)
85. Poologasundarampillai, G., Lee, P.D., Lam, C., Kourkouta, A.M., Jones, J.R.: Compressive strength of bioactive sol-gel glass foam scaffolds. Int. J. Appl. Glas. Sci. **9**, 1–9 (2016)
86. Jones, J.R., Ehrenfried, L.M., Hench, L.L.: Optimizing bioactive glass scaffolds for bone tissue engineering. Biomaterials **27**, 964–973 (2006)
87. Jones, J.R., Ehrenfried, L.M., Saravanapavan, P., Hench, L.L.: Controlling ion release from bioactive glass foam scaffolds with antibacterial properties. J. Mater. Sci. Mater. Med. **17**, 989–996 (2006)
88. Baino, F., et al.: Fe-doped bioactive glass-derived scaffolds produced by sol-gel foaming. Mater. Lett. **235**, 207–211 (2019)
89. Navarro, M., et al.: New macroporous calcium phosphate glass ceramic for guided bone regeneration. Biomaterials **25**, 4233–4241 (2004)
90. Fiume, E., et al.: Comprehensive assessment of bioactive glass and glass-ceramic scaffold permeability: experimental measurements by pressure wave drop, modelling and computed tomography-based analysis. Acta Biomater. **119**, 405–418 (2021)
91. Boccardi, E., et al.: Characterisation of Bioglass based foams developed via replication of natural marine sponges. Adv. Appl. Ceram. **114**, S56–S62 (2015)
92. Fiume, E., Serino, G., Bignardi, C., Verné, E., Baino, F.: Bread-derived bioactive porous scaffolds: an innovative and sustainable approach to bone tissue engineering. Molecules **24**, 2954 (2019)

93. Vitale-Brovarone, C., Vernè, E., Bosetti, M., Appendino, P., Cannas, M.: Microstructural and in vitro characterization of SiO2-Na2O-CaO-MgO glass-ceramic bioactive scaffolds for bone substitutes. J. Mater. Sci. Mater. Med. **16**, 909–917 (2005)
94. Baino, F., Verné, E., Vitale-Brovarone, C.: 3-D high strength glass-ceramic scaffolds containing fluoroapatite for load-bearing bone portions replacement. Mater. Sci. Eng., C **29**, 2055–2062 (2009)
95. Liang, W., et al.: Bioactive borate glass scaffold for Bone tissue engineering. J. Non-Cryst. Solids **354**, 1690–1696 (2008)
96. Fiume, E., Tulyaganov, D., Ubertalli, G., Verné, E., Baino, F.: Dolomite-foamed bioactive silicate scaffolds for bone tissue repair. Materials **13**, 628 (2020)
97. Minaberry, Y., Jobbágy, M.: Macroporous bioglass scaffolds prepared by coupling sol-gel with freeze drying. Chem. Mater. **23**, 2327–2332 (2011)
98. Maquet, V., Boccaccini, A.R., Pravata, L., Notingher, I., Jérôme, R.: Porous poly(α-hydroxyacid)/bioglass® composite scaffolds for bone tissue engineering. I: preparation and in vitro characterization. Biomaterials **25**, 4185–4194 (2004)
99. Gmeiner, R., et al.: Additive manufacturing of bioactive glasses and silicate bioceramics. J. Ceram. Sci. Technol. **6**, 75–86 (2015)
100. Liu, J., Hu, H., Li, P., Shuai, C., Peng, S.: Fabrication and characterization of porous 45S5 glass scaffolds via direct selective laser sintering. Mater. Manuf. Processes **28**, 610–615 (2013)
101. Tesavibul, P., et al.: Processing of 45S5 bioglass® by lithography-based additive manufacturing. Mater. Lett. **41**, 81–84 (2012)
102. Bose, S., Vahabzadeh, S., Bandyopadhyay, A.: Bone tissue engineering using 3D printing. Mater. Today **16**, 496–504 (2013)
103. Barberi, J., et al.: Robocasting of SiO2-based bioactive glass scaffolds with porosity gradient for bone regeneration and potential load-bearing applications. Materials (Basel) 12 (2019)
104. Yun, H.S., Kim, S.E., Park, E.K.: Bioactive glass-poly(epsilon-caprolactone) composite scaffolds with 3 dimensionally hierarchical pore networks. Mater. Sci. Eng., C **31**, 198–205 (2011)
105. McEntire, B., Bal, B., Rahaman, M., Chevalier, J., Pezzotti, G.: Ceramics and ceramic coatings in orthopaedics. J. Eur. Ceram. Soc. **35**, 4327–4369 (2015)
106. Sun, L., Berndt, C.C., Gross, K.A., Kucuk, A.: Material fundamentals and clinical performance of plasma-sprayed hydroxyapatite coatings: a review. J. Biomed. Mater. Res. Part A **58**, 570–592 (2001)
107. Sola, A., Bellucci, D., Cannillo, V., Cattini, A.: Bioactive glass coatings: a review. Surf. Eng. **27**, 560–572 (2011)
108. Lopez-Esteban, S., et al.: Bioactive glass coatings for orthopedic metallic implants. J. Eur. Ceram. Soc. **23**, 2921–2930 (2003)
109. Verne, E., Brovarone, C.V., Moisescu, C.: Glazing of alumina by a fluoroapatite-containing glass-ceramic. J. Mater. Sci. **40**, 1209–1215 (2005)
110. Bosetti, M., et al.: Fluoroapatite glass-ceramic coating on alumina: surface behaviour with biological fluids. J. Biomed. Mater. Res. **66A**, 615–621 (2003)
111. Verne, E., Fernandez-Valles, C., Brovarone, C.V., Spriano, S., Moisescu, C.: Double-layer glass-ceramic coatings on Ti6Al4V for dental implants. J. Eur. Ceram. Soc. **24**, 2699–2705 (2004)
112. Verné, E., Bona, E., Angelini, E., Rosalbino, F., Appendino, P.: Correlation between microstructure and properties of biocomposite coatings. J. Eur. Ceram. Soc. **22**, 2315–2323 (2002)
113. Vernè, E., et al.: Fluoroapatite glass-ceramic coatings on alumina: structural, mechanical and biological characterisation. Biomaterials **23**, 3395–3403 (2002)
114. Ferraris, M., et al.: Coatings on zirconia for medical applications. Biomaterials **21**, 765–773 (2000)
115. Verné, E., Ferraris, M., Jana, C., Paracchini, L.: Bioverit®I base glass/Ti par-ticulate biocomposite: "in situ" vacuum plasma spray deposition. J. Eur. Ceram. Soc. **20**, 473–479 (2000)

116. Verné, E., et al.: Sintering and plasma spray deposition of bioactive glass-matrix composites for biomedical applications. J. Eur. Ceramic Soc. **18**, 363–372 (1998)
117. Alonso-Barrio, J.A., et al.: Bioglass-coated femoral stem. J. Bone Joint Surg. **86-B**, 138 (2004)
118. Gomez-Vega, J.M., et al.: Novel bioactive functionally graded coatings on Ti6Al4V. Adv. Mater. **12**, 894–898 (2000)
119. Boccaccini, A.R., Keim, S., Ma, R., Li, Y., Zhitomirsky, I.: Electrophoretic deposition of biomaterials. J. Royal Soc. Interf. **7** (2010)
120. Baino, F., et al.: Novel antibacterial ocular prostheses: proof of concept and physico-chemical characterization. Mater. Sci. Eng., C **60**, 467–474 (2016)
121. Moskalewicz, T., Seuss, S., Boccaccini, A.R.: Microstructure and properties of composite poly-etheretherketone/Bioglass® coatings deposited on Ti-6Al-7Nb alloy for medical applications. Appl. Surf. Sci. **273**, 62–67 (2013)
122. Seuss, S., Heinloth, M., Boccaccini, A.R.: Development of bioactive composite coatings based on combination of PEEK, bioactive glass and ag nanoparticles with antimicrobial properties. Surf. Coat. Technol. **301**, 100–105 (2016)
123. Baino, F., Verné, E.: Glass-based coatings on biomedical implants: a state-of-the-art review. Biomed. Glass. **3**, 1–17 (2017)
124. Ferraris, M., et al.: Chemical, mechanical, and antibacterial properties of silver nanocluster–silica composite coatings obtained by sputtering. Adv. Eng. Mater. **12**, B276-282 (2010)
125. Ferraris, M., Chiaretta, D., Fokine, M., Miola, M., Verné, E.: Pellicole antibatteriche ottenute da sputtering e procedimento per conferire proprietà antibatteriche ad un substrato, TO2008A000098 (2008)
126. Ferraris, M., Balagna, C., Perero, S.: Method for the application of an antiviral coating to a substrate and relative coating, WO2019/082001 (2019)
127. Muzio, G., et al.: Biocompatibility versus peritoneal mesothelial cells of polypropylene prostheses for hernia repair, coated with a thin silica/silver layer. J. Biomed. Mater. Res. B Appl. Biomater. **105**, 1586–1593 (2017)
128. Irfan, M., et al.: Antimicrobial functionalization of cotton fabric with silver nanoclusters/silica composite coating via RF co-sputtering technique. Cellulose **24**, 2331–2345 (2017)
129. Balagna, C., et al.: Characterization of antibacterial silver nanocluster/silica composite coating on high performance Kevlar® textile. Surf. Coat. Technol. **321**, 438–447 (2017)
130. Miola, M., et al.: Silver nanocluster-silica composite antibacterial coatings for materials to be used in mobile telephones. Appl. Surf. Sci. **313**, 107–115 (2014)
131. Balagna, C., et al.: Antipathogen nanostructured coating for air filters. Appl. Surf. Sci. 508, 145283 (2020)
132. Balagna, C., et al.: Antibacterial coating on polymer for space application. Mater. Chem. Phys. **135**, 714–722 (2012)
133. Balagna, C., Francese, R., Perero, S., Lembo, D., Ferraris, M.: Nanostructured composite coating endowed with antiviral activity against human respiratory viruses deposited on fibre-based air filters. Surf. Coat. Technol. 409, 126873 (2021)
134. Balagna, C., Perero, S., Percivalle, E., Vecchio Nepita, E., Ferraris, M.: Virucidal effect against coronavirus SARS-CoV-2 of a silver nanocluster/silica composite sputtered coating. Open Ceram. **1**, 100006 (2020)
135. Verné, E., Vitale-Brovarone, C., Robiglio, I., Baino, F.: Single-Piece Ceramic Prosthesis Elements, EP2152328A2 (2008)
136. Baino, F., Vitale-Brovarone, C.: Feasibility of glass-ceramic coatings on alumina prosthetic implants by airbrush spraying method. Ceram. Int. **41**, 2150–2159 (2015)
137. Baino, F., et al.: Novel full-ceramic monoblock acetabular cup with a bioactive tra-becular coating: design. Fabr. Charact. Ceram. Int. **42**, 6833–6845 (2016)
138. Baino, F., Vitale-Brovarone, C.: Trabecular coating on curved alumina substrates using a novel bio-active and strong glass-ceramic. Biomed. Glass. **1**, 31–40 (2015)
139. Baino, F., Montealegre, M.A., Orlygsson, G., Novajra, G., Vitale-Brovarone, C.: Bioactive glass coatings fabricated by laser cladding on ceramic acetabular cups: a proof-of-concept study. J. Mater. Sci. **52**, 9115–9128 (2017)

140. Cevc, G., Blume, G.: Hydrocortisone and dexamethasone in very deformable drug carriers have increased biological potency, prolonged effect, and reduced ther-apeutic dosage. Biochem. Biophys. Acta. **1663**, 61–73 (2004)
141. Hum, J., Boccaccini, A.R.: Bioactive glasses as carriers for bioactive molecules and therapeutic drugs: a review. J. Mater. Sci. Mater. Med. **23**, 2317–2333 (2012)
142. Arcos, D., Vallet-Regí, M.: Bioceramics for drug delivery. Acta Mater. **61**, 890–911 (2013)
143. Yan, X., Yu, C., Zhou, X., Tang, J., Zhao, D.: Highly ordered mesoporous bioactive glasses with superior in vitro bone-forming bioactivities. Angewandte Chemie Int. Ed. Engl. **43**, 5980–5984 (2004)
144. Migneco, C., Fiume, E., Verné, E., Baino, F.: A Guided walk through the world of mesoporous bioactive glasses (MBGs): fundamentals, processing, and applications. Nanomaterials **10**, 2571 (2020)
145. Izquierdo-Barba, I., Vallet-Regi, M.: Mesoporous bioactive glasses: relevance of their porous structure compared to that of classical bioglasses. Biomed. Glass. **1**, 140–150 (2015)
146. Wu, C., et al.: Bioactive mesopore-glass microspheres with controllable protein-delivery properties by biomimetic surface modification. J. Biomed. Mater. Res. A **95**, 476–485 (2010)
147. Zhao, L.Z., et al.: Mesoporous bioactive glasses for controlled drug release. Micropor. Mesop. Mater. **109**, 210–215 (2008)
148. Xia, W., Chang, J.: Well-ordered mesoporous bioactive glasses (MBG): a promising bioac-tive drug delivery system. J. Control. Release **110**, 522–530 (2006)
149. Zhao, Y.F., Loo, S.C., Chen, Y.Z., Boey, F.Y., Ma, J.: In situ SAXRD study of sol-gel induced well-ordered mesoporous bioglasses for drug delivery. J. Biomed. Mater. Res. A **85**, 1032–1042 (2008)
150. López-Noriega, A., et al.: Ordered mesoporous bioactive glasses for bone tissue regeneration. Chem. Mater. **18**, 3137–3144 (2006)
151. Baino, F., Fiorilli, S., Vitale-Brovarone, C.: Bioactive glass-based materials with hierarchical porosity for medical applications: review of recent advances. Acta Biomater. **42**, 18–32 (2016)
152. Kargozar, S., Montazerian, M., Hamzehlou, S., Kim, H.W., Baino, F.: Mesoporous bioactive glasses: promising platforms for antibacterial strategies. Acta Biomater. **81**, 1–19 (2018)
153. Ahmed, A.A., Ali, A.A., Mahmoud, D.A.R., El-Fiqi, A.M.: Preparation and characterization of antibacterial P2O5–CaO–Na2O–Ag2O glasses. J. Biomed. Mater. Res. Part A **98A**, 132–142 (2011)
154. Bellantone, M., Coleman, N.J., Hench, L.L.: A novel sol-gel derived bioactive glass featuring antibacterial properties. Key Eng. Mater. **192–195**, 597–600 (2001)
155. Mulligan, A.M., Wilson, M., Knowles, J.C.: The effect of increasing copper 770 content in phosphate-based glasses on biofilms of *Streptococcus sanguis*, 771. Biomaterials **24**, 1797–1807 (2003)
156. Schuhladen, K., et al.: Cu, Zn doped borate bioactive glasses: antibacterial efficacy and dose-dependent in vitro modulation of murine dendritic cells. Biomater. Sci. **8**, 2143 (2020)
157. Vernè, E., et al.: Surface silver-doping of biocompatible glass to induce antibacterial properties. Part I: massive glass. J. Mater. Sci. Mater. Med. **20**, 2935–2942 (2009)
158. Miola, M., Vernè, E.: Bioactive and antibacterial glass powders doped with copper by ion-exchange in aqueous solutions. Materials **9**, 405 (2016)
159. Di Nunzio, S., et al.: Silver containing bioactive glasses prepared by molten salt ion-exchange. J. Eur. Ceram. Soc. **24**, 2935–2942 (2004)
160. Bellantone, M., Williams, H.D., Hench, L.L.: Broad-spectrum bactericidal activity of Ag (2)O-doped bioactive glass. Antimicrob. Agents Chemother. **46**, 1940–1945 (2002)
161. Diba, M., Boccaccini, A.R.: Silver-containing bioactive glasses for tissue engineering applications. In: Baltzer, N., Copponnex, T. (eds.) Precious Metals for Biomedical Applications, pp. 177–211. Woodhead Publishing (2014)
162. Neel, E.A., Ahmed, I., Pratten, J., Nazhat, S.N., Knowles, J.C.: Characterisation of antibacterial copper releasing degradable phosphate glass fibres. Biomaterials **26**, 2247–2254 (2005)

163. Baghbani, F., Moztarzadeh, F., Hajibaki, L., Mozafari, M.: Synthesis, characterization and evaluation of bioactivity and antibacterial activity of quinary glass system (SiO2–CaO–P2O5–MgO–ZnO): in vitro study. Bull. Mater. Sci. **36**, 1339–1346 (2013)
164. Valappil, S.P., et al.: Role of gallium and silver from phosphate-based glasses on in vitro dual species oral biofilm models of *Porphyromonas gingivalis* and *Streptococcus gordonii*. Acta Biomater. **8**, 1957–1965 (2012)
165. Sakthi, P.S., et al.: In vitro bioactivity and antibacterial properties of bismuth oxide modified bioactive glasses. J. Mater. Res. **33**, 178–190 (2018)
166. Liu, L., Pushalkar, S., Saxena, D., LeGeros, R.Z., Zhang, Y.: Antibacterial property expressed by a novel calcium phosphate glass. J. Biomed. Mater. Res. B Appl. Biomater. **102**, 423–429 (2014)
167. Marta, M., et al.: Tellurium: a new active element for innovative multifunctional bioactive glasses. Mater. Sci. Eng. C 123, 111957 (2021)
168. Gruian, K.M.C., et al.: Addressing the optimal silver content in bioactive glass systems in terms of BSA adsorption. J. Mater. Chem. B **2**, 5799–5808 (2014)
169. Verné, E., et al.: Synthesis and characterisation of bioactive and antibacterial glass–ceramic part 1—microstructure, properties and biological behaviour. Adv. Appl. Ceram. **107**, 234–244 (2008)
170. Vitale-Brovarone, C., Miola, M., Balagna, C., Verné, E.: 3D-glass–ceramic scaffolds with antibacterial properties for bone grafting. Chem. Eng. J. **137**, 129–136 (2008)
171. Gholipourmalekabadi, M., et al.: Detection and qualification of optimum antibacterial and cytotoxic activities of silver-doped bioactive glasses. ET Nanobiotechnol. **9**, 209–214 (2015)
172. Wilkinson, H.N., Iveson, S., Catherall, P., Hardman, M.J.: A novel silver bioactive glass elicits antimicrobial efficacy against *Pseudomonas aeruginosa* and *Staphylococcus aureus* in an ex vivo skin wound biofilm model. Front. Microbiol. **9**, 1450 (2018)
173. Miola, M., et al.: Antibiotic-free composite bone cements with antibacterial and bioactive proper-ties. A preliminary study. Mater. Sci. Eng. C **43**, 65–75 (2014)
174. Miola, M., Verné, E., Vitale-Brovarone, C., Baino, F.: Antibacterial bioglass-derived scaffolds: innovative synthesis approach and characterization. Int. J. Appl. Glas. Sci. **7**, 238–247 (2016)
175. Miola, M., Fucale, G., Maina, G., Verné, E.: Composites bone cements with different viscosities loaded with a bioactive and antibacterial glass. J. Mater. Sci. **52**, 5133–5146 (2017)
176. Miola, M., Fucale, G., Maina, G., Verné, E.: Antibacterial and bioactive composite bone cements containing surface silver-doped glass particles. Biomed. Mater. **10**, 055041 (2015)
177. Hosny, A.E.-D.M., Rasmy, S.A., Aboul-Magd, D.S., Kashef, M.T., El-Bazza, Z.E.: The increasing threat of silver-resistance in clinical isolates from wounds and burns. Inf. Drug Resist. **12**, 1985–2001 (2019)
178. Aina, V., et al.: Influence of the chemical composition on nature and activity of the surface layer of Zn-substituted sol-gel (Bioactive) glasses. J. Phys. Chem. C **115**, 2196–2210 (2011)
179. Raffi, M., et al.: Investigations into the antibacterial behavior of copper nanoparticles against *Escherichia coli*. Ann. Microbiol. **60**, 75–80 (2010)
180. Goh, Y.F., et al.: Bioactive glass: an in-vitro comparative study of doping with nanoscale copper and silver particles. Appl. Glass Sci. **5**, 255–266 (2014)
181. Srivastava, A.K., Pyare, R.: Characterization of CuO substituted 45S5 bioactive glasses and glass-ceramics. Int. J. Sci. Technol. Res. **1**, 28–41 (2012)
182. Miola, M., et al.: Copper-doped bioactive glass as filler for PMMA-based bone cements: morphological, mechanical, reactivity, and preliminary antibacterial characterization. Materials (Basel) **11**, 961 (2018)
183. Esteban-Tejeda, L., et al.: Calcium and zinc containing bactericidal glass coatings for biomedical metallic substrates. Int. J. Mol. Sci. **15**, 13030–13044 (2014)
184. Balasubramanian, P., Strobel, L.A., Kneser, U., Boccaccini, A.R.: Zinc-containing bioactive glasses for bone regeneration, dental and orthopedic applications. Biomed. Glass. **1**, 51–69 (2015)
185. Shahrbabak, M.S.N., Sharifianjazi, F., Rahban, D., Salimi, A.: A comparative investigation on bioactivity and antibacterial properties of sol-gel derived 58S bioactive glass substituted by Ag and Zn. Silicon **11**, 2741–2751 (2019)

186. Grandi, S., et al.: Bone reconstruction: Au nanocomposite bioglasses with antibacterial properties. Int. J. Artif. Organs **34**, 920–928 (2011)
187. Vulpoi, A., et al.: Bioactivity and protein attachment onto bioactive glasses containing silver nanoparticles. J. Biomed. Mater. Res. A **100**, 1179–1186 (2012)
188. Ferraris, S., et al.: In situ reduction of antibacterial silver ions to metallic silver nanoparticles on bioactive glasses functionalized with polyphenols. Appl. Surf. Sci. **396**, 461–470 (2017)
189. Miola, M., Bertone, E., Vernè, E.: In situ chemical and physical reduction of copper on bioactive glass surface. Appl. Surf. Sci. **495**, 143559 (2019)
190. Hamad, A., Khashan, K.S., Hadi, A.: Silver nanoparticles and silver ions as potential antibacterial agents. J. Inorg. Organomet. Polym. **30**, 4811–4828 (2020)
191. Miola, M., Vitale-Brovarone, C., Mattu, C., Verné, E.: Antibiotic loading on bioactive glasses and glass-ceramics: an approach to surface modification. J. Biomater. Appl. **28**, 308–319 (2012)
192. Arcos, D., Lopez-Noriega, A., Ruiz-Hernandez, E., Terasaki, O., Val-let-Regi, M.: Ordered mesoporous microspheres for bone grafting and drug delivery. Chem. Mater. **21**, 1000–1009 (2009)
193. Anand, A., Das, P., Nandi, S.K., Kundua, B.: Development of antibiotic loaded mesoporous bioactive glass and its drug release kinetics. Ceram. Int. **46**, 5477–5483 (2020)
194. Cheng, T., Qu, H., Zhang, G., Zhang, X.: Osteogenic and antibacterial properties of vancomycin-laden mesoporous bio-glass/PLGA composite scaffolds for bone regeneration in infected bone defects. Artif. Cells, Nanomed. Biotechnol. **46**, 1935–1947 (2018)
195. Bortuzzo, K.Z.J.A., et al.: Bio-templated bioactive glass particles with hierarchical macro–nano porous structure and drug delivery capability. Colloids Surf. B Biointerf. **135**, 825–832 (2015)
196. Zhang, X., et al.: Teicoplanin-loaded borate bioactive glass implants for treating chronic bone infection in a rabbit tibia osteomyelitis model. Biomaterials **31**, 5865–5874 (2010)
197. Drago, L., Toscano, M., Bottagisio, M.: Recent evidence on bioactive glass antimicrobial and antibiofilm activity: a mini-review. Materials **11**, 326 (2018)
198. Gorustovich, J., Rivadeneira, A.: Bioactive glasses as delivery systems for antimicrobial agents. J. Appl. Microbiol. **122**, 1424–1437 (2017)
199. Hu, S., Chang, J., Liu, M., Ning, C.: Study on antibacterial effect of 45S5 Bioglass. J. Mater. Sci. Mater. Med. **20**, 281–286 (2009)
200. Rivadeneira, J., Audisio, C.M., Boccaccini, A.R., Gorustovich, A.A.: In vitro antistaphylococcal effects of a novel 45S5 bioglass/agar-gelatin biocomposite films. J. Appl. Microbiol. **115**, 604–612 (2013)
201. Drago, L., et al.: In vitro antibiofilm activity of bioactive glass S53P4. Future Microbiol. **9**, 593–601 (2014)
202. Saima Begum, J.W.E., Worthingtonand, T., Martin, R.A.: The influence of pH and fluid dynamics on the antibacterial efficacy of 45S5 Bio-glass. Biomed. Mater. **11**, 015006 (2016)
203. Xie, Z.P., et al.: Failure of particulate bioglass to prevent experimental staphylococcal infection of open tibial fractures. J. Antimicrob. Chemother. **62**, 1162–1163 (2008)
204. Verné, E., Ferraris, S.: Surface functionalization of bioactive glasses: reactive groups, biomolecules and drugs on bioactive surfaces for smart and functional biomaterials. In: Boccaccini, l.R., Brauer, D.S., Hupa, L. (eds.) Bioactive Glasses: Fundamentals, Technology and Applications, pp. 221–235. Royal Society of Chemistry (2017)
205. Cao, W., Hench, L.L.: Bioactive materials. Ceram. Int. **22**, 493–507 (1996)
206. Verné, E., Vitale-Brovarone, C., Bui, E., Bianchi, C.L., Boccaccini, A.R.: Surface functionalization of bioactive glasses. J. Biomed. Mater. Res. **90A**, 981–992 (2009)
207. Kargozar, S., et al.: Functionalization and surface modifications of bioactive glasses (BGs): tailoring of the biological response working on the outermost surface layer. Materials (Basel) **12**, 3696 (2019)
208. Schuhladen, K., Roether, J.A., Boccaccini, A.R.: Bioactive glasses meet phytotherapeutics: the potential of natural herbal medicines to extend the functionality of bioactive glasses. Biomaterials **217**, 119288 (2019)

209. Chen, Q.Z., Rezwan, K., Armitage, D., Nazhat, S.N., Boccaccini, A.R.: The surface functionalization of 45S5 Bioglassr-based glass-ceramic scaffolds and its impact on bioactivity. J. Mater. Sci. Mater. Med. **17**, 979–987 (2006)
210. Verné, E., et al.: Surface functionalization of 3D glass–ceramic porous scaffolds for enhanced miner-alization in vitro. Appl. Surf. Sci. **271**, 412–420 (2013)
211. Verné, E., et al.: Alkaline phosphatase grafting on bioactive glasses and glass ceramics. Acta Biomater. **6**, 229–240 (2010)
212. Stanić, S.: Variation in properties of bioactive glasses after surface modification. In: Clinical Applications of Biomaterials, pp. 35–63. Springer, Cham (2017)
213. Sun, J., et al.: Functionalization and bioactivity in vitro of mesoporous bioactive glasses. J. Non-Cryst. Solids **354**, 3799–3805 (2008)
214. Gruian, C., Vanea, E., Simon, S., Simon, V.: FTIR and XPS studies of protein adsorption onto functionalized bioactive glass. Biochem. Biophys. Acta. **1824**, 873 (2012)
215. Toworfe, G.K., et al.: Effect of functional end groups of silane self-assembled monolayer surfaces on apa-tite formation, fibronectin adsorption and osteoblast cell function. J. Tissue Eng. Regen. Med. **3**, 26–36 (2009)
216. Magyari, K., et al.: Bioactivity evolution of the surface functionalized bioactive glasses. J. Biomed. Mater. Res. Part B **103**, 261–272 (2015)
217. Schickle, K., et al.: Synthesis of novel tricalcium phosphate-bioactive glass composite and functionalization with rhBMP-2. J. Mater. Sci. - Mater. Med. **22**, 763–771 (2011)
218. Massera, J., Mishra, A., Guastella, S., Ferraris, S., Verné, E.: Surface functionalization of phosphate-based bioactive glasses with 3-aminopropyltriethoxysilane (APTS). Biomed. Glass. **2**, 51–62 (2016)
219. Ferraris, S., Nommeots-Nomm, A., Spriano, S., Vernè, E., Massera, J.: Surface reactivity and silanization ability of borosilicate and Mg-Sr-based bioactive glasses. Appl. Surf. Sci. **475**, 43–55 (2019)
220. Zhang, Y., Luan, J., Jiang, S., Zhou, X., Li, M.: The effect of amino-functionalized mesoporous bioactive glass on MC3T3-E1 cells in vitro stimulation. Composites B **172**, 397–405 (2019)
221. Höhn, S., et al.: Effects of medium pH and preconditioning treatment on protein adsorption on 45S5 bioactive glass surfaces. Adv. Mater. Interf. **7**, 2000420 (2020)
222. Lou, T., Bai, X., He, X., Yuan, C.: Antifouling performance analysis of peptide-modified glass microstructural surfaces. Appl. Surf. Sci. **541**, 148384 (2021)
223. Zhang, X., et al.: Ultrahigh adhesion force between silica-binding peptide SB7 and glass substrate studied by single-molecule force spectroscopy and molecular dynamic simulation. Front. Chem. **8**, 600918 (2020)
224. Zhang, L., et al.: Facile surface modification of glass with zwitterionic polymers for improving the blood compatibility. Mater. Res. Exp. **5**, 065401 (2018)
225. Terpilowski, K., Rymuszka, D.: Surface properties of glass plates activated by air, oxygen, nitrogen and argon plasma. Glass Phys. Chem. **42**, 535–541 (2016)
226. Lopez-Noriega, A., Arcos, D., Vallet-Regí, M.: Functionalizing mesoporous bioglasses for long-term anti-osteoporotic drug delivery. Chem. Eur. J. **16**, 10879–10886 (2010)
227. Leonor, I.B., Alves, C.M., Azevedo, H.S., Reis, R.L.: Effects of protein incorporation on calcium phosphate coating. Mater. Sci. Eng., C **29**, 913–918 (2009)
228. Chen, Q.Z., et al.: Collagen release kinetics of surface functionalized 45S5 Bioglass®-based porous scaffolds. J. Biomed. Mater. Res. Part A **86A**, 987–995 (2007)
229. Zhang, X., Ferraris, S., Prenesti, E., Vernè, E.: Surface functionalization of bioactive glasses with natural molecules of biological significance, part I: gallic acid as model molecule. Appl. Surf. Sci. **287**, 329–340 (2013)
230. Ferraris, S., et al.: Gallic acid grafting to a ferrimagnetic bioactive glass-ceramic. J. Non-Cryst. Solids **432**, 167–175 (2016)
231. Corazzari, I., et al.: Gallic acid grafting modulates the oxidative potential of ferrimagnetic bioactive glass-ceramic SC-45. Colloids Surf. B Biointerf. **148**, 592–599 (2016)
232. Cazzola, M., et al.: Bioactive glasses functionalized with polyphenols: in vitro interactions with healthy and cancerous osteoblast cells. J. Mater. Sci. **52**, 9211–9223 (2017)

233. Cazzola, M., et al.: Bioactive glass coupling with natural polyphenols: surface modification, bioactivity and anti-oxidant ability. Appl. Surf. Sci. **367**, 237–248 (2016)
234. Zhang, X., Ferraris, S., Prenesti, E., Vernè, E.: Surface functionalization of bioactive glasses with natural molecules of biological significance, part II: grafting of polyphenols extracted from grape skin. Appl. Surf. Sci. **287**, 341–348 (2013)
235. Dziadek, M., Dziadek, K., Zagrajczuk, B., Menaszek, E., Cholewa-Kowalska, K.: Poly ("-caprolactone)/bioactive glass composites enriched with polyphenols extract-ed from sage (*Salvia officinalis* L.). Mater. Lett. **183**, 386–390 (2016)
236. Sayed Abdelgeliel, A., et al.: Surface functionalization of bioactive glasses with polyphenols from *Padina pavonica* Algae and in situ reduction of silver ions: physico-chemical characterization and biological response. Coatings **9**, 394 (2019)
237. Ferlenda, G., et al.: Surface functionalization of a silica-based bioactive glass with compounds from *Rosa canina* bud extracts. ACS Biomater. Sci. Eng. **7**, 96–104 (2021)
238. Malavasi, G., et al.: The role of coordination chemistry in the development of innovative gallium-based bioceramics: the case of curcumin. J. Mater. Chem. **21**, 5027–5037 (2011)
239. Verne, E., et al.: Surface activation of a ferrimagnetic glass-ceramic for antineoplastic drugs graftingy. Adv. Eng. Mater. **12**, B309–B319 (2010)
240. Zhu, M., Zhang, J., Tao, C., He, X., Zhu, Y.: Design of mesoporous bioactive glass/hydroxyapatite composites for controllable co-delivery of chemotherapeutic drugs and proteins. Mater. Lett. **115**, 194–197 (2014)
241. Boanini, E., et al.: Alendronate functionalized mesoporous bioactive glass nanospheres. Materials (Basel) **9**, 135 (2016)
242. Aina, V., et al.: New formulation of functionalized bioactive glasses to be used as carriers for the development of pH-stimuli responsive biomaterials for bone diseases. Langmuir **30**, 4703–4715 (2014)
243. Ferraris, S., et al.: Effects of sterilization and storage on the properties of ALP-grafted biomaterials for prosthetic and bone tissue engineering applications. Biomed. Mater. **7**, 054102 (2012)
244. Jones, J.R., Brauer, D.S., Greenspan, D.C.: Bioglass and bioactive glasses and their impact on healthcare. Int. J. Appl. Glas. Sci. **7**, 423–432 (2016)
245. Merwin, G.E.: Bioglass middle ear prosthesis: preliminary report. Ann. Otol. Rhinol. Laryngol. **95**, 78–82 (1986)
246. Bahmad, F., Merchant, S.N.: Histopathology of ossicular grafts and implants in chronic otitis media. Ann. Otol. Rhinol. Laryngol. **116**, 181–191 (2007)
247. Downing, M., et al.: A bone-anchored percutaneous connector system for neural prosthetic applications. Ear Nose Throat J. **76**, 328–332 (1997)
248. Stanley, H., Hall, M., Clark, A.: Using 45S5 bioglass cones as endosseous ridge maintenance implants to prevent alveolar ridge resorption: a 5-year evaluation. Int. J. Oral Maxillofac. Implants **12**, 1–19 (1997)
249. Baino, F.: Ceramics for bone replacement: commercial and clinical use. In: Palmero, P., Cambier, F., De Barra, E. (eds.) Advances in Ceramic Biomaterials, pp. 249–278. Woodhead Publishing, Elsevier (2017)
250. Chen, Q.Z., Thompson, I.D., Boccaccini, A.R.: 45S5 Bioglass®-derived glass–ceramic scaffolds for bone tissue engineering. Biomaterials **27**, 2414–2425 (2006)
251. Gillam, D.G., Tang, J.Y., Mordan, N.J., Newman, H.N.: The effects of a novel Bioglass® dentifrice on dentine sensitivity: a scanning electron microscopy investigation. J. Oral Rehabil. **29**, 305–313 (2002)
252. Vahid Golpayegani, M., Sohrabi, A., Biria, M., Ansari, G.: Remineralization effect of topical NovaMin versus sodium fluoride (1.1%) on caries-like lesions in permanent teeth. J. Dentistry (Tehran) **9**, 68–75 (2012)
253. Banerjee, A., Hajatdoost-Sani, M., Farrell, S., Thompson, I.: A clinical evaluation and comparison of bioactive glass and sodium bicarbonate air-polishing powders. J. Dent. **38**, 475–479 (2010)

254. Kargozar, S., Baino, F., Hamzehlou, S., Hill, R.G., Mozafari. M.: Bioactive glasses: sprouting angiogenesis in tissue engineering. Trends Biotechnol. **36**, 430–444 (2018)
255. Wray, P.: Cotton candy'that heals. Am. Ceram. Soc. Bull. **90**, 24–31 (2011)
256. Ma, X., Schou, K.R., Maloney-Schou, M., Harwin, F.M., Ng, J.D.: The porous polyethylene/bioglass spherical orbital implant: a retrospective study of 170 cases. Ophthalmic Plast. Reconstr. Surg. **27**, 21–27 (2011)
257. Miola, M., et al.: Glass-ceramics for cancer treatment: so close, or yet so far? Acta Biomater. **83**, 55–70 (2019)
258. Velasco, M.V., Souza, M.T., Crovace, M.C., Aparecido de Oliveira, A.J.A., Zanotto, E.D.: Bioactive magnetic glass-ceramics for cancer treatment. Biomed. Glass. **5**, 148–177 (2019)
259. Cochis, A., Miola, M., Bretcanu, O., Rimondini, L., Verné, E.: Magnetic bioactive glass ceramics for bone healing and hyperthermic treatment of solid tumors. In: Tiwari, A., Iyer, P. K., Kumar, V., Swart, H. (eds.) Advanced Magnetic and Optical Materials, pp. 81–112. Scrivener Publishing LLC (2016)
260. Miola, M., et al.: Composite bone cements for hyperthermia: modeling and characterization of magnetic, calorimetric and in vitro heating properties. Ceram. Int. **43**, 4831–4840 (2017)
261. Bretcanu, O., et al.: In vitro biocompatibility of a ferrimagnetic glass-ceramic for hyperthermia application. Mater. Sci. Eng., C **73**, 778–787 (2017)
262. Miola, M., Gerbaldo, R., Laviano, F., Bruno, M., Vernè, E.: Multifunctional ferrimagnetic glass–ceramic for the treatment of bone tumor and associated complications. J. Mater. Sci. **52**, 9192–9201 (2017)
263. Verné, E., et al.: Composite bone cements loaded with a bioactive and ferrimagnetic glass-ceramic: leaching, bioactivity and cytocompatibility. Mater. Sci. Eng., C **53**, 93–103 (2015)
264. Tiberto, P., et al.: Magnetic relaxation in ferrimagnetic glass-ceramics obtained by co-precipitation at different temperatures. IEEE Trans. Magn. **43**, 2471–2473 (2007)
265. Bretcanu, O., Verné, E., Coisson, M., Tiberto, P., Allia, P.: Temperature effect on the magnetic properties of the coprecipitation-derived ferrimagnetic glass-ceramics. J. Magn. Magn. Mater. **300**, 412–417 (2006)
266. Bretcanu, O., Verné, E., Coisson, M., Tiberto, P., Allia, P.: Magnetic properties of the ferromagnetic glass-ceramics for hyperthermia. J. Magn. Magn. Mater. **305**, 529–533 (2006)
267. Bretcanu, O., Spriano, S., Verné, E., Coisson, M., Tiberto, P.: The influence of crystallised Fe3O4 on the magnetic properties of coprecipita-tion-derived ferrimagnetic glass-ceramics. Acta Biomater. **1**, 421–429 (2005)
268. Bretcanu, O., Spriano, S., Vitale-Brovarone, C., Verné, E.: Synthesis and characterization of coprecipitation-derived ferrimagnetic glass-ceramics. J. Mater. Sci. **41**, 1029–1037 (2006)
269. Day, D.E.: Glasses for radiotherapy. In: Jones, J.R., Clarke, G.C. (eds.) Bio-Glasses: An Introduction, pp. 203–228. Wiley (2012)
270. Miguez-Pacheco, V., Hench, L.L., Boccaccini, A.R.: Bioactive glasses beyond bone and teeth: emerging applications in contact with soft tissues. Acta Biomater. **13**, 1–15 (2015)
271. Baino, F., Novajra, G., Miguez-pacheco, V., Boccaccini, A.R., Vitale-Brovarone, C.: Bioactive glasses: special applications outside the skeletal system. J. Non-Cryst. Solids **432**, 15–30 (2016)
272. Kargozar, S., Hamzehlou, S., Baino, F.: Potential of bioactive glasses for cardiac and pulmonary tissue engineering. Materials (Basel). **10**, 1429 (2017)
273. Gocha, A., McDonald, L.: Better bodies with biomaterials: how ceramic and glass contribute to the $110B global market for implantable bio-materials. Am. Ceram. Soc. Bull. **99**, 17–31 (2020)

Glass–Ceramic Sealants for Solid Oxide Cells Research at Politecnico di Torino: An Overview on Design, Sinter-Crystallization, Integration and Interfacial Issues

Antonio G. Sabato, Hassan Javed, Milena Salvo, Andreas Chrysanthou, and Federico Smeacetto

Abstract Solid oxide cells (SOCs) devices are of great interest for effective power generation and for highly efficient conversion of electricity to hydrogen, fuels and chemicals using high temperature electrolysis. Electrochemical energy conversion in SOCs can be achieved only by a reliable integration and joining technology. In this context, glass-based materials, with multiple functional compositions and properties play a key role in the performance and durability. In most SOC stack designs the metallic interconnect must be sealed both to the interconnect frame and to other cell ceramics components, thus presenting significant challenges. An overview and update about the research activity done at Politecnico di Torino (Italy) on design, sinter-crystallization, characterization and testing of glasses and glass–ceramics sealants is presented and discussed.

Keywords Glass · Glass–ceramics · Sintering · Crystallization · Solid oxide cells · Sealants

---

A. G. Sabato
Institut de Recerca en Energia de Catalunya (IREC), Jardins de les Dones de Negre 1, 2ª pl, 08930 Sant Adrià de Besòs, Barcelona, Spain
e-mail: gsabato@irec.cat

H. Javed
Sunfire GmbH, Gasanstaltstraße 2, 01237 Dresden, Germany
e-mail: Hassan.Javed@sunfire.de

M. Salvo · F. Smeacetto (✉)
Department of Applied Science and Technology, Politecnico di Torino, Corso Duca degli Abruzzi 24, 10129 Torino, Italy
e-mail: federico.smeacetto@polito.it

M. Salvo
e-mail: milena.salvo@polito.it

A. Chrysanthou
School of Physics, Engineering and Computer Science, University of Hertfordshire, College Lane, Hatfield AL10 9AB, Hertfordshire, UK
e-mail: a.chrysanthou@herts.ac.uk

F. Baino et al. (eds.), *Ceramics, Glass and Glass-Ceramics*, PoliTO Springer Series,
https://doi.org/10.1007/978-3-030-85776-9_6

1 Glass–Ceramic Sealants for Solid Oxide Cells (SOC) Applications

Glass and glass–ceramics play a fundamental role in the development of SOC technology. Indeed, thanks to their nature, glasses and glass–ceramics can be used as sealants in harsh conditions ensuring a very high gas tightness (leak rate < 10^{-9} mbar l s^{-1}). The use of glass to metal sealing technology is well known in different applications such as hermetic interconnectors, incandescent light bulb, glass-encapsulate diodes, pressure tight glass windows etc. [1, 2].

In SOC technology different materials need to be joined together for use in severe operating conditions. In order to obtain useful SOC performance, it is necessary to combine more cells in a stack connected by an interconnect which is typically made by high Cr-containing stainless steel (the most common being Crofer22APU), while ceramic materials are used for both the electrodes and the electrolyte. Furthermore, SOC stacks operate at high temperatures (700–900 °C) and are exposed to dual atmospheres (reducing and oxidizing) for long periods of time and their lifetime is expected to exceed 40 kh.

The use of sealants is necessary in these systems to ensure the separation between the atmospheres, otherwise their mixing, even due to micro leaks, can lead to fast reduction of the device efficiency as well as to side reactions that may damage the device itself (i.e. formation of hotspots due to strong exothermic reactions). In this scenario, the need for sealants that are able to withstand the high operating temperatures and thermal cycles as well as the aggressive atmospheres has emerged. In recent years, researchers have focused on different sealing approaches; these include the use of a variety of possible sealants including compressive sealants such as mica-based sealants [3–6], metallic-based sealants or rigid ones (like brazes or glass–ceramics). Out of these options, glass–ceramic sealants have been observed to be among the best-performing. The great advantage of the use of these materials as sealants is represented by the possibility of tuning their properties by tailoring their composition. It is possible to change the type or the amounts of the oxides contained in a parent glass in order to obtain favourable properties for the performance of the final glass–ceramic sealant. Properties that can be altered by variation in the glass–ceramic composition include the glass transition temperature, T_g, the softening temperature, T_s, the crystallization temperature, T_c, the crystalline phases and crystallization behaviour, the thermal expansion coefficient (CTE) which is fundamental in SOC applications due to the high operating temperatures as well as its chemical compatibility with the other materials involved in a SOC stack.

In such a challenging application, different factors need to be considered at the same time; a good sealant should not develop detrimental phases with the joined materials at the operating T and should withstand exposure to both oxidizing and reducing atmospheres and of course it must not degrade under these conditions. Furthermore, it has to show excellent thermo-mechanical compatibility with such materials and high electrical resistivity (> 10^4 Ω cm) in order to avoid current losses or short circuits [3–5]. Therefore, it is clear that the starting composition of a glass-based sealant has to be carefully tailored since an improvement on one of these factors may be concomitant with a worsening of other properties. The development of a suitable glass-based sealant during recent years has attracted the efforts of many players in the scientific community including: Pacific Northwest National Laboratories (PNNL) [7], Denmark Technical University (DTU) [8]. Furthermore, different compositions have been commercialized by SCHOTT and other companies [9].

There is a growing body of literature that recognizes the important and crucial role of sealants. Among these, a strong contribution in the last 10 years has come from Politecnico di Torino and its partners focusing the efforts on production, characterization and testing in real operating conditions of different glass compositions (either containing or in the absence of alkali oxides). In this section, their main results will be summarized.

1.1 Alkali-Containing Glass–Ceramic Sealants at POLITO

Part of the research activity at POLITO on glass–ceramic sealants focused on alkali-containing compositions. The work has addressed topics including their thermal characterization by means of differential thermal analyses (DTA), heating stage microscopy (HSM) and dilatometry (DIL) and their compatibility with typical materials involved in the joining of SOC stacks components including YSZ (electrolyte), Crofer22APU and AISI 441 stainless steels (both bare and coated with different protective coatings).

The introduction of alkali oxides in the glass composition can help the tuning of the CTE of the sealant after its sinter-crystallization, achieving a high thermomechanical compatibility with the sealed materials which typically have a CTE that ranges between 10×10^{-6} and 12×10^{-6} K^{-1}. Furthermore, thanks to their role as network modifiers in a glass, alkali oxides can contribute to the lowering of the characteristic temperatures of the material.

In the case of SOC applications, it is very important to understand the behaviour of a sealant not only in its starting glassy form but especially as a glass–ceramic obtained as a consequence of a sinter crystallization thermal treatment (carried out in order to consolidate the joining process). Indeed, the formation of crystalline phases can strongly influence the overall properties of the material and therefore the soundness of the joining as well as its compatibility in relevant operating conditions.

For this reason, Smeacetto and co-workers [10–13] carried out a variety of studies on the crystallization behaviour of different glasses by DTA measurements demonstrating how these systems tend to crystallize mainly by following a heterogeneous crystallization mechanism, starting from the surface of the glassy particles. It is possible to evaluate the crystallization behaviour of a glass by carrying out DTA measurements on a fixed particle size with different heating rates. Furthermore, DTA analyses on different glass particle sizes may provide understanding of the main crystallization behaviour. Indeed, if heterogeneous crystallization from the surface is the predominant mechanism, a finer powder will give a DTA thermogram with a more

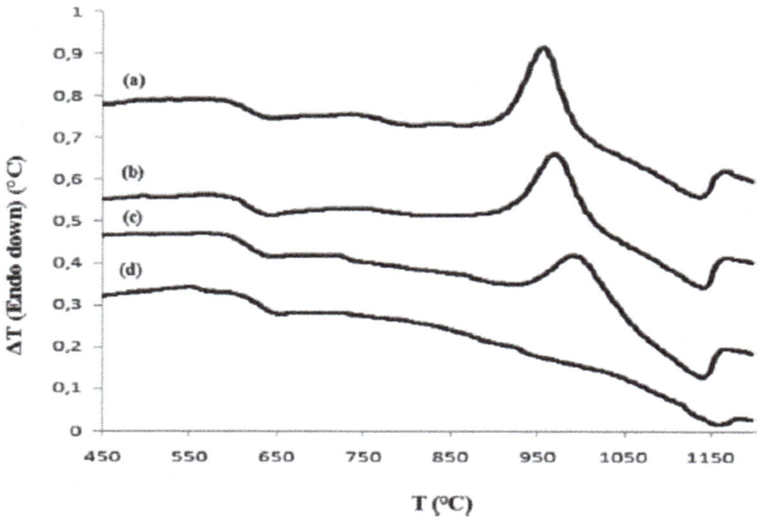

Fig. 1 DTA analyses carried out on alkali-containing glass with different powders dimension at the same heating rate (20 °C/min): **a** < 38 μm, **b** 38–75 μm, **c** > 75 μm, **d** bulk glass [12]

pronounced crystallization peak, in comparison with coarser powders (Fig. 1). A deep understanding of the crystallization mechanism and of the effect of the particle size is fundamental since it can affect the compatibility of a sealant with the interface material. An excessive surface crystallization can lead to high degree of shrinkage with consequent poor wetting on the substrate (as shown in the left image of Fig. 2).

Furthermore, the control of the amount of crystalline phases is also crucial. A certain degree of crystallization can be beneficial for a sealant since the crystalline phase improves the mechanical behaviour of the joint avoiding excessive flowing behaviour of the glass–ceramic at the operating conditions of a SOC stack (much higher of the typical glass transition temperatures). By contrast, a sufficiently residual glassy phase is able to accommodate thermal stresses during the operation at high T. Furthermore, the evolution of the crystalline phases needs to be taken into account since the long-term exposure of these materials $T > T_g$ can lead to changes in the amount and number of crystallized phases in comparison with the "as-joined" state.

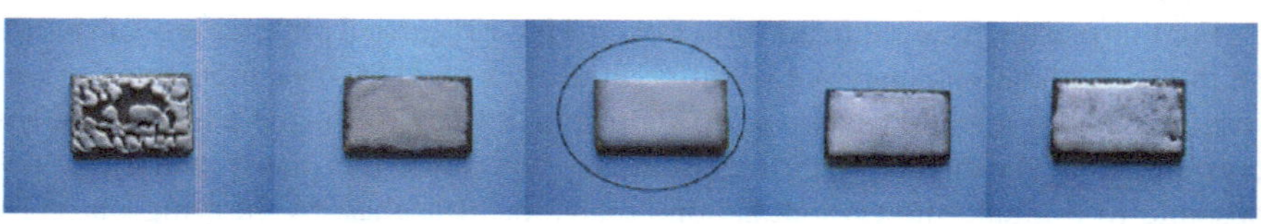

Fig. 2 Effect of different powder sizes on the compatibility between a glass–ceramic sealant (SACN) and a Crofer22APU stainless steel substrate. From left to right different SACN powders sizes: d < 20 μm, d = 20–38 μm, d = 38–75 μm, d = 75–106 μm and d > 106 μm [10]

An example is reported in Figs. 3 and 4 concerning an alkali-containing glass–ceramic labelled as V10. This material was studied after different steps of exposure at 800 °C in static air (100 and 300 h) and its microstructure compared with of its structure after the joining treatment (850 °C for 1 h in air). The XRD analysis (Fig. 3) revealed the formation of Al-containing diopside (with a CTE 11.6×10^{-6} K^{-1}) as the only crystalline phase after the joining treatment ($\approx$ 37 wt%) (Fig. 4b). This led to a beneficial increase in the CTE of the sealant from 7.6×10^{-6} K^{-1} (parent glass) to 11.5×10^{-6} K^{-1} (glass–ceramic), a value that was closer to those of the materials typically involved in a stack. In this case the further aging at 800 °C led to the formation of nepheline (CTE 9×10^{-6} K^{-1}) as a secondary crystalline phase ($\approx$8 wt%). The formation of this phase did not affect excessively the CTE of the glass–ceramic and led to a more rigid structure "delaying" the dilatometric softening point (Fig. 4a); the amount of the residual glassy phase was found to be about 50 wt%. This example demonstrates the importance of having knowledge of the evolution of a glass–ceramic sealant for SOC applications, especially when considering the long-term operations expected for these devices.

One of the first alkali-containing glass–ceramic compositions studied at POLITO was labelled SACN (already cited in Fig. 2). This system was extensively studied and exhibited excellent behaviour in terms of compatibility with both YSZ and Crofer22APU. Figure 5 shows the morphology of a YSZ/SACN/Crofer22APU sample in the "as-joined" state, as well as after aging for 400 h in air at 800 °C. Both the SEM pictures revealed good adherence with no evidence of cracks or delamination and there were no adverse reactions even after aging.

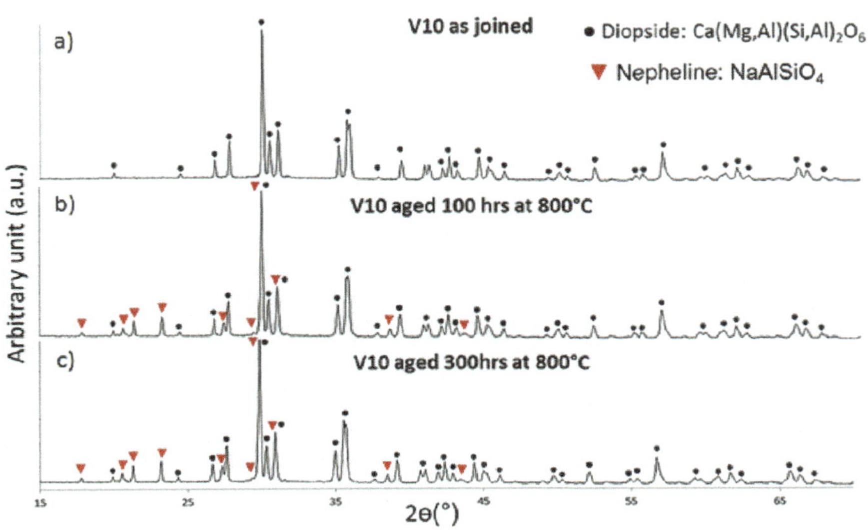

Fig. 3 XRD of V10 glass–ceramic after different aging at 800 °C in air [14]

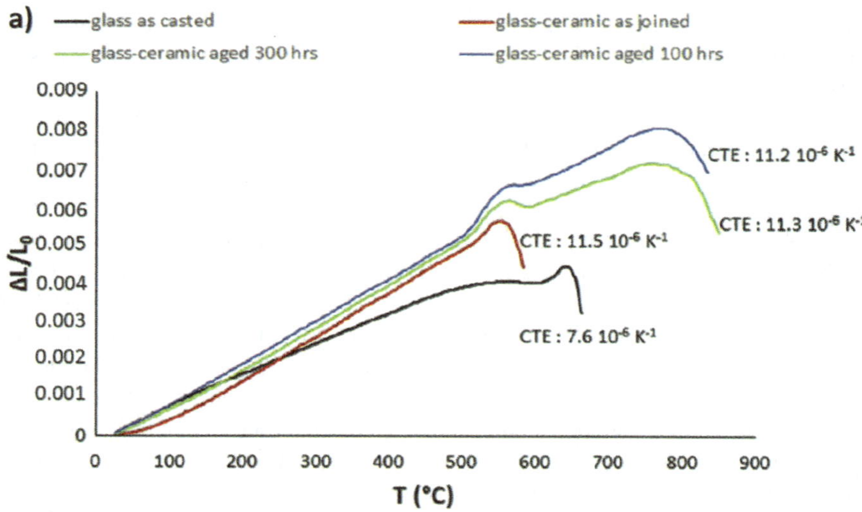

b)

Wt%	Amorphous	Diopside Ca(Mg,Al) $(Si,Al)_2O_6$	Nepheline $NaAlSiO_4$
As-joined	63	37	–
Aged 100 h	49	43	8
Aged 300 h	50	42	8

Fig. 4 Dilatometric measurements of V10 glass as casted and glass–ceramic after different aging at 800 °C in air (**a**) and results of Rietveld quantitative analyses based on XRD reported in Fig. 3 [14]

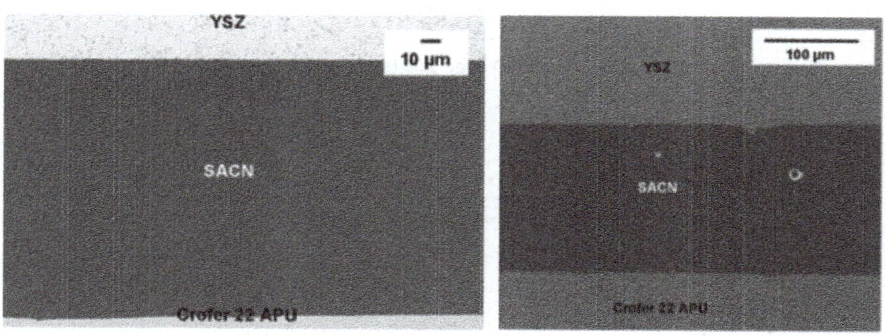

Fig. 5 SEM pictures of a joining Crofer22APU/SACN/YSZ after the joining treatment and after an aging of 400 h at 800 °C [10]

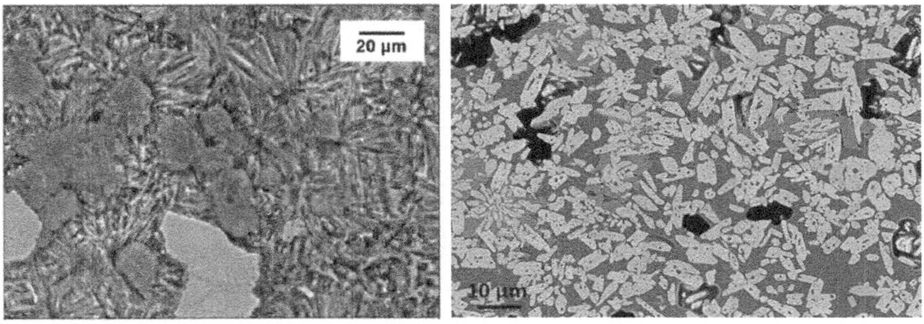

Fig. 6 Comparison between microstructure of a refractory sealant on the left Smeacetto et al. [10] and a compliant one on the right

However, the SACN glass–ceramic showed high crystallinity after the sinter-crystallization treatment. For this reason, the studies that followed focused on glassy systems that were more "compliant" and able to maintain a considerable amount of the residual amorphous phase thanks to a meticulous tailoring of the alkali content. A higher amount of residual glassy phase can be beneficial in real operating conditions ensuring a less rigid behaviour able to accommodate thermal stresses. Furthermore, the amorphous portion could provide self-healing during the thermal cycles. One example was represented by the V10 glass–ceramic. As reported in Fig. 4b after long aging at 800 °C, the residual glassy phase represents ≈ 50 wt% of the material. Figure 6 depicts a clear difference in the microstructure between the refractory glass–ceramic (SACN) and the compliant one (V10).

The behaviour of alkali-containing glasses has been the subject of several research investigations. Some investigations have shown that these glasses can exhibit satisfactory long-term behaviour when tested under harsh conditions [15–23]. There have also been studies in which there was evidence of reaction with Cr leading to formation of Cr-containing volatile species with consequent degradation of the sealant integrity and possible formation of conductive phases which may cause unwanted short circuits.

Sabato et al. [14] carried out an extensive study on the degradation phenomena of an alkali-containing glass–ceramic exposed to real-operating conditions under a dual atmosphere at 800 °C with application of a DC electrical voltage. The electrical resistivity was measured for the duration of the test. The work involved investigation of the behaviour of (i) the glass–ceramic itself with a 1.3 V applied voltage, (ii) a joined sample Crofer22APU/glass–ceramic/Crofer22APU with a 0.7 V applied voltage and (iii) a joined sample Crofer22APU/glass–ceramic/Crofer22APU with a 1.3 V applied voltage. The resulting resistivity measurements are shown in Fig. 7. While in the case of the first two samples the electrical resistivity was found to be stable exhibiting a value of above 10^4 Ω cm for the entire duration of the test, in the case of the sealant interfaced with the metallic counterpart under a higher voltage, fast degradation of the electrical resistivity was recorded.

A detailed post-mortem analysis was carried out on the degraded sample (Fig. 8)

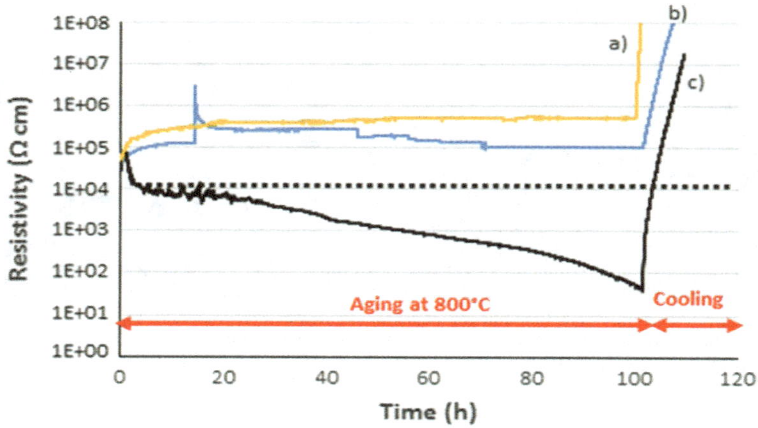

Fig. 7 Electrical resistivity measurements at 800 °C for 100 h of: **a** the glass–ceramic itself at 1.3 V, **b** joined sample Crofer22APU/glass–ceramic/Crofer22APU at 0.7 V and **c** joined sample Crofer22APU/glass–ceramic/Crofer22APU at 1.3 V [14]

showing the development of degradation phenomena at the air side of the sample. The integrity of the sealants seemed to be compromised close to its boundary with the air. Furthermore, the formation of a Cr_2O_3 bridge was detected starting from the cathodic (negative) polarized interface and growing towards the anodic (positive) polarized one. The good electronic conductivity of chromia explained the rapid decrease in the electrical resistivity of the system as reported in Fig. 7. On the other hand, the same phenomenon was not detected when a 0.7 V DC voltage was applied. This behaviour was explained starting with the well-known tendency of Cr-forming alloys to form Cr-containing volatile species in typical SOC operating conditions by the following the reaction:

$$Cr_2O_{3(s)} + 2H_2O_{(g)} + 3/2O_{2(g)} \leftrightarrow 2CrO_2(OH)_{2(g)} \quad (1)$$

$CrO_2(OH)_2$ then reacted with Na contained in the glass forming Na_2CrO_4 which is highly volatile at 800 °C and has a partial pressure much higher than $CrO_2(OH)_2$:

$$CrO_2(OH)_{2(g)} + Na_2O_{(s)} \leftrightarrow Na_2CrO_{4(g)} + H_2O_{(g)} \quad (2)$$

Na_2CrO_4 is reduced at the negative polarized interface forming again Cr_2O_3 under the electrochemical potential applied to this side of the sample:

$$2Na_2CrO_{4(g)} + 2H_2O_{(g)} \rightarrow Cr_2O_{3(s)} + 3/2\ O_2 + 4NaOH_{(g)} \quad (3)$$

Taking into account the fact that reaction 3 has a nominal Nerst potential of 1.19 V at 800 °C, it is clear how this degradation mechanism was triggered at an applied

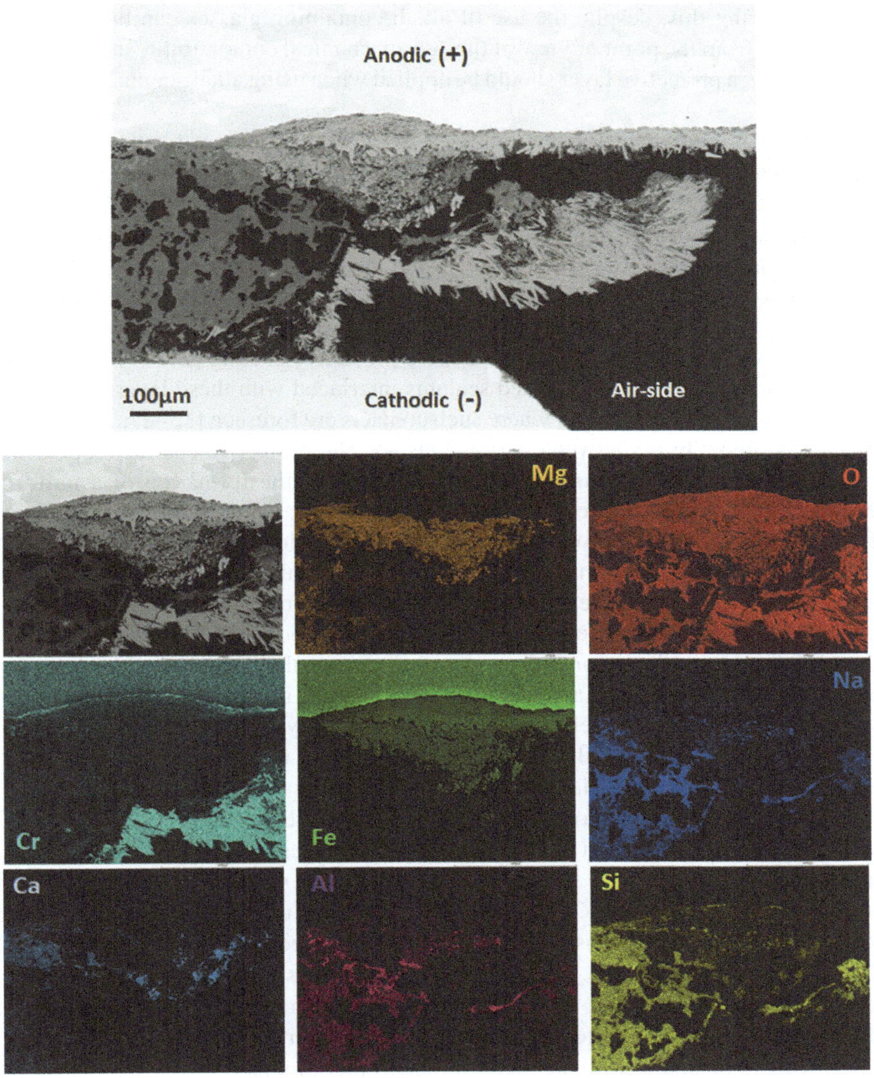

Fig. 8 SEM and EDS elemental mapping of alkali-containing glass–ceramic sealant interfaced with pre-oxidised (900 °C Crofer22APU in a sandwiched sample tested in dual atm. at 800 °C for 100 h under the application of a DC voltage of 1.3 V) [14]

voltage of 1.3 V, but did not take place in the case of the voltage of 0.7 V. Furthermore, the huge Cr reservoir represented by the steel and the re-formation of NaOH as a reaction product of Eq. 3 can lead to continuous catastrophic degradation of these materials under these conditions.

Considering this, despite the use of alkali-containing glasses can bring several advantages from the point of view of thermo-mechanical compatibility at the sealing temperature, a protective layer should be applied when using alkali-containing glass–ceramics.

Another important factor that emerged in the last few years for the successful development and operation of SOCs is the compatibility of glass-based sealants and protective coatings typically used to protect the metallic materials in a SOC stack. Furthermore, the deposition of protective coating can also prevent the development of detrimental reaction between the steels and the glass–ceramic sealants. The most common materials used as protective coatings in this application are represented by the family of spinels based on $(MnCo)_3O_4$ and their variants [24–26]. However, many studies focused also on the use of Al_2O_3 as a protective coating [27–29]. Therefore, the compatibility of the glass-based sealants interfaced with these layers is crucial, especially in the case of design where such contacts are foreseen [15–27, 27, 30, 31].

Good compatibility between the sealants and the protective coatings represents a strong advantage in industrial production allowing the formation of a uniform coating on the entire surface of interconnects without using masking techniques. Furthermore, the presence of a protective coating would block the development of detrimental reactions that may develop between the sealant and bare Cr-containing stainless steels. An example is reported in Fig. 9 that presents SEM and EDS analyses at the interface between a $Mn_{1.5}Co_{1.5}O_4$ coating obtained by electrophoretic deposition and an alkali-containing glass–ceramic sealant. The interface between the coating and the sealant and the underlying steel is in both cases excellent with no evidence of cracks or delamination. The EDS analyses have also shown excellent Cr-retention ability of the coating and the absence of any reaction between the protective layer and the sealant.

Another type of coating often applied to metallic interconnects is represented by a thin layer of alumina that acts as both a barrier layer to prevent Cr evaporation and as a barrier between the sealants and the underlying steel preventing detrimental interaction between the two. Ritucci et al. [27] successfully reported the interfacing of an alkali containing glass–ceramic sealants labelled V11 and alumina-coated Crofer22APU (see Fig. 10). The glass–ceramic appears to have excellent compatibility with the Al_2O_3 layer; the interface is free from cracks and reaction products. The EDS in Fig. 10 also shows that there were no inter-diffusion phenomena. The Cr appears to be confined to the metallic side of the sample.

1.2 Sr-Containing Glass–Ceramic Seals

As discussed in the previous section, alkali and alkaline earth oxides are commonly used as modifiers in glasses; however, alkali metal oxides have the disadvantage of having a high chemical reactivity and low electrical resistivity when in contact with metallic interconnects under applied voltage [32]. In spite of that, a large part of the research on glass–ceramic sealants has focused on the use of BaO as the

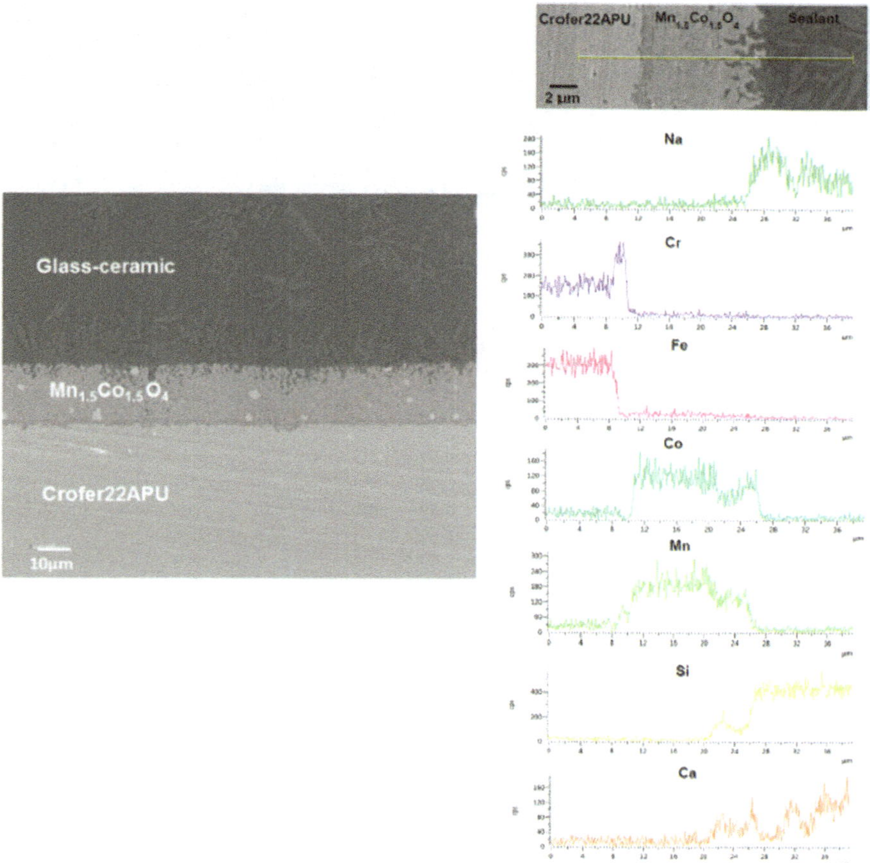

Fig. 9 SEM cross-section of Crofer22APU/protective coating/glass–ceramic sealant after the joining thermal treatment, together with EDS line-scan analyses carried out at the interface between the three materials. $Mn_{1.5}Co_{1.5}O_4$ coating was applied by electrophoretic deposition [17]

main modifier [5, 33–36], while in more recent years, the trend seems to be shifted towards the use SrO [37–44]. Similar to the behaviour of BaO, the addition of SrO also reduces viscosity and adjusts the CTE, but SrO is less prone to form chromates in comparison to BaO. While several studies have assessed the efficacy of SrO-based glasses [40, 42–46], the long-term durability of SrO-containing glasses under SOFC/SOEC conditions is still unclear. Another topic of interest is the fact that the addition of small concentrations of components like B_2O_3, Y_2O_3, Al_2O_3, etc. has been shown to improve the wettability of glasses and adjust the CTE and to control the crystallization behaviour [6, 47].

At POLITO, an extensive research has been conducted to investigate the use of Sr-containing glasses for SOC applications. One of the most promising glass

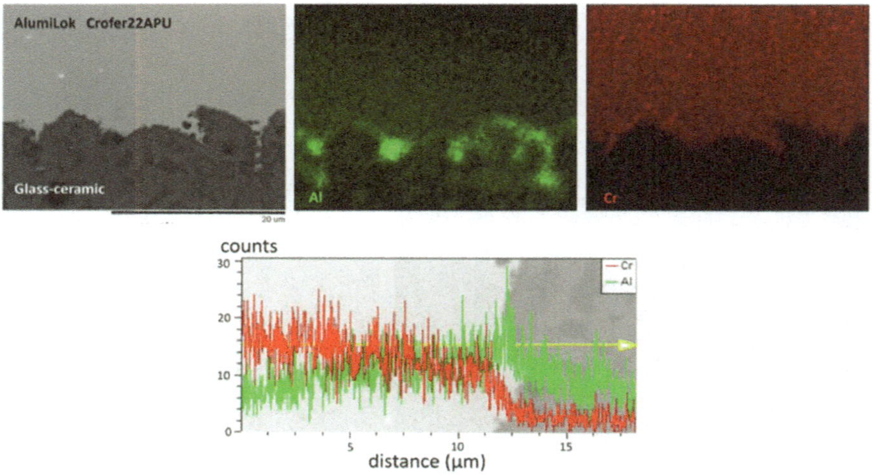

Fig. 10 SEM cross-section and EDS maps and line-scan of interface aluminized Crofer22APU/glass–ceramic sealant as-joined [27]

Table 1 Glass composition of HJ4 glass in mol% [48]

mol%	SiO_2	B_2O_3	SrO	Al_2O_3	Y_2O_3
HJ4	57.6	5.65	28.84	6.17	1.74

compositions that has been developed by the POLITO group is termed as HJ4 glass and the corresponding composition is given in Table 1.

In a SiO_2–SrO based system, a SiO_2/SrO of 1 is required in order to obtain the high CTE $SrSiO_3$ phase ($10.9 \times 10^{-6}\ K^{-1}$). However, for SOC sealant applications a high amount of SrO increases the possibility of formation of the unwanted high CTE $SrCrO_4$ phase ($18–20 \times 10^{-6}\ K^{-1}$). Therefore, the SiO_2/SrO ratio in the HJ4 glass system was kept to ~ 2 with an aim to minimize the possibility of $SrCrO_4$ formation. On the other hand, the high silica content would also be beneficial considering the high-operating temperatures of the SOECs of 850 °C.

1.2.1 Thermal and Microstructural Analysis

The thermal properties glass transition temperature (T_g), crystallization temperature (T_p) and sintering temperatures of as-cast glass were analysed by differential thermal analysis (DTA) and heating stage microscopy (HSM). The corresponding DTA analysis and the shrinkage behaviour vs temperature, found from HSM, are shown in Fig. 11.

The results have shown that there was no sharp exothermic crystallization peak, thus indicating that the crystallization in HJ4 glass was not sufficient to be identified

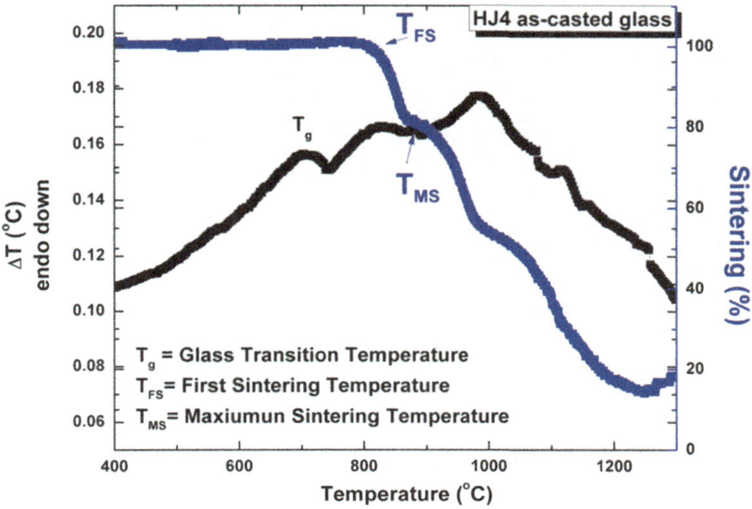

Fig. 11 DTA and HSM analysis of HJ4 as-cast glass; the analysis was performed at a heating rate of 5 °C/min [48]

during the DTA analysis. The corresponding HSM curve shows that the sintering process starts at (T_{FS}) 809 °C, while complete sintering was obtained at (T_{MS}) 875 °C followed by continuous viscous flow. To achieve a dense and leakage-free sealant, it is necessary to complete the sintering before the start of crystallization, thus avoiding the formation of porosity in the glass–ceramic due to increased viscosity triggered by crystal growth [3, 6]. From the DTA and HSM analyses, it is clear that the sintering process was completed at 875 °C and no evidence of crystallization was detected by DTA, thus ensuring the formation of a dense sealant. Moreover, the presence of a low degree of crystallization (higher residual glass) further promoted the formation of a dense sealant due to the viscous flow above the glass transition temperature.

From the data obtained from DTA and HSM, a heat treatment of 950 °C with a dwelling time of 5 h at a heating rate of 2 °C/min was used to process an optimal joining treatment. The as-joined HJ4 glass–ceramic showed a CTE of 9.3 10^{-6} K^{-1} which is a value within the desired range (9–12 × 10^{-6} K^{-1}), taking into consideration the CTEs of the other cell components (CTEs for Crofer22APU and 3YSZ are 12 × 10^{-6} K^{-1} and 10.5 × 10^{-6} K^{-1} respectively) [24, 49]. Further details about the crystalline phases that were observed are discussed in the later sections of this chapter.

A SEM cross-section of the HJ4 glass–ceramic in contact with Crofer22APU and 3YSZ substrates is shown in Fig. 12. The joining area was found to be crack-free with no evidence of delamination at the two interfaces, thus indicating very good thermomechanical compatibility. The microstructure is dense with a negligible concentration of closed pores. Various crystals are uniformly distributed into the residual glassy phase as shown in the magnified image (Fig. 12). The EDS point analysis carried out at different areas of the HJ4 as-joined glass–ceramic is shown

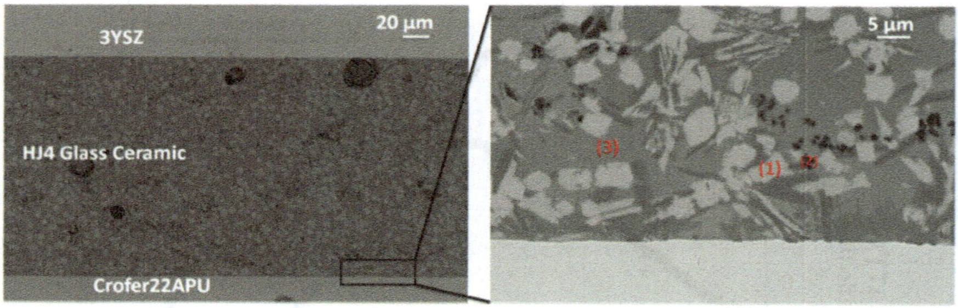

Fig. 12 SEM cross-section of Crofer22APU/HJ4 glass–ceramic/3YSZ joined sample [48]

Table 2 EDS point analysis (at.%) carried out on HJ4 as-joined glass–ceramic, shown in Fig. 12 [48]

Elements	Point 1	Point 2	Point 3
O	68.6	74.3	72.4
Si	15.6	16.4	15.8
Sr	14.9	5.5	8.2
Al	0.4	3.8	2.2
Y	0.5	0.0	1.4

in Table 2. The EDS analysis at the bright crystals (point 1) shows a phase based on silicon, strontium and oxygen and from the EDS composition, this was identified to be $SrSiO_3$, while the black crystals (point 2) represent a silica-rich phase. The EDS point analysis conducted at the residual glassy phase (point 3) is also shown in Table 2; the presence of 8.2 at.% Sr is beneficial to have good wettability of residual glassy phase.

The XRD patterns of the as-joined HJ4 glass–ceramics as well as after ageing are shown in Fig. 13. The XRD analysis confirmed the presence of $SrSiO_3$ as the main crystalline phase, in addition to SiO_2 (cristobalite) as the secondary phase. The XRD pattern for the HJ4 glass–ceramic after ageing at 850 °C for 1000 h is also shown in Fig. 13. The identical patterns after joining and ageing indicates that the HJ4 glass–ceramic is stable and no new phase formation was detected after ageing.

The SEM and XRD analysis have confirmed that a suitable SiO_2/SrO ratio has resulted in the formation of the high CTE $SrSiO_3$ phase. However, a SiO_2/SrO of ~ 2 also resulted in the formation of cristobalite phase. Due to phase transformation and consequently volume change, in cristobalite around 270 °C [50], its presence might determine thermomechanical stresses with possible crack formation, thus affecting the reliability of sealant during thermal cycles.

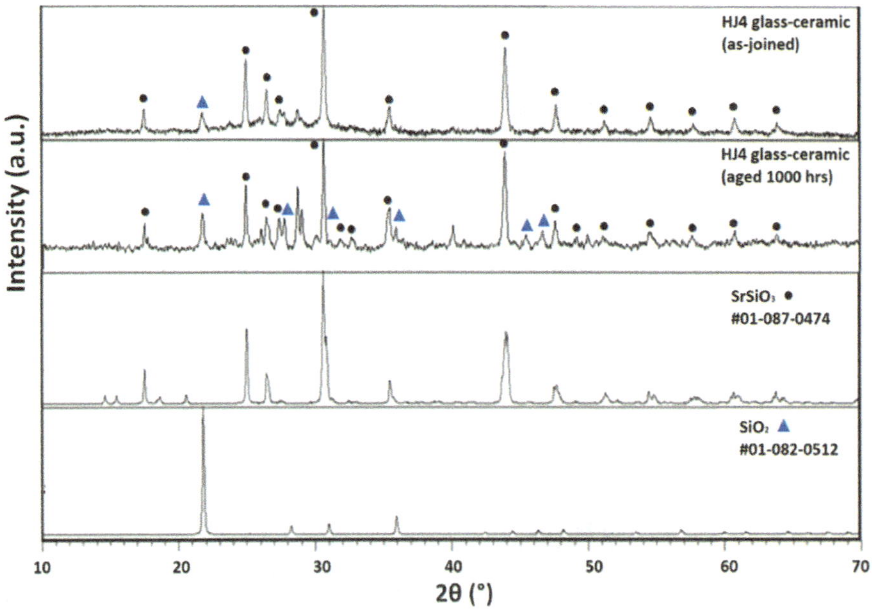

Fig. 13 XRD analysis of as-joined and thermally aged (1000 h, 850 °C) HJ4 glass–ceramic [48]

1.2.2 Electrical Properties and Post-Mortem Analysis

For SOC applications, the sealants should have high resistivity in order to avoid electrical short circuit. For long-term operation, any possible microstructural changes of the sealing glasses must be considered. This is of particular importance especially considering that interfacial reactions with the interconnect may be boosted by an applied voltage. To analyse and address this issue, the electrical resistivity of Crofer/HJ4 glass–ceramic/Crofer joined samples was measured at 850 °C under an applied voltage of 1.6 V. Figure 14 shows a schematic diagram of the test as well as the corresponding electrical resistivity curve of the joined sample. The resistivity is in the range of (10^6–10^7 Ω cm) and is higher than the minimum value (10^4 Ω cm) required to ensure electrical insulation and to work effectively in the SOC working conditions. Moreover, the electrical resistivity of HJ4 glass–ceramic is similar or slightly higher than the resistivity values reported in literature [15, 32, 51].

Figure 14 shows that after an initial reduction in the electrical resistivity during the first 200 h of testing, a constant resistivity was measured until 800 h. The continuous reduction at start of resistivity analysis is due to the fact that the as-joined HJ4 glass–ceramic has significantly high level of residual glass, where the presence of different free ions (such as Si, Sr, O, Sr, Al) can move under applied voltage and can cause reduction in resistivity. Around 800 h, another discontinuity caused further reduction in electrical resistivity. This second discontinuity is most likely due to polarization

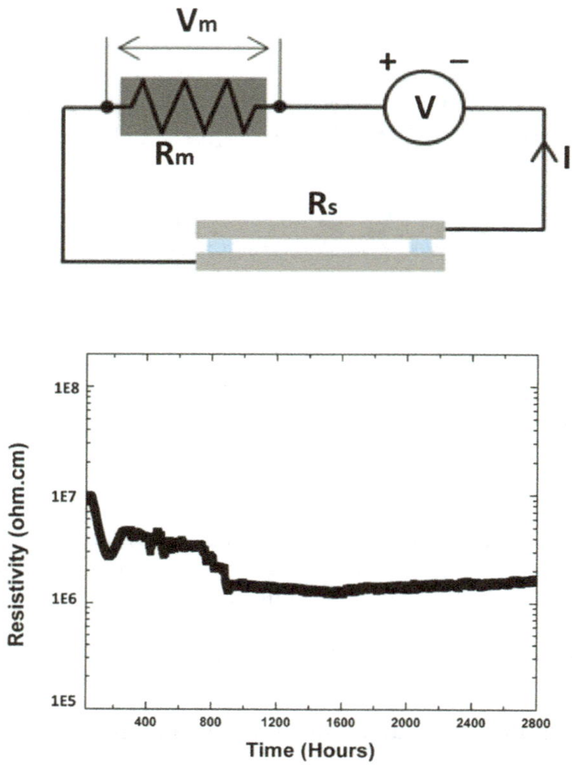

Fig. 14 Electrical resistivity of Crofer22APU/glass–ceramics/Crofer22APU joined samples for HJ4 glass–ceramics. These measurements were carried out at 850 °C with 1.6 V applied [52]

effects. Nevertheless, after 800 h the resistivity values for the HJ4-based joint became constant and no abnormalities or gaps were observed until the end of test.

After the electrical resistivity analysis, SEM–EDS post-mortem analyses were carried out to investigate the compatibility of the HJ4 glass–ceramic with an anodic polarized Crofer22APU plate and the results are shown in Fig. 15. The microstructure of the HJ4 glass–ceramic is homogenous, dense with apparently no signs of porosity, showing a good adhesion and an excellent compatibility with the Crofer22APU substrate. The microstructure is similar to that of as-joined glass–ceramic where a bright crystalline phase of $SrSiO_3$ is distributed within the residual amorphous matrix together with the presence of black cristobalite (SiO_2) crystals. Although a thin (~ 2 μm) MnCr-rich oxide scale is present at the interface, Cr is confined within the oxide scale without any diffusion into the glass–ceramic. Nevertheless, some traces of Mn diffusion into the residual glass are observed. Anyway, no further elements diffused or segregate from the glass–ceramic and the Crofer22APU. These results demonstrated that HJ4 glass–ceramic is chemically stable after long-term ageing at high temperature and under electric load.

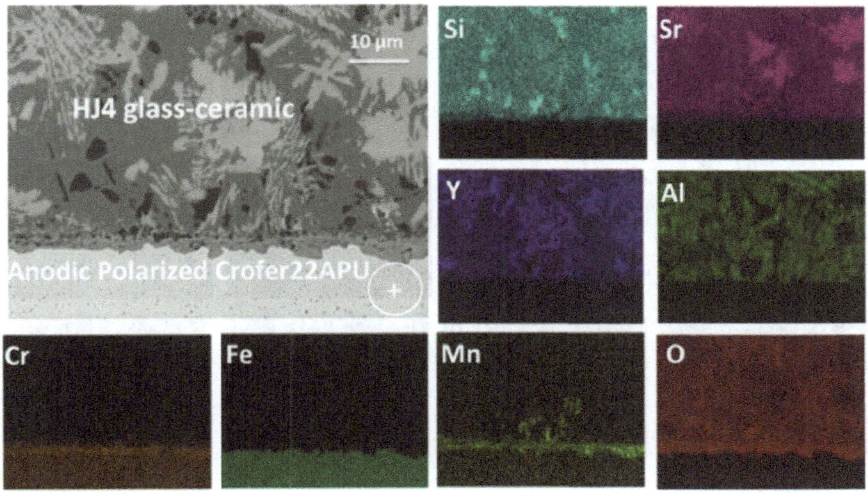

Fig. 15 EDS mapping of the SEM image corresponding to the post mortem analyses of anodic polarized Crofer22APU/HJ4 glass–ceramic interface, after the electrical resistivity test for 2800 h in static air, under a voltage of 1.6 V [52]

The formation of $SrCrO_4$ has frequently been observed in Sr-containing glasses and can harmfully affect the compatibility of sealants with Cr-based metallic interconnects due to its high CTE ($18–20 \times 10^{-6}$ K^{-1}) [53, 54]. However, despite the presence of a Cr-rich oxide scale, the Crofer22APU/sealant interface did not show any indication of the $SrCrO_4$ phase, thus maintaining an excellent adhesion.

The cathodic Crofer22APU/sealant interface also showed no segregation of elements from the glass–ceramic. The EDS mapping (Fig. 16) detected a small concentration of Cr and Mn that has been diffused from the Crofer22APU to the sealant. Sabato et al. [14] reported similar data, and observed outward diffusion of Cr and Mn close to the cathodic polarized interface. The results reported by Sabato et al. [14] after 100 h of testing, showed that around 4 at.% Cr was detected at approx. 10 μm from the interface. However, the diffusion in case of HJ4 glass–ceramic is limited, likely due to a slight lower amount of residual glassy phase. Finally, no interaction or signs of formation of the $SrCrO_4$ phase were found from SEM–EDS analyses, thus demonstrating an excellent compatibility with the Crofer22APU substrate.

1.2.3 Mechanical Properties and Post Mortem Analysis

Reliable operation of SOC devices can be critically determined on the mechanical behaviour of the sealants when in contact with the metallic interconnect. Mechanical issues can have a key impact on the performance and degradation rate of the SOFC/SOEC device and there is a relatively small body of literature that is concerned

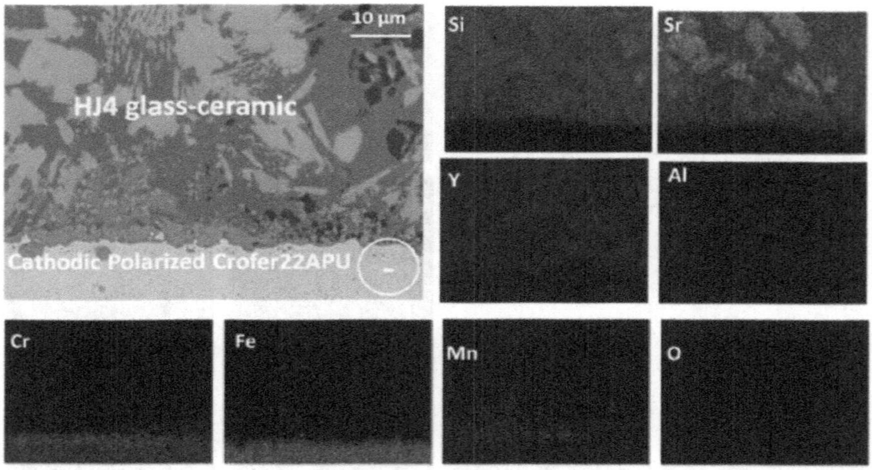

Fig. 16 EDS mapping of SEM image corresponding to post mortem analyses of cathodic polarized Crofer22APU/HJ4 glass–ceramic interface, after the electrical resistivity test for 2800 h in static air, under a voltage of 1.6 V [52]

with the mechanical properties of sealants both at RT and at operating conditions [8, 32–36, 55–58].

In order to evaluate the mechanical stability and shear resistance of the sealant, Crofer22APU/HJ4 glass–ceramic/Crofer22APU joints were prepared to investigate their behaviour under shear load conditions. The mechanical stability and shear strength of joined samples were measured at room temperature, 650 °C (below T_g) and at 850 °C (working temperature of SOC). Figure 17 illustrates the sample configuration (including dimensions) and setup used for mechanical testing of the joined samples under the shear load.

Figure 18a shows the shear stress versus load displacement data for the HJ4 joint samples, tested at RT, 650 °C and 850 °C respectively. It is worth mentioning that the slope of the plot can be slightly different due to possible inexact value of the joint area that was measured during the post mortem analysis of the fractured surfaces. The stress versus load curves show the effect of the test temperature on the resulting shear stress. When the Crofer22APU/HJ4 glass–ceramic/Crofer22APU joined samples were tested below T_g (at RT and at 650 °C), the stress versus displacement curve reflects almost linear elastic deformation until fracture and a brittle behaviour of the glass–ceramic joint. On the other hand, for shear testing performed at 850 °C ($T > T_g$), an enhanced displacement was observed for Crofer22APU/HJ4 glass–ceramic/Crofer22APU joint under applied shear load. This effect was as expected and is due to the softening of the residual glassy phase above T_g, that resulted in glass viscous flow. The softening of HJ4 glass–ceramic sealant at 850 °C can promote the stress relaxation phenomenon that can reduce thermal stresses at high temperature. Besides that, viscous flow of the residual glassy phase can lead to self-healing and improved long-term integrity of the sealant. Such stress relaxation phenomenon is

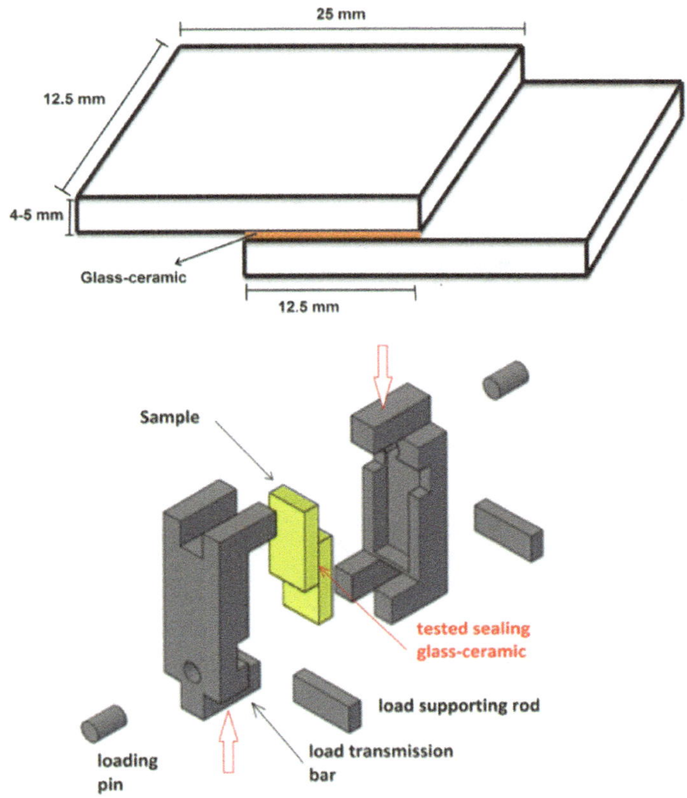

Fig. 17 Illustration of Crofer22APU/glass–ceramic/Crofer22APU samples with glass ceramic joint for shear testing (left) and setup for testing the sample under shear load (right) [59]

also reported in literature. For instance, according to Chang et al. [60], the viscoelastic behaviour of the residual glassy phase in GC-9 glass–ceramic system showed the stress relaxation at the temperature range of 650–750 °C.

The shear strength of the joined samples as calculated from the stress displacement curves are given in Fig. 18b. The average shear strength obtained after testing three samples at each test temperature is shown in Fig. 18b. The shear strength of the joined samples was reduced with the increase in the testing temperatures i.e., the shear strength of 13.9 MPa was measured at RT, however, testing at 850 °C showed shear strength of 1.8 MPa. As mentioned earlier, the residual glassy phase becomes viscous above T_g, therefore, the observed reduction in the strength with increasing temperature was due to the viscous flow occurring at the softening of the glassy phase at high temperature.

Figure 18c shows the elastic modulus of the HJ4 glass–ceramic measured by vibrational method from room temperature to 650 °C. The measurements were performed both during the heating and cooling stages as shown in Fig. 18c The elastic modulus

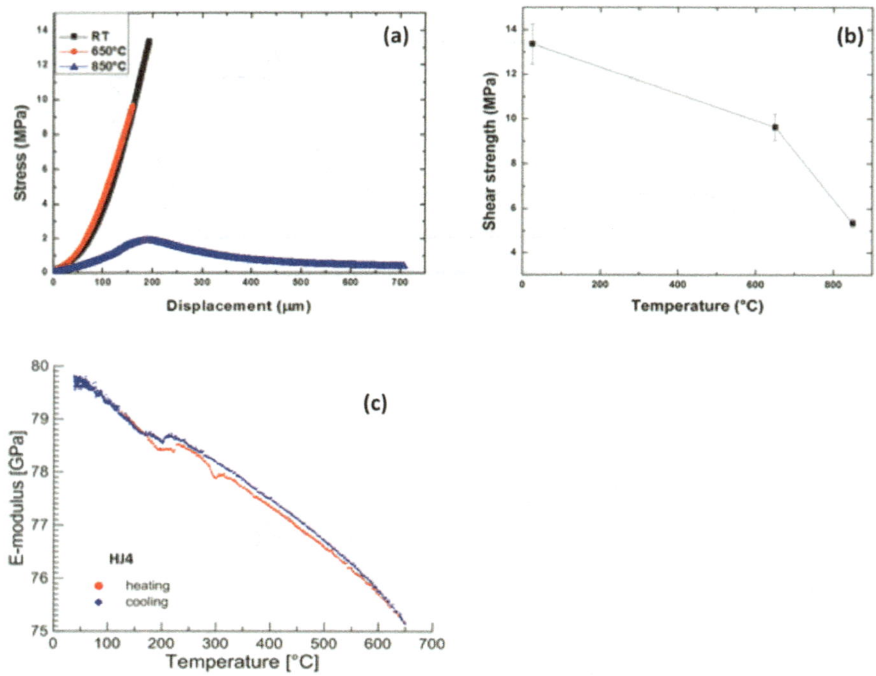

Fig. 18 **a** Shear stress versus displacement curves for HJ4 joints, **b** shear strength versus temperature for HJ4 joints, and**c** elastic modulus of HJ4 glass–ceramic [59]

curves for the HJ4 glass–ceramic during heating and cooling show overlapping, thus indicating that no microstructural changes occurred within the glass–ceramic. On the other hand, some discontinuities are visible in Fig. 18c around 230 °C during the heating cycle and around 230–270 °C during cooling. These are due to the presence of the cristobalite (SiO_2) phase in the HJ4 glass–ceramic that shows volume expansion around 230–270 °C. Nevertheless, the obtained elastic modulus value is comparable with measurements reported in the literature (50–80 GPa) [39, 62].

In order to better understand the mechanical behaviour as described above, it is important to investigate the fracture surfaces as well as cross-sections of samples after fracture. Figure 19 shows the SEM cross section images of the Crofer22APU/HJ4 glass–ceramic fractured interfaces. The SEM images showed a uniform microstructure of glass and strong bonding with the Crofer22APU at room temperature as well as at higher testing temperatures, with no evidence of any crack or delamination. The Crofer22APU/HJ4 glass–ceramic interface showed comparable morphology. The HJ4 glass–ceramic microstructure after mechanical testing at different temperatures is also similar to the as-joined HJ4 glass–ceramic as shown in Fig. 12.

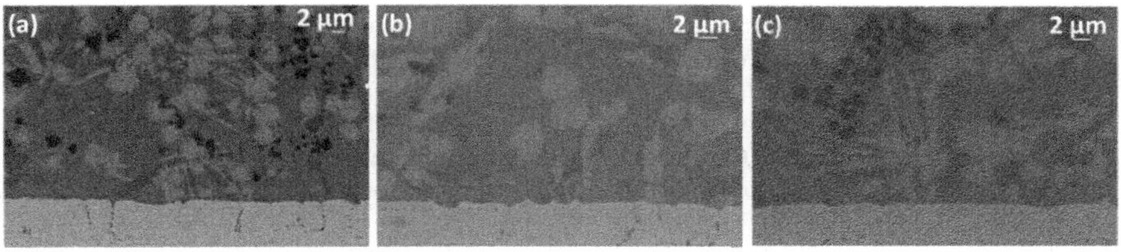

Fig. 19 SEM images of interface of Crofer22APU with **a** HJ4 tested at RT, **b** HJ4 tested at 650 °C, **c** HJ4 tested at 850 °C [59]

2 Conclusions

The purpose of the current chapter was to present and discuss different approaches in the design and specific tests of glass-based materials as sealants for SOCs.

One of the more significant findings to emerge from this comparison is that a single and ideal sealant does not exists, since the performances are strictly related to the SOC design and the operating conditions.

The findings of this study suggest that Na-containing glass sealants suffer from the concurrent high voltage application and contact with chromia forming alloys.

The current data highlight the importance of protective coatings needed to overcome this issue.

Main results on research activity done in the past 10 years at the Politecnico di Torino on both Na-and Sr-containing sealants have been presented and discussed for SOC applications in the T range 750–850 °C. Different glass compositions have been designed and successfully tested, thus providing a deeper insight into relationships between glass compositions, characteristic temperature and SOC operating conditions.

The right balance of sealants composition plays a key role in obtaining the desired high-CTE phases in the glass–ceramic and a dense sealant with sufficient quantity of the residual glassy phase that is beneficial for stress relaxation below T_g.

The detailed studies conducted at Politecnico di Torino, Italy, showed that both Na and Sr-based glass–ceramics can be promising candidates to use for SOC sealant applications.

Bibliography

1. Holland, W., Beall, G.: Glass-Ceramic Technology. The American Ceramic Society, Westerville (2002)
2. Donald, I.W., Mallinson, P.M., Metcalfe, B.L., Gerrard, L.A., Fernie, J.A.: Recent developments in the preparation, characterization and applications of glass-and glass–ceramic-to-metal seals and coatings. J. Mater. Sci. **46**, 1975–2000 (2011). https://doi.org/10.1007/s10853-010-5095-y
3. Mahapatra, M.K., Lu, K.: Seal glass for solid oxide fuel cells. J. Power Sources **195**, 7129–7139 (2010). https://doi.org/10.1016/j.jpowsour.2010.06.003

4. Schilm, J., Rost, A., Poenicke, A., Kusnezoff, M., Michaelis, A.: Ceramic integration technologies for solid oxide fuel cells. Int. J. Appl. Ceram. Technol. **9**, 688–699 (2012). https://doi.org/10.1111/j.1744-7402.2012.02789.x
5. Mahapatra, M.K., Lu, K.: Glass-based seals for solid oxide fuel and electrolyzer cells—a review. Mater. Sci. Eng. R Reports. **67**, 65–85 (2010). https://doi.org/10.1016/j.mser.2009.12.002
6. Fergus, J.W.: Sealants for solid oxide fuel cells. J. Power Sources **147**, 46–57 (2005). https://doi.org/10.1016/j.jpowsour.2005.05.002
7. Chou, Y.S., Stevenson, J.W., Choi, J.P.: Long-term evaluation of solid oxide fuel cell candidate materials in a 3-cell generic short stack fixture, part II: sealing glass stability, microstructure and interfacial reactions. J. Power Sources **250**, 166–173 (2014). https://doi.org/10.1016/j.jpowsour.2013.09.148
8. Ritucci, I., Kiebach, R., Talic, B., et al.: Improving the interface adherence at sealings in solid oxide cell stacks. J. Mater. Res. **34**, 1167–1178 (2019). https://doi.org/10.1557/jmr.2018.459
9. Gödeke, D., Dahlmann, U.: Study on the crystallization behaviour and thermal stability of glass-ceramics used as solid oxide fuel cell-sealing materials. J. Power Sources **196**, 9046–9050 (2011). https://doi.org/10.1016/j.jpowsour.2010.12.054
10. Smeacetto, F., Salvo, M., Ferraris, M., Cho, J., Boccaccini, A.R.: Glass–ceramic seal to join Crofer 22 APU alloy to YSZ ceramic in planar SOFCs. J. Eur. Ceram. Soc. **28**, 61–68 (2008). https://doi.org/10.1016/j.jeurceramsoc.2007.05.006
11. Sabato, A.G., Salvo, M., De Miranda, A., Smeacetto, F.: Crystallization behaviour of glass–ceramic sealant for solid oxide fuel cells. Mater. Lett. **141**, 284–287 (2015). https://doi.org/10.1016/j.matlet.2014.11.128
12. Smeacetto, F., De Miranda, A., Chrysanthou, A., Bernardo, E., Secco, M., Bindi, M., Salvo, M., Sabato, A.G., Ferraris, M.: Novel glass-ceramic composition as sealant for SOFCs. J. Am. Ceram. Soc. **97**, 3835–3842 (2014). https://doi.org/10.1111/jace.13219
13. Sabato, A.G., Cempura, G., Montinaro, D., Chrysanthou, A., Salvo, M., Bernardo, E., Secco, M., Smeacetto, F.: Glass–ceramic sealant for solid oxide fuel cells application: characterization and performance in dual atmosphere. J. Power Sources **328**, 262–270 (2016). https://doi.org/10.1016/j.jpowsour.2016.08.010
14. Sabato, A.G., Rost, A., Schilm, J., Kusnezoff, M., Salvo, M., Chrysanthou, A., Smeacetto, F.: Effect of electric load and dual atmosphere on the properties of an alkali containing diopside-based glass sealant for solid oxide cells. J. Power Sources **415** (2019). https://doi.org/10.1016/j.jpowsour.2019.01.051
15. Chou, Y.S., Stevenson, J.W., Xia, G.G., Yang, Z.G.: Electrical stability of a novel sealing glass with (Mn, Co)-spinel coated Crofer22APU in a simulated SOFC dual environment. J. Power Sources **195**, 5666–5673 (2010). https://doi.org/10.1016/j.jpowsour.2010.03.052
16. Smeacetto, F., De Miranda, A., Sandra Cabanas Polo, S.M., Boccaccini, D., Salvo, M., Boccaccini, A.R.: Electrophoretic deposition of $Mn_{1.5}Co_{1.5}O_4$ on metallic interconnect and interaction with glass–ceramic sealant for solid oxide fuel cells application. J. Power Sources **280**, 379–386 (2015). https://doi.org/10.1016/j.ijhydene.2013.04.106
17. Smeacetto, F., De Miranda, A., Cabanas Polo, S., Molin, S., Boccaccini, D., Salvo, M., Boccaccini, A.R.: Electrophoretic deposition of $Mn_{1.5}Co_{1.5}O_4$ on metallic interconnect and interaction with glass–ceramic sealant for solid oxide fuel cells application, J. Power Sources **280**, 379–386 (2015). https://doi.org/10.1016/j.jpowsour.2015.01.120
18. Smeacetto, F., Salvo, M., Leone, P., Santarelli, M., Ferraris, M.: Performance and testing of joined Crofer22APU-glass–ceramic sealant-anode supported cell in SOFC relevant conditions. Mater. Lett. **65**, 1048–1052 (2011). https://doi.org/10.1016/j.matlet.2010.12.050
19. Ogasawara, K., Kameda, H., Matsuzaki, Y., Sakurai, T.: Chemical stability of ferritic alloy interconnect for SOFCs. J. Electrochem. Soc. **154**, B657–B663 (2007). https://doi.org/10.1149/1.2735919
20. Coillot, D., Me, O.: New viscous sealing glasses for electrochemical cells. Int. J. Hydrogen Energy **37**, 9351–9358 (2012). https://doi.org/10.1016/j.ijhydene.2012.02.194
21. Shyam, A., Trejo, R., McClurg, D., Ladouceur, A., Kirkham, M., Song, X., Howe, J., Lara-Curzio, E.: Microstructural evolution in two alkali multicomponent silicate glasses as a result of

long-term exposure to solid oxide fuel cell environments. J. Mater. Sci. **48**, 5880–5898 (2013). https://doi.org/10.1007/s10853-013-7384-8
22. Smeacetto, F., Salvo, M., Ferraris, M., Casalegno, V., Asinari, P.: Glass and composite seals for the joining of YSZ to metallic interconnect in solid oxide fuel cells. J. Eur. Ceram. Soc. **28**, 611–616 (2008). https://doi.org/10.1016/j.jeurceramsoc.2007.07.008
23. Smeacetto, F., Chrysanthou, A., Salvo, M., Moskalewicz, T., D'Herin Bytner, F., Ajitdoss, L.C., Ferraris, M.: Thermal cycling and ageing of a glass-ceramic sealant for planar SOFCs. Int. J. Hydrogen Energy **36,** 11895–11903 (2011). https://doi.org/10.1016/j.ijhydene.2011.04.083
24. Mah, J.C.W., Muchtar, A., Somalu, M.R., Ghazali, M.J.: Metallic interconnects for solid oxide fuel cell: a review on protective coating and deposition techniques, Int. J. Hydrogen Energy pp. 1–11 (2015). https://doi.org/10.1016/j.ijhydene.2016.03.195
25. Shaigan, N., Qu, W., Ivey, D.G., Chen, W.: A review of recent progress in coatings, surface modifications and alloy developments for solid oxide fuel cell ferritic stainless steel interconnects. J. Power Sources **195**, 1529–1542 (2010). https://doi.org/10.1016/j.jpowsour.2009.09.069
26. Zanchi, E., Sabato, A.G., Molin, S., Cempura, G., Boccaccini, A.R., Smeacetto, F.: Recent advances on spinel-based protective coatings for solid oxide cell metallic interconnects produced by electrophoretic deposition. Mater. Lett. **286**, 129229 (2021). https://doi.org/10.1016/j.matlet.2020.129229
27. Ritucci, I., Agersted, K., Zielke, P., Wulff, A.C., Khajavi, P., Smeacetto, F., Sabato, A.G., Kiebach, R.: A Ba-free sealing glass with a high coefficient of thermal expansion and excellent interface stability optimized for SOFC/SOEC stack applications. Int. J. Appl. Ceram. Technol. **15**, 1011–1022 (2018). https://doi.org/10.1111/ijac.12853
28. Qi, H.B., Lees, D.G.: The effects of surface-applied oxide films containing varying amounts of yttria, chromia, or alumina on the high-temperature oxidation behavior of chromia-forming and alumina-forming alloys. Oxid. Met. **53**, 507–527 (2000). https://doi.org/10.1023/A:1004633026323
29. Chou, Y.S., Choi, J.P., Stevenson, J.W.: Compliant alkali silicate sealing glass for solid oxide fuel cell applications: the effect of protective alumina coating on electrical stability in dual environment. Int. J. Hydrogen Energy **37**, 18372–18380 (2012). https://doi.org/10.1016/j.ijhydene.2012.08.084
30. Chou, Y., Thomsen, E.C., Choi, J., Stevenson, J.W.: Compliant alkali silicate sealing glass for solid oxide fuel cell applications: the effect of protective YSZ coating on electrical stability in dual environment. J. Power Sources **202**, 149–156 (2012). https://doi.org/10.1016/j.jpowsour.2011.11.017
31. Sabato, A.G., Chrysanthou, A., Salvo, M., Cempura, G., Smeacetto, F.: Interface stability between bare, Mn–Co spinel coated AISI 441 stainless steel and a diopside-based glass–ceramic sealant. Int. J. Hydrogen Energy **43**, 1824–1834 (2018). https://doi.org/10.1016/j.ijhydene.2017.11.150
32. Chou, Y.-S.S., Stevenson, J.W., Choi, J.-P.: Alkali effect on the electrical stability of a solid oxide fuel cell sealing glass. J. Electrochem. Soc. **157**, B348–B353 (2010). https://doi.org/10.1111/j.1551-2916.2009.03466.x
33. Reddy, A.A., Tulyaganov, D.U., Goel, A., Pascual, M.J., Kharton, V.V., Tsipis, E.V., Ferreira, J.M.F.: Diopside-Mg orthosilicate and diopside-Ba disilicate glass-ceramics for sealing applications in SOFC: sintering and chemical interactions studies. Int. J. Hydrogen Energy **37**, 12528–12539 (2012). https://doi.org/10.1016/j.ijhydene.2012.05.130
34. Yang, Z., Meinhardt, K.D., Stevenson, J.W.: Chemical compatibility of barium–calcium–aluminosilicate-based sealing glasses with the ferritic stainless steel interconnect in SOFCs. J. Electrochem. Soc. **150**, A1095 (2003). https://doi.org/10.1149/1.1590325
35. Javed, H., Sabato, A.G., Mansourkiaei, M., Ferrero, D., Santarelli, M., Herbrig, K., Walter, C., Smeacetto, F.: Glass–ceramic sealants for SOEC: thermal characterization and electrical resistivity in dual atmosphere. Energies **13**, 3682 (2020). https://doi.org/10.3390/en13143682
36. Ferraris, M., De la Pierre, S., Sabato, A.G., Smeacetto, F., Javed, H., Walter, C., Malzbender, J.: Torsional shear strength behavior of advanced glass–ceramic sealants for SOFC/SOEC

applications. J. Eur. Ceram. Soc. **40**, 4067–4075 (2020). https://doi.org/10.1016/j.jeurceramsoc.2020.04.034
37. Kothiyal, G.P., Goswami, M., Tiwari, B., Sharma, K., Ananthanarayanan, A., Montagne, L.: Some recent studies on glass/glass–ceramics for use as sealants with special emphasis for high temperature applications. J. Adv. Ceram. **1**, 110–129 (2012). https://doi.org/10.1007/s40145-012-0009-x
38. Kaur, M., Kaur, G., Pandey, O.P., Singh, K., Kumar, V.: Influence of CaO/MgO ratio on the crystallization kinetics and interfacial compatibility with crofer 22APU and YSZ of strontium based alumino-borosilicate glasses for SOFC applications. Int. J. Hydrogen Energy **42**, 16244–16257 (2017). https://doi.org/10.1016/j.ijhydene.2017.05.026
39. Fakouri Hasanabadi, M., Faghihi-Sani, M.A., Kokabi, A.H., Groß-Barsnick, S.M., Malzbender, J.: Room-and high-temperature flexural strength of a stable solid oxide fuel/electrolysis cell sealing material. Ceram. Int. 1–7 (2018). https://doi.org/10.1016/j.ceramint.2018.09.236
40. Luo, Z., Lei, W., Liang, H., Xu, W., Liu, X., Qin, C., Lu, A.: Improving sealing properties of $CaO–SrO–Al_2O_3–SiO_2$ glass and glass–ceramics for solid oxide fuel cells: effect of La_2O_3 addition. Ceram. Int. **46**, 17698–17706 (2020). https://doi.org/10.1016/j.ceramint.2020.04.074
41. Wang, X., Ou, D.R., Zhao, Z., Cheng, M.: Stability of $SrO–La_2O_3–Al_2O_3–SiO_2$ glass sealants in high-temperature air and steam. Ceram. Int. **42**, 7514–7523 (2016). https://doi.org/10.1016/j.ceramint.2016.01.158
42. Mahapatra, M.K., Lu, K., Reynolds, W.T.: Thermophysical properties and devitrification of $SrO-La_2O_3–Al_2O_3–B_2O_3–SiO_2$-based glass sealant for solid oxide fuel/electrolyzer cells. J. Power Sources **179**, 106–112 (2008). https://doi.org/10.1016/j.jpowsour.2007.12.101
43. Zhang, Q., Yang, H., Zeng, F., Wang, S., Tang, D., Zhang, T.: Development of the $CaO–SrO–ZrO_2–B_2O_3–SiO_2$ sealing glasses for solid oxide fuel cell applications: structure. RSC Adv. **5**, 41772–41779 (2015). https://doi.org/10.1039/C5RA04781A
44. Ojha, P.K., Chongdar, T.K., Gokhale, N.M., Kulkarni, A.R.: Investigation of crystallization kinetic of $SrO–La_2O_3–Al_2O_3–B_2O_3–SiO_2$ glass and its suitability for SOFC sealant. Int. J. Hydrogen Energy **36**, 14996–15001 (2011). https://doi.org/10.1016/j.ijhydene.2010.12.120
45. Reddy, A.A., Goel, A., Tulyaganov, D.U., Sardo, M., Mafra, L., Pascual, M.J., Kharton, V.V., Tsipis, E.V., Kolotygin, V.A., Ferreira, J.M.F.: Thermal and mechanical stability of lanthanide-containing glass–ceramic sealants for solid oxide fuel cells. J. Mater. Chem. A **2**, 1834 (2014). https://doi.org/10.1039/c3ta13196c
46. Mahapatra, M.K., Lu, K.: Effect of atmosphere on interconnect-seal glass interaction for solid oxide fuel/electrolyzer cells. J. Am. Ceram. Soc. **94**, 875–885 (2011). https://doi.org/10.1111/j.1551-2916.2010.04020.x
47. Elsayed, H., Javed, H., Sabato, A.G., Smeacetto, F., Bernardo, E.: Novel glass-ceramic SOFC sealants from glass powders and a reactive silicone binder. J. Eur. Ceram. Soc. **38**, 4245–4251 (2018). https://doi.org/10.1016/j.jeurceramsoc.2018.05.024
48. Javed, H., Sabato, A.G., Herbrig, K., Ferrero, D., Walter, C., Salvo, M., Smeacetto, F.: Design and characterization of novel glass-ceramic sealants for solid oxide electrolysis cell (SOEC) applications. Int. J. Appl. Ceram. Technol. **15**, 999–1010 (2018). https://doi.org/10.1111/ijac.12889
49. Kaur, G.: Solid Oxide Fuel Cell Components: Seal Glass for Solid Oxide Fuel Cells (2006). https://doi.org/10.1007/s11837-006-0052-6
50. Beals, M.D., Zerfoss, S.: Volume change attending low-to-high inversion of cristobalite. J. Am. Ceram. Soc. **27**, 285–292 (1944). https://doi.org/10.1111/j.1151-2916.1944.tb14471.x
51. Ghosh, S., Das Sharma, A., Kundu, P., Basu, R.N.: Glass-ceramic sealants for planar IT-SOFC: a bilayered approach for joining electrolyte and metallic interconnect. J. Electrochem. Soc. **155**, B473–B478 (2008). https://doi.org/10.1149/1.2883732
52. Javed, H., Herbrig, K., Sabato, A.G., Ferrero, D., Santarelli, M., Walter, C., Smeacetto, F.: Electrical characterization of glass-ceramic sealant-metallic interconnect joined samples under solid oxide electrolysis cell conditions; influence on the microstructure and composition at the different polarized interfaces. Ceram. Int. (2020). https://doi.org/10.1016/j.ceramint.2020.11.176

53. Chou, Y.S., Stevenson, J.W., Singh, P.: Effect of pre-oxidation and environmental aging on the seal strength of a novel high-temperature solid oxide fuel cell (SOFC) sealing glass with metallic interconnect. J. Power Sources **184**, 238–244 (2008). https://doi.org/10.1016/j.jpowsour.2008.06.020
54. Zhang, Q., Du, X., Tan, S., Tang, D., Chen, K., Zhang, T.: Effect of Nb_2O5 Doping on Improving the Thermo-mechanical Stability of Sealing Interfaces for Solid Oxide Fuel Cells, pp. 1–8 (2017). https://doi.org/10.1038/s41598-017-05725-y
55. Lin, C.K., Liu, Y.A., Wu, S.H., Liu, C.K., Lee, R.Y.: Joint strength of a solid oxide fuel cell glass–ceramic sealant with metallic interconnect in a reducing environment. J. Power Sources **280**, 272–288 (2015). https://doi.org/10.1016/j.jpowsour.2015.01.126
56. Fakouri Hasanabadi, M., Faghihi-Sani, M.A., Kokabi, A.H., Malzbender, J.: The analysis of torsional shear strength test of sealants for solid oxide fuel cells. Ceram. Int. **43**, 12546–12550 (2017). https://doi.org/10.1016/j.ceramint.2017.06.128
57. Smeacetto, F., De Miranda, A., Ventrella, A., Salvo, M., Ferraris, M.: Shear strength tests of glass ceramic sealant for solid oxide fuel cells applications. Adv. Appl. Ceram. **114**, S70–S75 (2015). https://doi.org/10.1179/1743676115Y.0000000042
58. Osipova, T., Wei, J., Pecanac, G., Malzbender, J.: Room and elevated temperature shear strength of sealants for solid oxide fuel cells. Ceram. Int. **42**, 12932–12936 (2016). https://doi.org/10.1016/j.ceramint.2016.05.064
59. Javed, H., Sabato, A.G., Dlouhy, I., Halasova, M., Bernardo, E., Salvo, M., Herbrig, K., Walter, C., Smeacetto, F.: Shear performance at room and high temperatures of glass–ceramic sealants for solid oxide electrolysis cell technology. Mater. **12**, 298 (2019). https://doi.org/10.3390/ma12020298
60. Chang, H.T., Lin, C.K., Liu, C.K.: Effects of crystallization on the high-temperature mechanical properties of a glass sealant for solid oxide fuel cell. J. Power Sources **195**, 3159–3165 (2010). https://doi.org/10.1016/j.jpowsour.2009.12.008
61. Zhao, Y., Malzbender, J., Gross, S.M.: The effect of room temperature and high temperature exposure on the elastic modulus, hardness and fracture toughness of glass ceramic sealants for solid oxide fuel cells. J. Eur. Ceram. Soc. **31**, 541–548 (2011). https://doi.org/10.1016/j.jeurceramsoc.2010.10.032
62. Wei, J., Osipova, T., Malzbender, J., Krüger, M.: Mechanical characterization of SOFC/SOEC cells. Ceram. Int. (2018). https://doi.org/10.1016/j.ceramint.2018.03.103

Optical Quality Resorbable Calcium-Phosphate Glasses for Biophotonic Applications

Diego Pugliese, Nadia Giovanna Boetti, Davide Janner, and Daniel Milanese

Abstract Recently developed calcium-phosphate glass formulations are proposed in this chapter as a new class of materials for biomedical optics and photonics. The glasses have been designed and carefully prepared in our laboratory to be dissolvable in biological fluids while being optically transparent, mechanically reliable both in dry and humid environments, and suitable for both preform extrusion and fiber drawing. Optical fibers have been drawn from these glasses using our custom-made induction heated drawing tower and showed attenuation loss values from one to two orders of magnitude lower than the counterpart polymeric-based bioresorbable devices reported in literature. In addition, the optical fibers have been implanted in living rats for several weeks and no clinical signs of any adverse effect have been found. Results on the inscription and characterization of different types of fiber Bragg grating-based optical filters will be also shown, together with the demonstration of the suitability of the above-mentioned bioresorbable optical fibers for time-domain diffuse optical spectroscopy.

Keywords Calcium-phosphate glass · Bioresorbable optical fiber · In-vivo testing · Extrusion · Fiber Bragg grating · Time-domain diffuse optics

1 Introduction

Among oxide glasses, phosphate glasses have attracted an increasing interest in the scientific community, thanks to the specific functionalities provided by proper design of their composition.

D. Pugliese (✉) · D. Janner
DISAT—Politecnico di Torino and RU INSTM, 10129 Torino, Italy
e-mail: diego.pugliese@polito.it

N. G. Boetti
LINKS Foundation—Leading Innovation and Knowledge for Society, 10138 Torino, Italy

D. Milanese
DIA—Università di Parma and RU INSTM, 43124 Parma, Italy

F. Baino et al. (eds.), *Ceramics, Glass and Glass-Ceramics*, PoliTO Springer Series,
https://doi.org/10.1007/978-3-030-85776-9_7

Initially phosphate glasses were studied thanks to their ability to resist to hydrofluoric acid and for other applications, such as incorporation of radioactive wastes [1, 2]. Further interest for phosphate glasses arose due to their high transparency in the Ultraviolet (UV) wavelength range [3] and for the development of near-infrared wavelength high power lasers [4], and optical fiber-based lasers and amplifiers [4–6].

More recently, suitable engineering of the composition of phosphate glasses allowed obtaining dissolution in aqueous media, opening the way toward the realization of resorbable materials for their use in biomedicine [7]. The base composition includes calcium and sodium oxides as network modifiers and was exploited to develop fibers to be used as fillers for biopolymers to obtain a resorbable composite for bone repair [8]. Fiber drawing of calcium-phosphate glasses could lead to the realization of scaffolds which played a key role in neuronal polarization and axonal growth direction [9].

In this chapter, the fabrication and characterization of a phosphate glass composition able to combine resorbability with high optical quality is reported. This formulation was engineered based on previous recipes in order to fabricate specialty optical fibers suitable for biophotonic applications, which range from fiber Bragg grating biosensors to diffuse optics and measurement of pH.

2 Phosphate Glass Optical Fibers

2.1 Phosphate Glass Synthesis and Characterization

Phosphate glasses were based on the composition (in mol%) 50 P_2O_5 – (30 – x) CaO – (3 + x) MgO – 11.5 Na_2O – 2.5 B_2O_3 – 3 SiO_2, with x = 0, 5, 12, 20. The glasses were fabricated by melting chemicals of high purity level (99+%) in alumina crucibles inside a chamber furnace at the temperature of 1200 °C for 1 h, followed by casting into a preheated brass mold [10]. Subsequently, the glasses were annealed at a temperature around the glass transition temperature, T_g, for 5 h to relieve internal stresses, and finally cooled down slowly to room temperature.

Differential thermal analysis (DTA) was performed using a Netzsch DTA 404 PC Eos differential thermal analyzer up to 1200 °C with a heating rate of 5 °C/min in sealed Pt/Rh pans.

Density of the glasses was measured at room temperature by the Archimede's method using distilled water as immersion fluid, while coefficient of thermal expansion (CTE) was measured by dilatometry using a Netzsch DIL 402 PC horizontal alumina dilatometer operating at 5 °C/min on 5 mm-long specimens.

The values of density, T_g, onset crystallization temperature (T_x), and CTE of the prepared glasses are reported in Table 1.

The specimens were cut and polished to an optical quality for optical and spectroscopic characterization, in order to assess the possible matching of two of the prepared glasses to constitute the core and cladding components of an optical fiber.

Table 1 Physical and thermal properties of the synthesized calcium-phosphate glasses

Glass name	MgO (mol%)	Density ± 0.005 (g/cm^3)	T_g ± 3 (°C)	T_x ± 3 (°C)	CTE ± 0.1 (10^{-6} °C^{-1})
G1	3	2.606	435	658	12.6
G2	8	2.600	435	628	12.2
G3	15	2.598	442	632	12.0
G4	23	2.589	444	625	12.2

The refractive index (R.I.) of the glasses was measured by means of the prism coupling technique, using a Metricon Model 2010, at the wavelengths of 633, 825, 1061, 1312, and 1533 nm. Ultraviolet-Visible/Near-infrared (UV-Vis/NIR) spectroscopy was carried out with a Varian Cary 500 spectrometer in the wavelength range between 190 and 3000 nm.

As an example, Fig. 1 shows the transmittance of the G1 glass as a function of the wavelength. The inset reports the refractive index values of all the prepared glasses at the wavelength of 633 nm.

From Fig. 1, it is clearly possible to observe that G1 glass shows a noticeably broad transparency window from the NIR down to the near UV, while the refractive index values appear to decrease linearly with increasing MgO molar content in substitution to CaO in the composition of the glass [10].

Degradation tests were carried out on 150 ± 6 μm diameter single-material fibers of the four different glass compositions, drawn using an in-house developed induction heated drawing tower and subsequently soaked in phosphate buffered saline solution (PBS, pH = 7.4) at a temperature of 37 °C for 21 days. The solution volume to sample exposed area ratio was kept fixed to 0.1 ml/mm^2, and the refreshing of the medium was performed twice a week to simulate physiological fluid exchange.

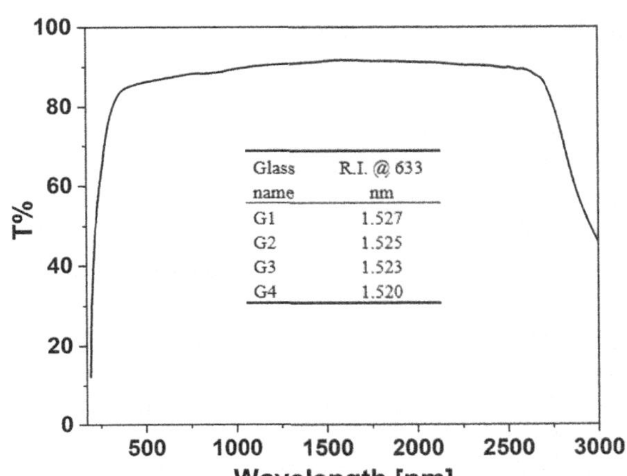

Fig. 1 Transmittance spectrum of G1 glass; the inset reports the refractive index values of all the synthesized calcium-phosphate glasses at 633 nm

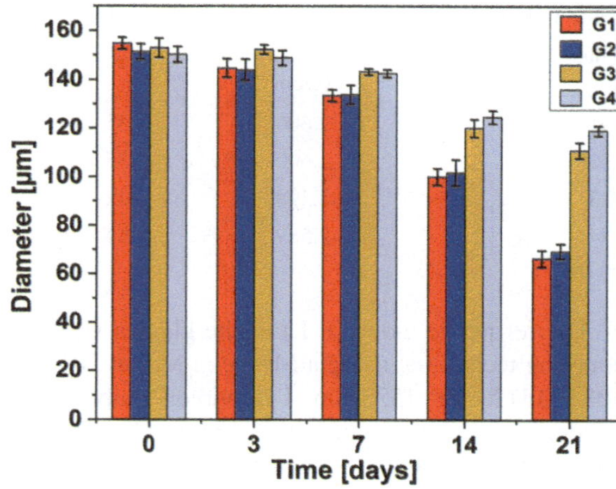

Fig. 2 Diameter values of the prepared glass fibers after 3, 7, 14, and 21 days of soaking in PBS solution

The decrease in diameter of the G1, G2, G3, and G4 fibers was periodically monitored over the whole dissolution time (see Fig. 2). Overall, the results show a complete dissolution within 30 days in PBS for the G1 and G2 compositions, while G3 and G4 exhibited slower dissolution rates. However, it is of high interest observing how the dissolution kinetics can be tailored by modifying the glass composition. In particular, a decrease in glass solubility occurred when replacing MgO with CaO, thus indicating a higher strengthening effect of MgO with respect to CaO in the glass network [10].

2.2 *Preform Fabrication*

In the development of the optical fibers, suitable glass pairs were selected. In the following paragraphs, details of the prepared optical fibers will include the selected glass pairs which suited best the requirements of the applications.

Glasses constituting the core of the optical fibers were shaped into rods by casting the melt into a pre-heated cylindrical mold. The so-obtained samples were polished to an optical quality using SiC disks of different grains. The cladding glass tubes were realized by means of extrusion or rotational casting techniques aiming at a central hole large enough to fit the core rod. Examples of core glass rod and cladding glass tubes constituting the optical fibers are reported in Fig. 3. The extrusion of the cladding preform is a very convenient process compared with other techniques such as rotational casting and allows to fabricate even very complex preforms [11].

The results obtained for rotational casting and extrusion of the bioresorbable phosphate glass cladding are compared in Fig. 3. It can be clearly observed that the surface of the preform obtained with rotational casting needs polishing to be used for fiber drawing. The surface of the extruded tube, instead, is readily usable for fiber

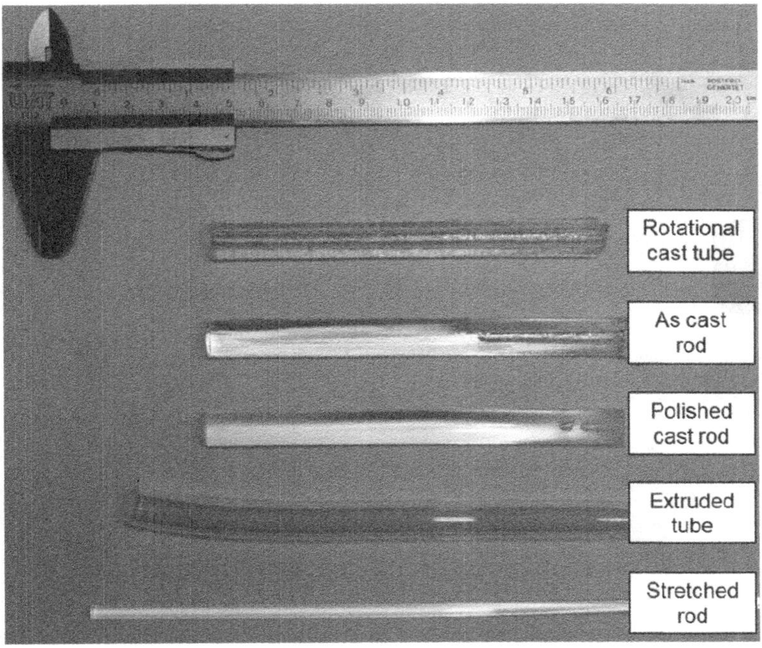

Fig. 3 Examples of core and cladding glass preforms used for the fabrication of the optical fibers: rotational cast cladding tube, as-cast and polished core rod, extruded cladding tube, and stretched core rod

drawing since it shows a smooth and even surface. With such technique there are no limitations given by the rotation speed, that restricts the achievable ratio between the outer and the inner diameter of the tube being at most 2:1. Indeed, with extrusion thin wall capillaries can be easily obtained achieving a ratio of outer to inner diameter ranging between 10:1 and 10:9. For the sake of completeness, it must be said that the main disadvantage of the extrusion process is the presence of a deformed initial and final part that must be cut off prior to the insertion of the core rod into the cladding tube. This fact leads to a reduction in the used material even though the discarded material can be remelted in future batches.

2.3 *Optical Fiber Fabrication*

The bioresorbable optical fibers can be fabricated in different configurations, however for the main applications as light guides the most common is the single core/cladding optical fiber. The preform, assembled as previously described, is fed into a drawing tower system that is reported in Fig. 4. The preform is progressively introduced into a susceptor under controlled nitrogen atmosphere with a radio frequency (RF)

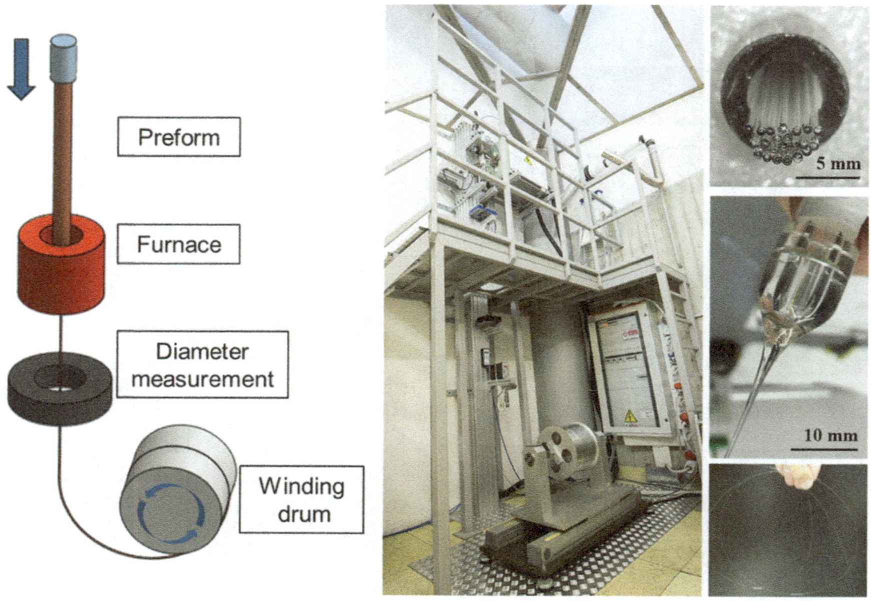

Fig. 4 Left panel: schematic illustration of a fiber drawing tower. Middle panel: optical fiber drawing tower located at Politecnico di Torino-LINKS. The preform is fed into a heated susceptor that brings it to the drawing temperature producing a fiber of controllable diameter. Right panel (from top to bottom): microstructured calcium-phosphate glass optical fiber preform, typical neck-down region of the optical fiber preform after the drawing process, example of optical fiber specimen

induction heating system (SAET, Torino, Italy) that can reach up to 800 °C. After reaching the drawing temperature, the preform starts forming a neck and starts the drawing process under its own weight. After reaching the winding drum at the bottom of the drawing tower, the fiber is attached to it and is rolled onto the drum upon drawing. The feed rate of the preform and the rotation speed of the winding drum determine and control the diameter of the fiber, and for this reason are used as process parameters.

The drawing process allowed to produce different types of fibers from multi-mode (MM) to single-mode (SM). For biological applications, both these types are of great interest since they can be used for imaging or greater collection of light in the case of MM fibers or for critical signal processing for spectroscopy in the case of SM ones. A typical example of bioresorbable MM and SM fibers is reported in Fig. 5. Their cross-sectional view shows a very good core/cladding interface and a circular and centered core.

After fiber fabrication, a full optical and mechanical characterization has been performed to verify if the requirements for potential biomedical applications could be met.

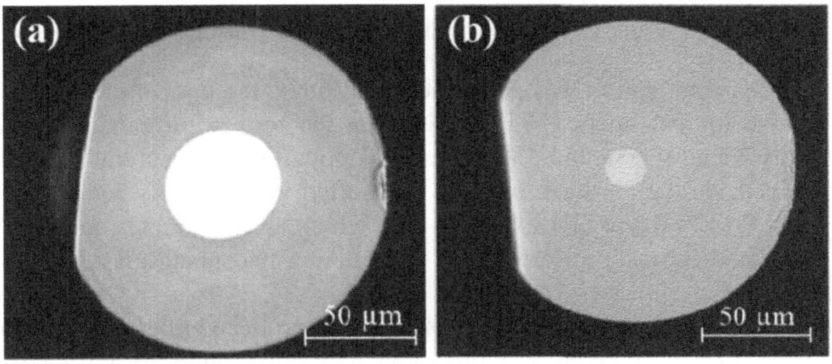

Fig. 5 Cross-sectional view of the drawn bioresorbable fibers showing a MM (**a**) and a SM (**b**) configuration. Adapted from Ceci-Ginistrelli et al. [10] under the Creative Common license CC BY and from Boetti et al. [12]

2.4 Characterization of the Prepared Optical Fibers

The two main requirements for the application of optical fibers in a biomedical context are centered around their mechanical strength and optical transmission properties. Indeed, a certain mechanical strength is required to withstand the possible stresses that might be generated by handling or implantation in the case of bioresorbable fibers. At the same time, a desired property is a reduced loss in light propagation and possibility to convey light in a part of the spectrum that is useful for e.g. fluorescence or diffusion spectroscopy.

The fabricated fibers were submitted to mechanical tensile tests as reported in Fig. 6 and adapted from [13]. The figure shows typical values of $\Delta L/F$ versusL_0/A, from which an estimation of the elastic modulus of 48 GPa can be derived. Such value is in line with the original glass materials and implies a good rigidity of the

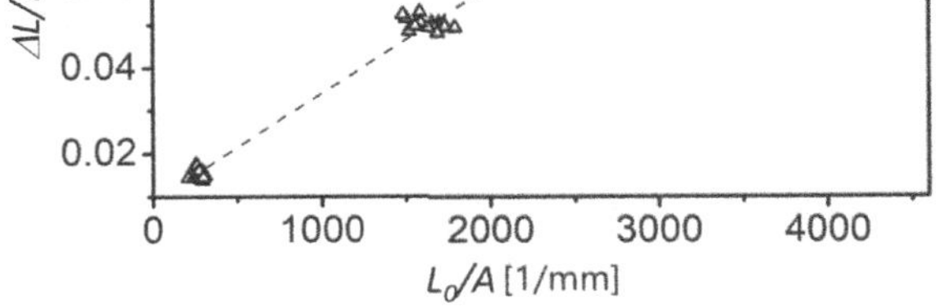

Fig. 6 A typical $\Delta L/F$ versusL_0/A diagram for the bioresorbable optical fiber. The fitting line used for the elastic modulus estimate is also shown. Adapted from [13]. Copyright 2019, with permission from Elsevier

fibers in a scenario of implantation inside the body, thus opening new interesting possibilities.

Further characterization was performed to evaluate the optical properties of the bioresorbable optical fibers [10]. In this case, the tests were carried out at two distinct wavelengths: 633 and 1300 nm. The optical loss measurement of the SM fiber obtained via the cut-back method is reported in Fig. 7. As can be seen, the losses are 1.9 (4.7) dB/m at 1300 (633) nm, which are low values and very close to the estimated intrinsic absorption of the calcium-phosphate glass used for the fabrication.

The mechanical and optical properties make the produced fiber a very promising candidate for in-vivo biomedical applications both in a scenario of intravital monitoring or body implantation for theranostics purposes. Before reviewing the different applications, a thorough characterization of the in-vivo bioresorbability properties will be detailed in the next Section.

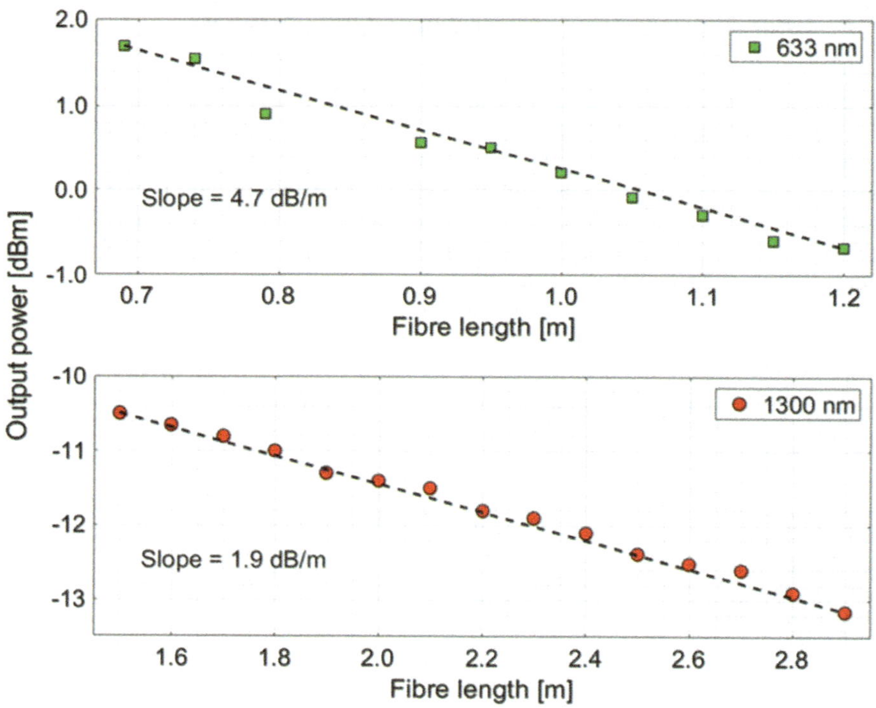

Fig. 7 Typical optical losses at 633 and 1300 nm of a SM bioresorbable optical fiber measured with the cut-back method. The square and circular symbols are the experimental points, while the dashed lines are the linear fitting curves. Adapted from [10] under the Creative Common license CC BY

2.5 *In-Vivo Tests of the Biodegradable Optical Fibers*

To assess the viability of the usage of the bioresorbable optical fibers, the response to implantation has been checked through in-vivo tests [14]. The latters were performed on adult male rats in the animal facility of the University of Defense, Faculty of Military Health Sciences and in the laboratories of the Biomedical Research Center, University Hospital in Hradec Králové. The administration of the phosphate fibers was realized by subcutaneous implantation of 1 cm-long phosphate fiber bundles to evaluate their biodegradability and potential toxicology risk. The bundle was secured by two knots of surgical thread as can be seen in Fig. 8a.

Control (CTRL) animals underwent the same operation without the administration of fibers. Tissue sampling was performed 2, 4 and 5 weeks after fibers administration, and histopathological tissues analysis and biochemical and hematological analysis of blood were performed. The implanted fiber bundle extracted after 4 weeks and its microphotograph are shown in Fig. 8b, c, respectively. The postmortem examination included the site of implantation, external surfaces of the body, all orifices of the body and the cranial, thoracic, abdominal and pelvic cavities and their contents.

The blood tests were realized for eight parameters, namely urea, creatinine, sodium, potassium, chlorides, alanine aminotransferase, aspartate aminotransferase and C-reactive protein (CRP). Histological slides were prepared by common paraffin technique and stained with hematoxylin and eosin. One-way analysis of variance (ANOVA) and Kruskal-Wallis' test with Dunn's multiple comparison test were

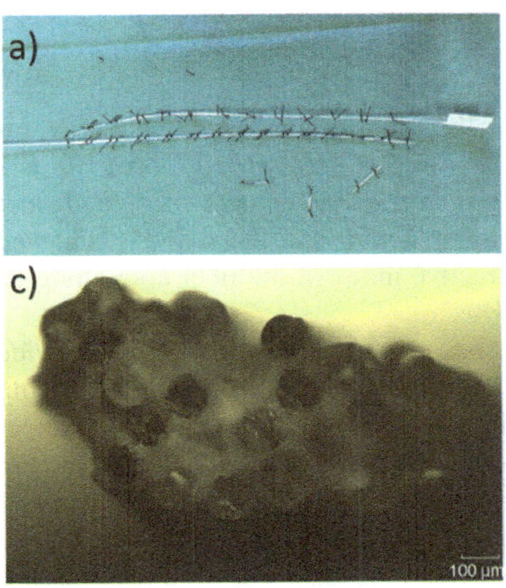

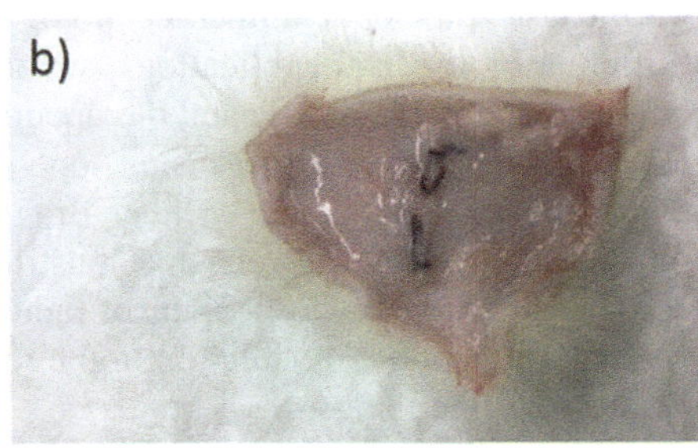

d)

	Urea	Creatinine	Na	K	CRP
CTRL	8.1 ± 0.0	15.0 ± 0.0	140.0 ± 0.0	6.1 ± 0.0	0.3 ± 0.0
Group 1	7.0 ± 1.2	18.0 ± 1.5	143.0 ± 1.9	6.0 ± 0.9	0.1 ± 0.1
CTRL	7.0 ± 0.0	20.0 ± 0.0	144.0 ± 0.0	6.2 ± 0.0	0.1 ± 0.0
Group 2	8.3 ± 1.0	21.0 ± 3.1	145.1 ± 0.7	5. 7 ± 0.7	0.2 ± 0.1
CTRL	7.1 ± 0.0	26.0 ± 0.0	143.0 ± 0.0	5.7 ± 0.0	0.1 ± 0.0
Group 3	7.4 ± 0.8	22.6 ± 3.1	143.4 ± 0.8	6.2 ± 1.5	0.2 ± 0.1

Fig. 8 In-vivo test of the bioresorbable optical fibers was realized by the implantation of a (**a**) fiber bundle in (**b**) the subcutis of rats. The bundle removed after test is shown in (**c**), while the blood test showing no signs of inflammation is reported in (**d**). Adapted from [14]. Copyright ©2019 WILEY–VCH Verlag GmbH and Co

selected for statistical evaluation of the blood tests. The skin from the implantation site was histopathologically evaluated on day 14, day 28 and day 35.

As confirmed in the Table of Fig. 8d, no clinical signs of adverse reactions were found in tested animals after implantation of the bioresorbable fibers. Indeed, the CRP, whose levels rise in response to an inflammation, showed normal values.

Mild to minimal fibrosis accompanied by some bands of newly formed capillaries were found in the subcutis of both test and control animals for all the time intervals considered. In the interval of 28 and 35 days, the intensity of these lesions got milder. Furthermore, occurrence of a mild focal chronic inflammation in the subcutis was detected in some administered and one control rat. In summary, there were no significant clinical differences between test and control animals. No signs of hepatotoxicity or nephrotoxicity were found in the liver and kidneys parenchyma of the experimental animals after implantation of the bioresorbable fibers.

The results of the in-vivo tests show a good biocompatibility of the fibers and open up very interesting scenarios for their application in biomedicine. In the following Sections, a few of these potential applications will be reviewed.

3 Applications

3.1 Bioresorbable Optical Fiber Bragg Grating Sensors

Over the last years, optical fiber technology has attracted an ever-increasing interest in different biomedical applications, such as endoscopic imaging, localized tissue modification, and diagnostic and therapeutic sensing [15].

Focusing on biomedical sensing, one of the most promising technologies involving optical fibers is that of fiber Bragg gratings (FBGs) [16, 17]. A FBG is a type of distributed Bragg reflector constructed in a short segment of optical fiber that reflects particular wavelengths of light and transmits all others, and is achieved by creating a periodic modulation in the refractive index of the fiber core along its longitudinal direction.

During the last two decades, FBG sensors have been efficiently applied in a wide selection of biosensing applications, including articular joint pressure monitoring [18], plantar pressure and human–machine interface sensing pads [19], DNA detection [20], measuring mechanical strain in bones [21], pressure detection inside the bloodstream, and local in-vivo temperature probes [22].

Thermally durable Bragg gratings have been successfully inscribed in phosphate glass optical fibers using a variety of laser sources [23–25]. The combination of a bioresorbable phosphate glass optical fiber with Bragg grating reflectors paves the way toward new soluble photonic sensing probes with the potential characteristic of in-body operation for the efficient monitoring of vital mechanical or chemical parameters. In the following Subsections, results on the inscription of Bragg gratings into our custom-made bioresorbable phosphate glass optical fiber, using both an UV

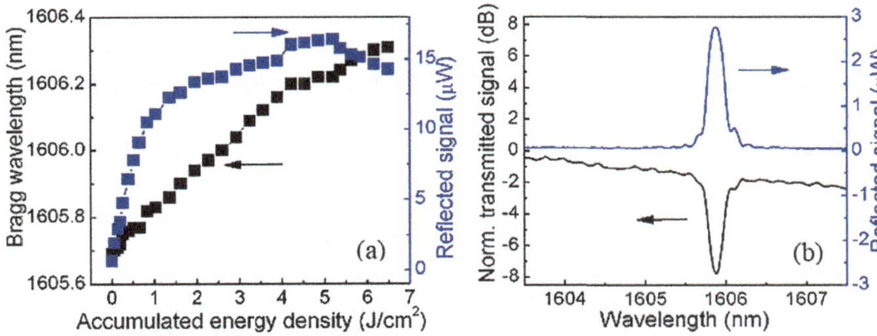

Fig. 9 (**a**) Bragg wavelength and reflected signal strength evolution during a 100 min inscription of a 5 mm FBG with an UV laser beam pulse energy density of 108 mJ/cm^2 and 10 Hz repetition rate. (**b**) Normalized transmission and reflection signals of a standard FBG after an accumulated energy density dose of 4 J/cm^2. Adapted from [26] under the Creative Common license CC BY

193 nm excimer laser [26] and a visible 517 nm femtosecond (fs) laser [27], will be presented and discussed.

UV Laser Written FBGs Through the Phase-Mask Technique The Bragg gratings were inscribed onto 10 cm-long bioresorbable phosphate fiber sections, previously fusion spliced to a standard SM fiber (SMF-28e, Corning Inc.). The inscription set-up consisted of a standard 1064.7 nm period phase-mask and a 193 nm high spatial coherence and 10 ns pulse duration excimer laser (Braggstar, TUI laser). The energy density of the inscription beam was set to 108 mJ/cm^2, as higher energy densities revealed to induce surface damage to the fiber cladding [26]. The gratings transmission and reflection spectra were real time monitored during inscription using a broadband super luminescent diode source and an optical spectrum analyzer (OSA).

A 100 min irradiation at 10 Hz resulted in the monotonous shift of the inscribed Bragg wavelength to higher values, as shown in Fig. 9a. The reflected signal strength for the first 20 min of irradiation (accumulated energy density of 1 J/cm^2) showed a linear increase, while a longer exposure to deep UV laser radiation led to saturation. A 6 dB strong notch was formed for an accumulated energy density dose of approximately 4 J/cm^2 (Fig. 9b) [26]. This grating inscription behavior is considered to be of Type I nature [24]. Based on the Bragg wavelength shift illustrated in Fig. 9a, the induced average refractive index change at the end of the inscription was found equal to 5.8×10^{-4} [16].

In addition to the standard FBGs, tilted FBGs [28] with tilting angles equal to 1 and 0.8° were also successfully recorded [26, 29]. These gratings, taking advantage of a certain tilting angle between the grating plane and the longitudinal fiber axis, allow not only the coupling of the counter-propagating core mode, but also the coupling of the core mode to higher-order cladding modes with azimuthal field component.

For the inscription of the tilted FBGs, the phase-mask holder was tilted with respect to the horizontal axis of the optical fiber. Initial grating recording attempts at a tilting angle of 2° and an exposure energy dose of 4 J/cm^2 did not result in the simultaneous inscription of measurable Bragg and cladding mode resonance signals due to the low numerical aperture (NA) of the optical fiber, thus the tilting angle was lowered to 1° or even less.

Figure 10a, b shows the normalized transmission signal for FBGs inscribed with tilting angles of 1° and 0.8°, respectively. The tilted FBGs were inscribed with the same UV laser beam inscription conditions employed earlier. The inscription time was 80 min for the case of 1° tilting angle and 60 min for the 0.8° one, yielding a corresponding accumulated energy density of 5.2 and 3.9 J/cm^2, respectively.

In addition to the fundamental Bragg peak attenuation signal, which is diminished in strength in comparison to the standard inscription but still present at 1606 nm, the tilted FBG spectra show the so-called "ghost" peak at 1605.3 nm. This peak has similar characteristics with the Bragg line and is ascribable to the superposition of several tightly confined lower-order cladding modes [30]. The increase of the tilting angle from 0.8 up to 1° proved to have a strong effect on the relative strength of

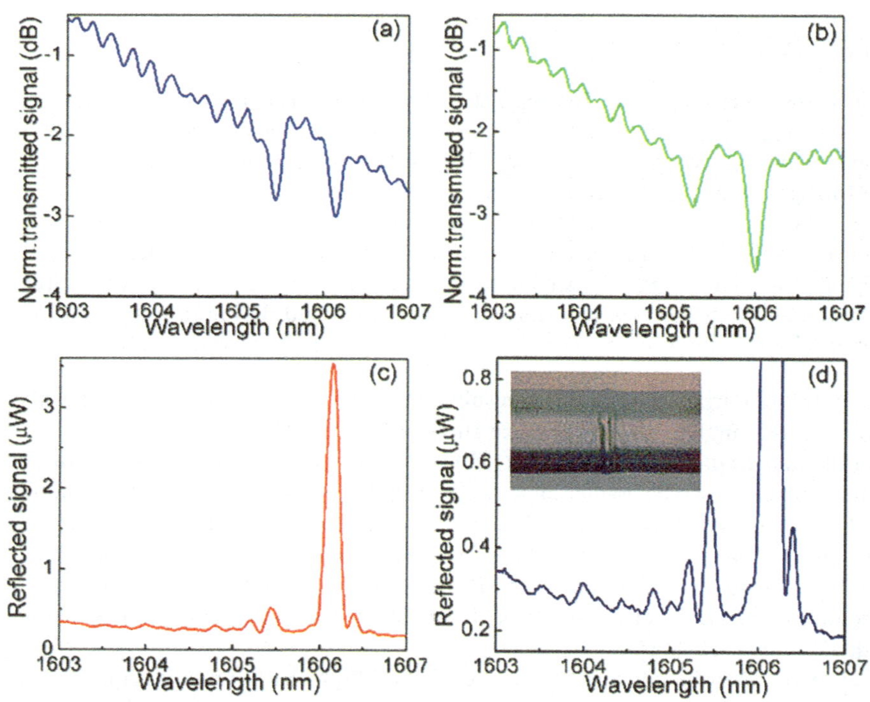

Fig. 10 Normalized transmission signal for FBGs inscribed with a phase-mask tilting angle of (**a**) 1° and (**b**) 0.8°. (**c**) Reflected signal of the 1° tilted FBG and (**d**) vertical magnification of (**c**). The inset of (**d**) shows an image of the splice joint of the phosphate and SMF-28e fibers. Adapted from [26] under the Creative Common license CC BY and from [29] with permission from the Institute of Electrical and Electronics Engineers

the fundamental and "ghost" mode peaks. For the case of the 1° inscription, the suppression of the fundamental mode and the corresponding enhancement of the "ghost" one rendered the strength of the two lines almost equal. As highlighted in Fig. 10a, b, the spectra do not display any additional noticeable higher order cladding mode resonances at shorter wavelengths. For the case of the 1° tilting angle, an indication of smaller peaks is visible at around 1603.8 nm, which is however masked by noise features of the transmission signal.

The corresponding reflected signal of the 1° tilted FBG and its vertical zoom-up are shown in Fig. 10c, d; similar data were recorded for the 0.8° case (not shown here). Figure 10d highlights the presence of a number of cladding mode peaks of smaller strength, compared to the Bragg one, at shorter wavelengths. The occurrence of those cladding modes is facilitated by the fringe recording along the whole fiber volume, which is responsible for the recoupling of a certain amount of light in the core in back propagating mode, as well as by the geometry of the splice joint (shown in the inset of Fig. 10d), that partially acts as a mirror [31].

To allow online monitoring of the optical fiber dissolution, the 1° tilted FBG was immersed into a PBS solution at 37 °C and pH = 7.4. The evolution of the grating signal was monitored regularly up to 56 h [26], and the spectra recorded for the reflected signal are shown in Fig. 11.

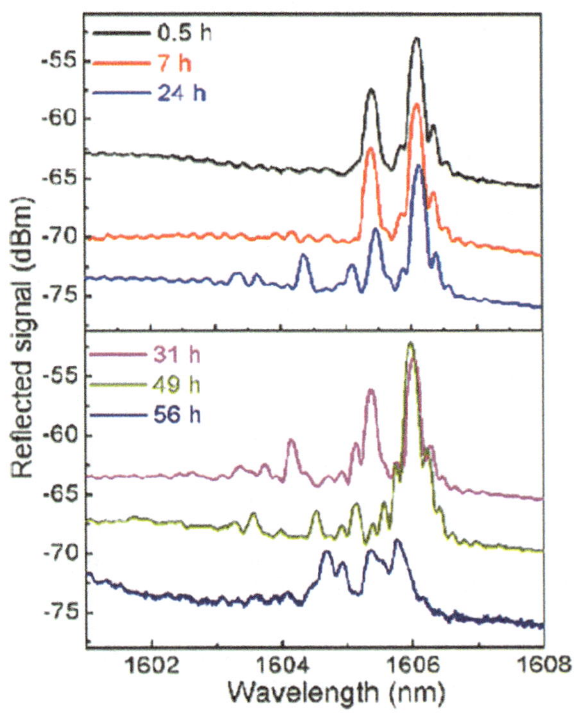

Fig. 11 Reflected signal evolution of an 1° tilted FBG immersed in PBS solution for up to 56 h. The baseline of the different spectra has been adjusted vertically to assist visualization. Adapted from [26] under the Creative Common license CC BY

It is worthwhile highlighting that changes in the reflected signal are evident just after a few hours in the solution, with the appearance of additional cladding mode peaks in the region between 1603 and 1605 nm. For prolonged immersion periods, i.e. 49 h onward, the fundamental FBG peak exhibited modifications also in wavelength, shape, and strength. In particular, a strength reduction from an initial value of 0.78 dB down to 0.25 dB was assessed after an immersion time of 56 h. This behavior is not surprising, since the "ghost" mode corresponds to low-order cladding modes characterized by a modal dispersion response that is highly similar to that of the core guiding mode manifested at the fundamental Bragg peak [28].

Femtosecond Laser Written FBGs Through the Plane-by-Plane Technique Our in-house developed bioresorbable phosphate glass optical fiber exhibited not only a good photosensitivity to deep UV light [26], but it also proved to be suitable for fs laser processing [27]. More specifically, different types of Bragg grating-based optical filters, such as uniform FBGs, chirped gratings, and fiber Fabry-Perot (FFP) cavities, were successfully inscribed into it using a 517 nm fs laser. The direct-write plane-by-plane inscription method was selected, as it offers excellent flexibility in sensors development [32].

The Bragg grating inscription set-up consisted of a fs laser system (femtoRegen HighQ) operating at 517 nm, with 220-fs pulse duration and a nm-accuracy air-bearing translation stage system (Aerotech) for controlled and accurate two-axis motion during the inscription [33, 34]. The fiber samples were carefully mounted on the translation stage and the laser beam was focused from above using a third stage and a long working distance microscope objective (Mitutoyo, $\times$ 50, NA 0.42). The pulse energy at the exit of the laser was set to 100 nJ for a laser repetition rate of 2 kHz, controlled using a pulse picker.

For the inscription of the uniform FBG operating at 1563.08 nm, the grating period (Λ) was close to 2 μm (4th order FBG) and was selected to avoid any overlapping index change. The laser was swept perpendicularly to the fiber axis generating a plane across the fiber core, and the inscription width was set to 15 μm (equal to the core diameter). The total number of periods was fixed to 1000, resulting in a $\sim$ 2 mm grating length (L) [27].

Following the inscription, the bioresorbable phosphate fiber sample was fusion spliced to a standard telecom SM fiber (SMF-28e, Corning Inc.) by setting the intensity of the electric arc at a very low value and by shifting the relative position of the splicing location 700 μm towards the silica fiber.

The FBG reflection spectrum (see Fig. 12a) was then recovered using an OSA with optical resolution of 10 pm through an optical circulator, with illumination from a broadband light-source (Thorlabs ASE730). From the reflection spectrum, the bandwidth full width at half maximum (FWHM) and the effective refractive index (n_{eff}) of the grating were found to be 0.5 nm and $\sim$ 1.515, respectively. A bright field microscope image of the inscribed grating, as captured with a charge-coupled device (CCD) camera during the inscription, is shown in Fig. 12b. The image shows the extent of the fiber diameter and the grating location, which has been designed to completely cover the core region.

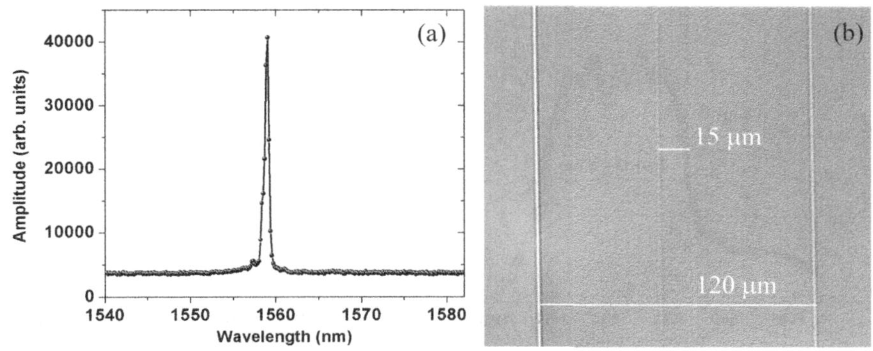

Fig. 12 (**a**) Reflection spectrum of a 4th order FBG inscribed in the bioresorbable phosphate fiber using the fs laser. (**b**) Microscope picture of the laser-induced refractive index change. Adapted from [27] with permission from the Institute of Electrical and Electronics Engineers

The uniform 4th order FBG response to temperature and relative humidity (RH) changes was initially evaluated. To this aim, the FBG was placed in a climate chamber (Memmet 108HCP) and interrogated in reflection.

For the measurement of the temperature sensitivity, the RH of the chamber was kept constant at 35%, whereas the temperature was set to increase from 23 to 73 °C in a one-hour cycle. The temperature response of the grating was found to be 12.7 pm/°C [27], in agreement with the 10.5 pm/°C typically measured for standard SM optical glass fibers [16]. For the humidity tests, instead, the climate chamber temperature was kept constant at 36 °C and the RH was increased from 35 to 95%. After 4 h, the FBG wavelength shifted by only 0.025 nm, thus leading to a very low response to RH of 0.416 pm/%RH [27], as expected when dealing with a glass-based fiber.

Subsequently, the grating sample was kept for up to 52 h into the climate chamber at 95% RH; the grating exhibited a significant profile distortion and its amplitude progressively decreased, whereas the resonance wavelength of the grating shifted towards shorter wavelengths, indicating erasure.

As a further step in this research, the inscription of a chirped grating was carried out. In this case, the period of the grating was linearly increased with a chirp coefficient for each period as follows:

$$\Lambda_z = \Lambda_0 + k_z \tag{1}$$

where Λ_0 is the initial period of the grating and k_z is the chirp coefficient for $0 < z < L$. For the 4th order FBG at 1561 nm, the initial and the final periods were ~ 2.060 and ~ 2.067 μm, respectively, corresponding to a reflection wavelength of 1566 nm and resulting in a 5 nm chirped grating (see Fig. 13a) [27]. Ripples of up to 1 dB of the maximum reflection value were observed, however they could potentially be reduced by using a smaller chirp coefficient and a higher number of periods.

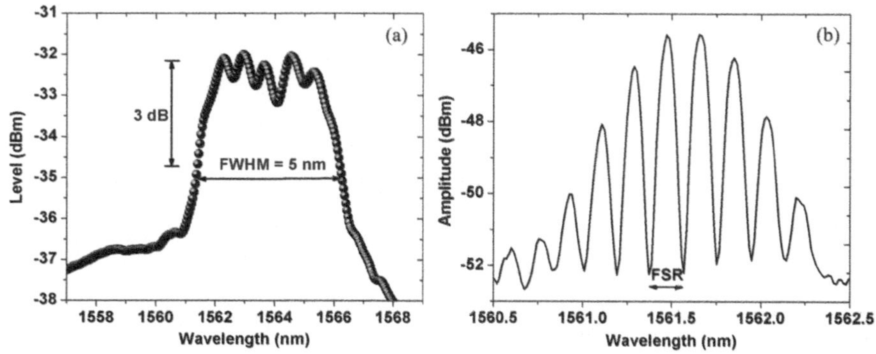

Fig. 13 Reflection spectrum of a (**a**) 5 nm chirped grating and a (**b**) Fabry-Perot cavity inscribed in the bioresorbable phosphate glass optical fiber. Adapted from [27] with permission from the Institute of Electrical and Electronics Engineers

Finally, an FFP cavity was inscribed using two identical FBGs with a physical separation of 3.4 mm [27]. The total cavity length must consider the light penetration depth through each grating, and so is estimated as the sum of the cavity length and approximately half of each grating length [35]. For gratings of length ~ 0.8 mm, the effective optical length of the cavity (L_{cavity}) is ~ 4.2 mm. The cavity free spectral range (FSR) is given by:

$$\lambda_{FSR} = \frac{\lambda^2}{2n_{eff}L_{cavity}}, \tag{2}$$

where λ is the light wavelength and n_{eff} is the effective refractive index. According to the FBG inscription, n_{eff} was ~ 1.515 and for gratings operating at ~ 1561 nm the FSR is approximately equal to 190 pm (see Fig. 13b). The finesse of the FFP cavity is given by:

$$finesse = \frac{\lambda_{FSR}}{\delta\lambda} = \frac{4\sqrt{R_1R_2}}{\left(1-\sqrt{R_1R_2}\right)^2}, \tag{3}$$

where $\delta\lambda$ is the FWHM of the fringes and R_1 and R_2 are the maximum reflectivity of each grating. From Eq. (3), the calculated cavity finesse is 3.16, thus indicating an average grating reflectivity of ~ 53%. The finesse and the reflectivity of the cavity could possibly be enhanced by inscribing Bragg gratings with longer length.

3.2 Time-Domain Diffuse Optics Using Bioresorbable Optical Fibers

Diffuse optics (DO) is an attractive optical technique which aims to investigate the optical properties of a diffusive medium using light in the near-infrared spectral region, where the water absorption is negligible [36]. The most widespread application of DO is on biological tissues such as brain, muscles, breast, bones and others [37–41], for which quantitative information on tissue components (e.g. lipid, collage, water) or metabolic and oxidative state indicators (e.g. oxygen, saturation, hemoglobin and flux) can be derived. Such information can be related to the tissue regeneration, to the healing process, or to a harmful evolution.

The time-domain (TD) approach, based on the use of a picosecond pulsed laser source and time-resolved single-photon-detection, improves DO capabilities in terms of depth resolution and sensitivity [42].

The use of a phosphate bioresorbable optical fiber in diffuse optics is particularly attractive in applications where there is the need to implant an optical fiber, like photodynamic therapy (PDT) or monitoring of a post-operative healing process. In these applications the possibility to leave the fiber in place after its function is completed without the need for an explant surgery, reduces the impact on the patient and on clinical procedures.

In the following paragraphs, a study of suitability of the phosphate bioresorbable optical fiber for TD diffuse optical spectroscopy (TD-DOS) deep into a biological tissue will be presented. The use of the fibers was validated first on tissue-mimicking liquid phantoms and then on a more realistic environment using a chicken breast.

Suitability of Bioresorbable Fibers for TD Measurements: Phantom Measurements The experimental set-up used for this research is shown in Fig. 14. Two different pulsed laser sources were employed for the different experiments: a four-wave mixing prototype (Fianium Ltd., UK), emitting optical pulses at 820 nm at a rate of 40 MHz, and a supercontinuum laser (SuperK Extreme, NKT Photonics, Denmark) coupled to a dual-channel Acousto-Optical Tunable Filter able to provide optical pulses at 80 MHz with a wavelength tunable in the 500–1100 nm range [43].

The laser light was then attenuated using a variable optical attenuator (VOA) and delivered to the sample through a bioresorbable optical fiber (depicted in grey in Fig. 14), that was coupled to a commercial silica fiber (depicted in blue).

A second bioresorbable fiber was used to collect light from the sample and was coupled to a detector (100 μm active area diameter single-photon avalanche diode (SPAD), Micro Photon Devices Srl, Italy) through a doublet of lenses with a demagnification factor of 2.

The phosphate fibers were immersed by about 30 mm into the sample to mimic an interstitial measurement and they were spaced by 20 mm (source-detector distance, ρ).

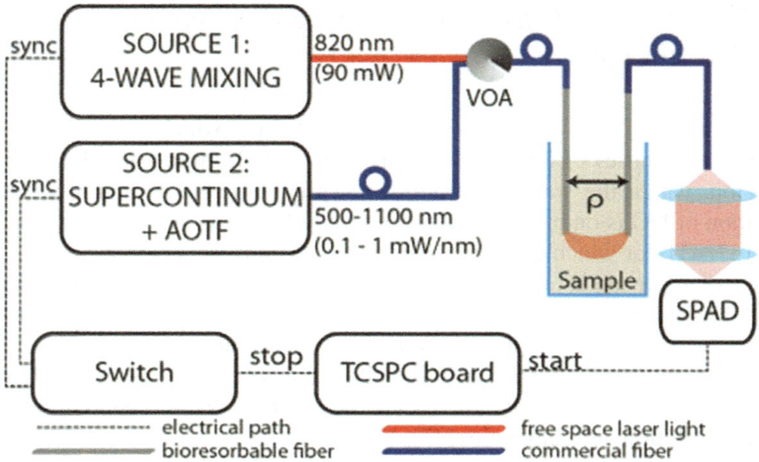

Fig. 14 Schematic of the experimental set-up. The sources 1 and 2 were used for the linearity and spectroscopy measurements, respectively. Adapted from [43]

The phosphate fibers, tailored in their composition to increase the numerical aperture (NA = 0.17 at 633 nm) so as to enhance the light harvesting, featured three different core diameters: 50, 100 and 200 μm.

Complete details of the bioresorbable fibers and of all the components of the set-up can be found in [43].

To study the aptness of phosphate bioresorbable fibers for TD-DOS, a standardized protocol for the objective performance assessment of diffuse optics instruments was followed. At first the instrument response function (IRF), described in the basic instrumental performance (BIP) protocol [44], was measured to characterize the overall time resolution of the instrument and to assess the intrinsic suitability of the set-up for DOS measurement.

Subsequently, the linearity test, contained in the MEDPHOT protocol [45], was carried out to determine the performance of the system in retrieving the absorption (μ_a) and reduced scattering coefficients (μ'_s) of a reference homogeneous diffusive medium [46].

Figure 15 displays the IRF of the realized set-up. In Fig. 15a the IRF at 820 nm acquired using all the fibers of the set (50, 100, 200 μm) is shown, while the Fig. 15b reports the IRF at three randomly chosen wavelengths (650, 820 and 1000 nm) using the 100 μm core phosphate fiber.

The FWHM is always lower than 75 ps for all wavelengths and fiber dimensions; the dynamic range is larger than 4 decades at 820 nm, while decreases to 3 for 1000 nm. No fluorescence or backscattering from all the bioresorbable fibers tested was detected at all wavelengths, while the only exponential decay present is the diffusion tail of the SPAD, thus confirming the suitability of bioresorbable fibers for TD-DOS [46].

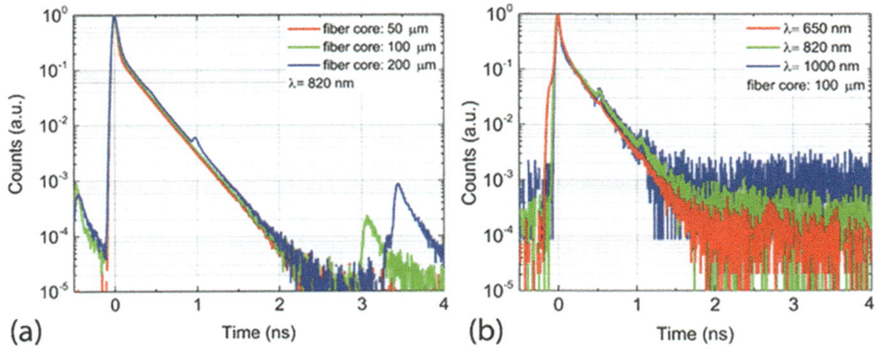

Fig. 15 (**a**) IRFs acquired at 820 nm using phosphate fibers with 50, 100 and 200 μm core diameter. (**b**) IRFs at three randomly selected wavelengths (650, 820 and 1000 nm) acquired using the 100 μm core fiber. Reproduced from [46]. Copyright ©2017 WILEY–VCH Verlag GmbH and Co.

The linearity measurement was checked by using a liquid phantom composed of water, intralipid (providing scattering properties) and ink (acting as absorber).

By properly mixing of the above components, the desired ("true" in Fig. 16) optical properties of the phantom were obtained [47]. For this research, phantoms combining 4 values for the scattering coefficient (μ'_s = 5, 10, 15 and 20 cm^{-1} at 820 nm) with 8 values for the absorption one (μ_a ranging from 0.023 to 0.690 cm^{-1}) were synthesized. The optical properties were then recovered by fitting the experimental measurements to an analytical model obtained under the diffusion approximation, which describes the photon transport in an infinite diffusive medium [48].

Figure 16 shows the results of linearity measurements for all set of fibers (columns) [46]. The first row reports the linearity in absorption, i.e. measured values of μ_a against the true ones. The system is linear till 0.690 cm^{-1} for the 50 μm core fiber, while the 100 and 200 μm ones show a linear behavior till 0.370 cm^{-1}. This difference is explained by the presence of a higher number of reflections in the larger core fibers, that can impair the fitting procedure.

The second and third rows of Fig. 16 show the coupling between absorption and scattering. The plots indicate a good independency between the two parameters, except for high scattering values (15 and 20 cm^{-1}) and high absorption (0.690 cm^{-1}), where the fit was not reliable as stated for the first row of the graph.

The last row of Fig. 16 displays the linearity in scattering, i.e. measured values of μ'_s against the true ones.

The behavior of the system is almost linear (relative error always lower than 20%) for the 50 and 100 μm core fibers, except for the scattering value of 20 cm^{-1}. For the 200 μm core fiber the linearity is good for all phantoms with absorption values lower than 0.690 cm^{-1} (maximum relative error of 10.3% at $\mu'_s = 5$ cm^{-1} and μ_a = 0.370 cm^{-1}, otherwise always lower than 4%).

Ex-Vivo Spectroscopy Measurements In order to demonstrate that bioresorbable fibers can be used to recover the absorption and the scattering coefficients of a living

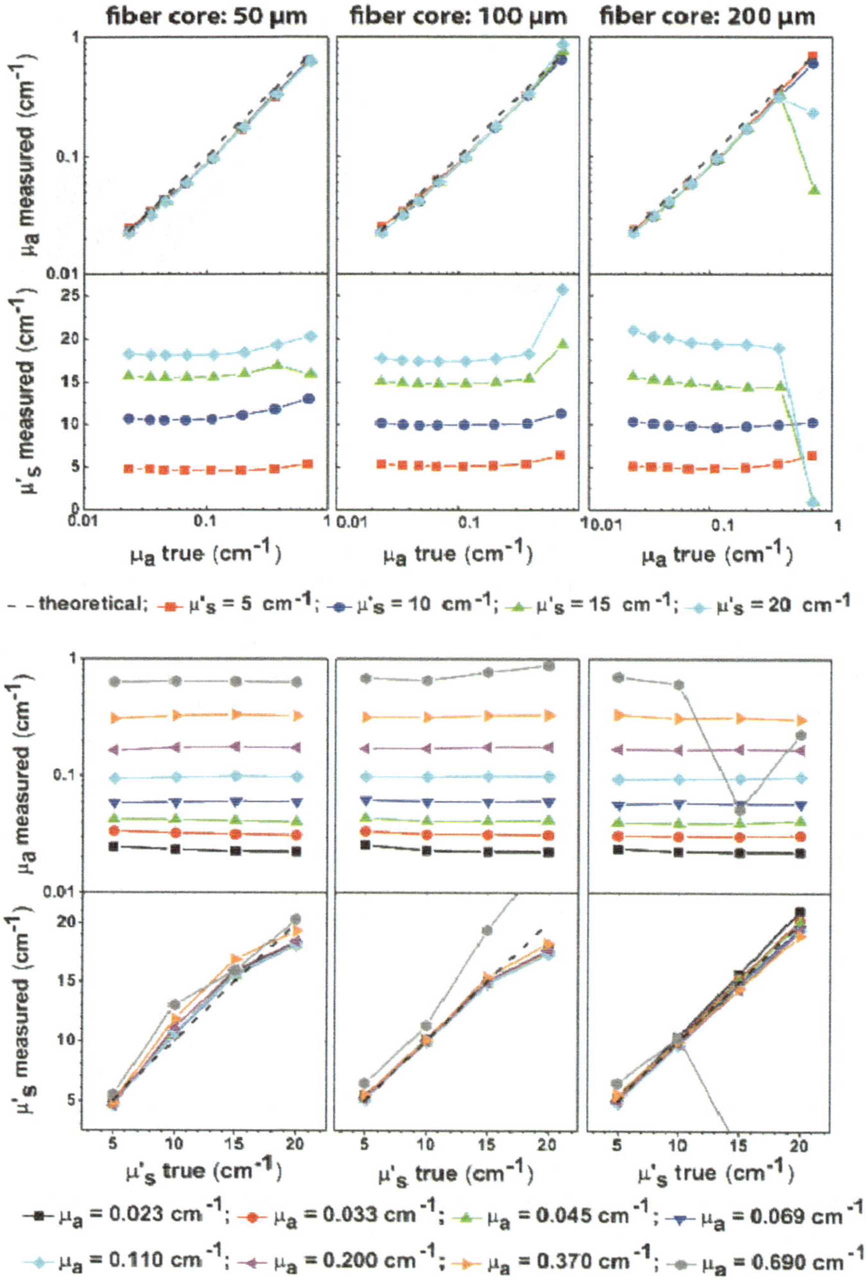

Fig. 16 Results of the linearity measurements for the three fibers under test. Reproduced from [46]. Copyright ©2017 WILEY–VCH Verlag GmbH and Co.

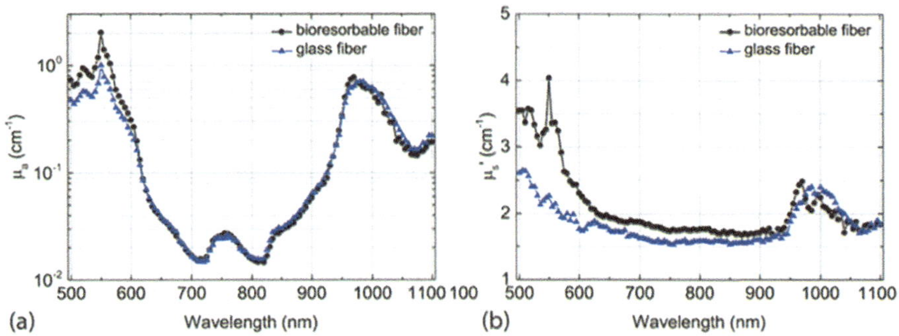

Fig. 17 Measured spectra of absorption coefficient (**a**) and reduced scattering (**b**) of a chicken breast obtained using a system based on bioresorbable fibers (black circles) and on standard glass fibers (blue triangles). Reproduced from [46]. Copyright ©2017 WILEY–VCH Verlag GmbH and Co.

tissue, spectroscopic measurement on a chicken breast in the range 500–1100 nm was performed [46].

For comparison, the measurement was repeated by substituting the bioresorbable fibers with standard silica glass-based ones (100 μm core, NA = 0.22 from Thorlabs GmbH, Germany), thus using a system close to a state-of-the-art clinical prototype.

Figure 17 reports the measured spectra obtained using the two different systems. The results show similar peaks and trends and the same absorption-to-scattering couplings.

The absorption spectrum is noisier around the water peak (at 970 nm) for the bioresorbable fiber, while tends to be overestimated below 600 nm, where the deoxy-hemoglobin peak is located (550 nm). In both spectral regions, a low signal using the bioresorbable fiber was obtained due to high absorption.

The retrieved scattering coefficient was slightly higher (around 10%) in the "safe" 600–900 nm region for the bioresorbable fibers. The difference between the reduced scattering exhibited by the bioresorbable and standard fibers was more pronounced below 600 nm, thus in a region where low signal and absorption coupling makes the scattering measurement quite unreliable.

All these measurements validated the use of phosphate bioresorbable fibers for TD-DOS and pave the way for a new generation of minimally invasive biomedical devices.

4 Future Developments

The development of optically transparent resorbable phosphate glasses paves the way toward the realization of a new generation of optical fiber-based devices for biophotonics.

Beyond the examples reported in this study, bioresorbable devices might be used in other specific applications, where minimally invasive procedures including drug delivery and light assisted processes are required. These may consist for instance in developing customized probes for metronomic photodynamic therapy [49], where the bioresorbable multifunctional fiber device could help minimizing the impact of the therapy on the patient and resulting in no need of explant surgery for removing the probe.

References

1. Vogel, W.: Glass Chemistry, 1st edn. Springer, Berlin (1994)
2. Sales, B.C., Boatner, L.A.: Lead–iron phosphate glass: a stable storage medium for high–level nuclear waste. Science **226**(4670), 45–48 (1984)
3. Kreidl, N.J., Weyl, W.A.: Phosphates in ceramic ware: IV, phosphate glasses. J. Am. Ceram. Soc. **24**(11), 372–378 (1941)
4. Campbell, J.H., Hayden, J.S., Marker, A.: High-power solid-state lasers: a laser glass perspective. Int. J. Appl. Glas. Sci. **2**(1), 3–29 (2011)
5. Boetti, N.G., Pugliese, D., Ceci-Ginistrelli, E., Lousteau, J., Janner, D., Milanese, D.: Highly doped phosphate glass fibers for compact lasers and amplifiers: a review. Appl. Sci. **7**(12), 1295 (2017)
6. http://www.npphotonics.com/products. Last Accessed 12 June 2020
7. Abou Neel, E.A., Chrzanowski, W., Knowles, J.C.: Effect of increasing titanium dioxide content on bulk and surface properties of phosphate-based glasses. Acta Biomater. **4**(3), 523–534 (2008)
8. Ahmed, I., Jones, I.A., Parsons, A.J., Bernard, J., Farmer, J., Scotchford, C.A., Walker, G.S., Rudd, C.D.: Composites for bone repair: phosphate glass fibre reinforced PLA with varying fibre architecture. J. Mater. Sci.-Mater. Med. **22**, 1825–1834 (2011)
9. Vitale-Brovarone, C., Novajra, G., Lousteau, J., Milanese, D., Raimondo, S., Fornaro, M.: Phosphate glass fibres and their role in neuronal polarization and axonal growth direction. Acta Biomater. **8**(3), 1125–1136 (2012)
10. Ceci-Ginistrelli, E., Pugliese, D., Boetti, N.G., Novajra, G., Ambrosone, A., Lousteau, J., Vitale-Brovarone, C., Abrate, S., Milanese, D.: Novel biocompatible and resorbable UV-transparent phosphate glass based optical fiber. Opt. Mater. Exp. **6**(6), 2040–2051 (2016)
11. Gallichi-Nottiani, D., Pugliese, D., Boetti, N.G., Milanese, D., Janner, D.: Toward the fabrication of extruded microstructured bioresorbable phosphate glass optical fibers. Int. J. Appl. Glas. Sci. **11**(4), 632–640 (2020)
12. Boetti, N.G., Ceci-Ginistrelli, E., Pugliese, D., Novajra, G., Vitale-Brovarone, C., Lousteau, J., Abrate, S., Milanese, D.: Bioresorbable calcium-phosphate optical fiber. In: Proceedings Advanced Photonics, Optical Society of America, Vancouver, Canada, p. 1 (2016)
13. Sglavo, V.M., Pugliese, D., Sartori, F., Boetti, N.G., Ceci-Ginistrelli, E., Franco, G., Milanese, D.: Mechanical properties of resorbable calcium-phosphate glass optical fiber and capillaries. J. Alloy. Compd. **778**, 410–417 (2019)
14. Podrazký, O., Peterka, P., Kašík, I., Vytykáčová, S., Proboštová, J., Mrázek, J., Kuneš, M., Závalová, V., Radochová, V., Lyutakov, O., Ceci-Ginistrelli, E., Pugliese, D., Boetti, N.G., Janner, D., Milanese, D.: In vivo testing of a bioresorbable phosphate-based optical fiber. J. Biophotonics **12**(7), e201800397 (2019)
15. Chin, L.C.L., Whelan, W.M., Vitkin, I.A.: Optical-Thermal Response of Laser-Irradiated Tissue, 2nd edn. Springer, Heidelberg (2011)
16. Othonos, A., Kalli, K.: Fiber Bragg Gratings: Fundamentals and Applications in Telecommunications and Sensing, 1st edn. Artech House, Boston (1999)

17. Berghmans, F., Geernaert, T., Baghdasaryan, T., Thienpont, H.: Challenges in the fabrication of fibre Bragg gratings in silica and polymer microstructured optical fibres. Laser Photon. Rev. **8**(1), 27–52 (2014)
18. Dennison, C.R., Wild, P.M., Wilson, D.R., Gilbart, M.K.: An in-fiber Bragg grating sensor for contact force and stress measurements in articular joints. Meas. Sci. Technol. **21**(11), 115803 (2010)
19. Candiani, A., Konstantaki, M., Pamvouxoglou, A., Pissadakis, S.: A shear sensing pad, based on ferrofluidic actuation in a microstructured optical fiber. IEEE J. Sel. Top. Quantum Electron. **23**(2), 210–216 (2017)
20. Bertucci, A., Manicardi, A., Candiani, A., Giannetti, S., Cucinotta, A., Spoto, G., Konstantaki, M., Pissadakis, S., Selleri, S., Corradini, R.: Detection of unamplified genomic DNA by a PNA-based microstructured optical fiber (MOF) Bragg-grating optofluidic system. Biosens. Bioelectron. **63**, 248–254 (2015)
21. Carvalho, L., Alberto, N.J., Gomes, P.S., Nogueira, R.N., Pinto, J.L., Fernandes, M.H.: In the trail of a new bio-sensor for measuring strain in bone: osteoblastic biocompatibility. Biosens. Bioelectron. **26**(10), 4046–4052 (2011)
22. Favero, F.C., Pruneri, V., Villatoro, J.: Microstructured opt fiber interferometric breathing sensor. J. Biomed. Opt. **17**(3), 037006 (2012)
23. Grobnic, D., Mihailov, S.J., Walker, R.B., Smelser, C.W., Lafond, C., Croteau, A.: Bragg gratings made with a femtosecond laser in heavily doped Er–Yb phosphate glass fiber. IEEE Photon. Technol. Lett. **19**(12), 943–945 (2007)
24. Albert, J., Schülzgen, A., Temyanko, V.L., Honkanen, S., Peyghambarian, N.: Strong Bragg gratings in phosphate glass single mode fiber. Appl. Phys. Lett. **89**(10), 101127 (2006)
25. Sozzi, M., Rahman, A., Pissadakis, S.: Non-monotonous refractive index changes recorded in a phosphate glass optical fibre using 248 nm, 500 fs laser radiation. Opt. Mater. Exp. **1**(1), 121–127 (2011)
26. Pugliese, D., Konstantaki, M., Konidakis, I., Ceci-Ginistrelli, E., Boetti, N.G., Milanese, D., Pissadakis, S.: Bioresorbable optical fiber Bragg gratings. Opt. Lett. **43**(4), 671–674 (2018)
27. Theodosiou, A., Pugliese, D., Ceci-Ginistrelli, E., Boetti, N.G., Janner, D., Milanese, D., Kalli, K.: Femtosecond laser written plane-by-plane Bragg grating sensors in bioresorbable phosphate optical fibres. J. Lightwave Technol. **37**(10), 2363–2369 (2019)
28. Albert, J., Shao, L.-Y., Caucheteur, C.: Tilted fiber Bragg grating sensors. Laser Photon. Rev. **7**(1), 83–108 (2013)
29. Konstantaki, M., Pissadakis, S., Pugliese, D., Ceci-Ginistrelli, E., Boetti, N.G., Milanese, D.: Bragg grating UV inscription in a bioresorbable phosphate glass optical fiber. In: 18th International Conference on Transparent Optical Networks (ICTON), Institute of Electrical and Electronics Engineers, Trento, Italy, pp. 1–4 (2016)
30. Poulsen, T., Berendt, O., Bjarklev, A., Gruner-Nielsen, L., Soccolich, C.E.: Bragg grating induced cladding mode coupling caused by ultra-violet light absorption. Electron. Lett. **34**(10), 1007–1009 (1998)
31. Chan, C.-F., Chen, C., Jafari, A., Laronche, A., Thomson, D.J., Albert, J.: Optical fiber refractometer using narrowband cladding-mode resonance shifts. Appl. Opt. **46**(7), 1142–1149 (2007)
32. Theodosiou, A., Lacraz, A., Polis, M., Kalli, K., Tsangari, M., Stassis, A., Komodromos, M.: Modified fs-laser inscribed FBG array for rapid mode shape capture of free-free vibrating beams. IEEE Photon. Technol. Lett. **28**(14), 1509–1512 (2016)
33. Fokine, M., Theodosiou, A., Song, S., Hawkins, T., Ballato, J., Kalli, K., Gibson, U.J.: Laser structuring, stress modification and Bragg grating inscription in silicon-core glass fibers. Optic. Mater. Exp. **7**(5), 1589–1597 (2017)
34. Theodosiou, A., Hu, X., Caucheteur, C., Kalli, K.: Bragg gratings and Fabry-Perot cavities in low-loss multimode CYTOP polymer fiber. IEEE Photon. Technol. Lett. **30**(9), 857–860 (2018)
35. Barmenkov, Y.O., Zalvidea, D., Torres-Peiró, S., Cruz, J.L., Andrés, M.V.: Effective length of short Fabry-Perot cavity formed by uniform fiber Bragg gratings. Opt. Exp. **14**(14), 6394–6399 (2006)

36. Gibson, A.P., Hebden, J.C., Arridge, S.R.: Recent advances in diffuse optical imaging. Phys. Med. Biol. **50**(4), R1–R43 (2015)
37. Durduran, T., Choe, R., Baker, W.B., Yodh, A.G.: Diffuse optics for tissue monitoring and tomography. Rep. Prog. Phys. **73**(7), 076701 (2010)
38. Bashkatov, A.N., Genina, E.A., Tuchin, V.V.: Optical properties of skin, subcutaneous, and muscle tissues: a review. J. Innov. Optic. Health Sci. **4**(1), 9–38 (2011)
39. Torricelli, A., Contini, D., Dalla Mora, A., Pifferi, A., Re, R., Zucchelli, L., Cani, M., Farina, A., Spinelli, L.: Neurophotonics: non-invasive optical techniques for monitoring brain functions. Funct. Neurol. **29**(4), 223–230 (2014)
40. Hoshi, Y., Yamada, Y.: Overview of diffuse optical tomography and its clinical applications. J. Biomed. Optic. **21**(9), 091312 (2016)
41. Lange, F., Tachtsidis, I.: Clinical brain monitoring with time domain NIRS: a review and future perspectives. Appl. Sci. **9**(8), 1612 (2019)
42. Dalla Mora, A., Di Sieno, L., Re, R., Pifferi, A., Contini, D.: Time-gated single-photon detection in time-domain diffuse optics: a review. Appl. Sci. **10**(3), 1101 (2020)
43. Di Sieno, L., Boetti, N.G., Dalla Mora, A., Pugliese, D., Farina, A., Konugolu Venkata Sekar, S., Ceci-Ginistrelli, E., Janner, D., Pifferi, A., Milanese, D.: Time-domain diffuse optics using bioresorbable fibers: a proof-of-principle study. In: Amelink, A., Vitkin, I.A. (eds.) European Conference on Biomedical Optics, Novel Biophotonics Techniques and Applications IV, Optical Society of America-Society of Photo-Optical Instrumentation Engineers, vol. 10413, pp. 1–5, Washington-Bellingham, USA (2017)
44. Wabnitz, H., Taubert, D.R., Mazurenka, M., Steinkellner, O., Jelzow, A., Macdonald, R., Milej, D., Sawosz, P., Kacprzak, M., Liebert, A., Cooper, R., Hebden, J., Pifferi, A., Farina, A., Bargigia, I., Contini, D., Caffini, M., Zucchelli, L., Spinelli, L., Cubeddu, R., Torricelli, A.: Performance assessment of time-domain optical brain imagers, part 1: basic instrumental performance protocol. J. Biomed. Opt. **19**(8), 086010 (2014)
45. Pifferi, A., Torricelli, A., Bassi, A., Taroni, P., Cubeddu, R., Wabnitz, H., Grosenick, D., Möller, M., Macdonald, R., Swartling, J., Svensson, T., Andersson-Engels, S., van Veen, R.L.P., Sterenborg, H.J.C.M., Tualle, J.-M., Nghiem, H.L., Avrillier, S., Whelan, M., Stamm, H.: Performance assessment of photon migration instruments: the MEDPHOT protocol. Appl. Opt. **44**(11), 2104–2114 (2005)
46. Di Sieno, L., Boetti, N.G., Dalla Mora, A., Pugliese, D., Farina, A., Konugolu Venkata Sekar, S., Ceci-Ginistrelli, E., Janner, D., Pifferi, A., Milanese, D.: Towards the use of bioresorbable fibers in time-domain diffuse optics. J. Biophoton. **11**(1), e201600275 (2018)
47. Spinelli, L., Botwicz, M., Zolek, N., Kacprzak, M., Milej, D., Sawosz, P., Liebert, A., Weigel, U., Durduran, T., Foschum, F., Kienle, A., Baribeau, F., Leclair, S., Bouchard, J.-P., Noiseux, I., Gallant, P., Mermut, O., Farina, A., Pifferi, A., Torricelli, A., Cubeddu, R., Ho, H.-C., Mazurenka, M., Wabnitz, H., Klauenberg, K., Bodnar, O., Elster, C., Bénazech-Lavoué, M., Bérubé-Lauzière, Y., Lesage, F., Khoptyar, D., Subash, A.A., Andersson-Engels, S., Di Ninni, P., Martelli, F., Zaccanti, G.: Determination of reference values for optical properties of liquid phantoms based on Intralipid and India ink. Biomed. Opt. Exp. **5**(7), 2037–2053 (2014)
48. Contini, D., Martelli, F., Zaccanti, G.: Photon migration through a turbid slab described by a model based on diffusion approximation. I. Theory. Appl. Opt. **36**(19), 4587–4599 (1997)
49. Bisland, S.K., Lilge, L., Lin, A., Rusnov, R., Wilson, B.C.: Metronomic photodynamic therapy as a new paradigm for photodynamic therapy: rationale and preclinical evaluation of technical feasibility for treating malignant brain tumors. Photochem. Photobiol. **80**(1), 22–30 (2004)

Glass and Glass–Ceramic Photonic Materials for Sensors

Matteo Giardino, Diego Pugliese, and Davide Janner

Abstract Recent developments in sensors are pushing for optimized materials that can increase their usage, bolster their sensitivity and enable new and more efficient transduction mechanisms. We hereby review some of the most relevant applications of glasses and glass–ceramics for photonic sensors considering the recent trends and innovative approaches. This review covers from bulk glasses to thin films and from fiber optics to nanocrystal-based and their applications in sensing.

Keywords Sensors · Nanoparticles · Photonics · Thin films · Waveguides

1 Introduction

With the development of technology and its pervasive implementation in all aspects of modern life, almost every research field is experiencing a "smart" revolution where (artificial) intelligence and feedback system are required for an ever-increasing number of products and services. The extension of this revolution covers public services and natural resources, e.g., smart cities, capillary monitoring devices, and low-cost implementation of control points. That drives the development of sensors able to measure many different parameters that later can be used for the algorithms/control systems as feedback. In this scenario, sensors are developing steadily, and particularly, photonics-based sensors are attracting interest because of their high sensitivity, immunity to electromagnetic radiations, and small form factor. A key role in photonic sensors development is assumed by selecting the more suitable optical materials to be employed. Indeed, if traditional bulk materials are well known and widely used, new materials and nano-/micro-structured ones can boost the performance of photonic sensors contributing to a broader adoption and more capillary network sensors implementation. Among the different photonic materials used, glasses and glass–ceramics are of great interest because of their optical properties (particularly for oxides). Although a vast knowledge base is available for these different materials [1–5], there

M. Giardino · D. Pugliese · D. Janner (✉)
Dipartimento di Scienza Applicata e Tecnologia (DISAT) and RU INSTM, Politecnico di Torino, C.so Duca degli Abruzzi, 24, 10129 Torino, Italy
e-mail: davide.janner@polito.it

F. Baino et al. (eds.), *Ceramics, Glass and Glass-Ceramics*, PoliTO Springer Series,
https://doi.org/10.1007/978-3-030-85776-9_8

is still room for improvement. This work will give an overview of glass/glass–ceramic materials of interest for sensing and their application. The following Sections will cover three main topics: bulk materials and thin films, fiber optics and fiber Bragg grating sensors, application of nanomaterials as sensing platforms.

2 Bulk Materials and Thin Film-Based Sensors

Photonic bulk materials can be used for sensing and detectors either as macroscopic pieces or in the form of thin films where waveguiding mechanisms can also be exploited. In the following, two main applications of these types of materials are considered: scintillation in bulk materials and thin film-based sensors.

2.1 *Scintillation*

Scintillators are defined as a class of materials that emit light when excited by ionizing radiations. Their applications include the detection of neutrons, X-rays, gamma-rays, alpha-particles, and beta-particles.

The physical process of scintillation can be essentially divided into three steps, as sketched in Fig. 1 [6]:

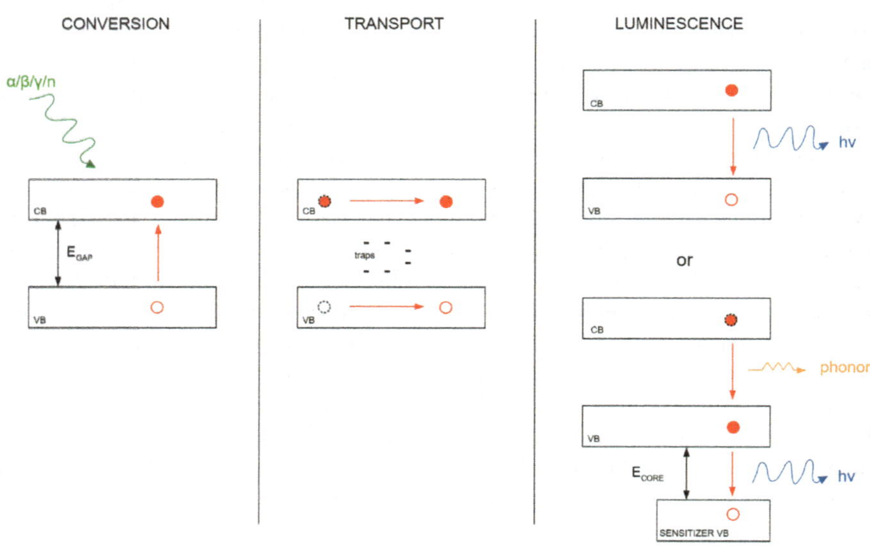

Fig. 1 Schematization of the three steps involved in the scintillation process

- Conversion: the energy of the radiation impacting the material can excite electrons and generate a significant number of electron–hole pairs. Since the ionization energy of an atom is usually in the range 5–20 eV, whereas the photon of the ionizing radiations can have an energy ranging from 10 keV (X-rays) to more than 1 MeV (gamma rays), after ionization, there is an energy transfer to the electron in the form of kinetic energy.
- Transport: the electron–hole pairs can move through the material, and, during this step, they can be trapped in the defects of the materials. This trapping leads to a significant loss in their energy and a delay in the emission process.
- Luminescence: the electrons and holes can also be trapped in a luminescence center. These luminescence centers can originate from some intrinsic defects of the material or an included element added for that aim. Light emission originates from the radiative recombination of the electron–hole pair. In some cases, the emission process may also involve the energy transfer from a sensitizer.

The excited electron may also decay in a nonradiative mode by transferring its energy to the bulk matrix through vibrations (phonons). If the energy surplus of the excited electron is much greater than the maximum phonon energy of the material, the system relaxation requires a multiphonon decay, and the scintillating efficiency of the glass is reduced [7].

Rare-earth halides, such as CeBr, $LaCl_3Ce$, and LuI_3Ce, are among the compounds with higher scintillation yield but are hygroscopic, exhibit fair mechanical properties, and require expensive production processes [8].

For these reasons, the scintillating properties of halides nano-/micro-crystals embedded in glass–ceramic matrices have been extensively explored in the last two decades and give a promising perspective to the field.

The material obtained by the inclusion of halides nano/micro-crystals in a glass matrix shows all the properties of a glass–ceramic system such as mechanical strength, chemical resistance, and thermal stability with low phonon energy, which result in a low probability of nonradiative recombination while retaining a good efficiency in the conversion. Moreover, impurity elements in the glass matrix are usually a fundamental requirement to obtain a scintillating behavior, and some dopants, e.g., Ce^{3+}, Pr^{3+}, Gd^{3+}, and Tb^{3+}, may also be added to increase the scintillation efficiency of the glass system.

Scintillating glasses are usually prepared by melt-quenching or by sol–gel processes.

Struebing et al. [9] reported on a neutron scintillator glass–ceramic system by incorporating CaF_2:Eu nanocrystals in a glass matrix containing some ^6Li, whose role is to increase the neutron absorption cross-section of the glass–ceramic scintillator (GCS). This system demonstrated a much higher scintillation efficiency than the glass matrix, and this enhancement is likely due to the more efficient energy transfer from the CaF_2 nanocrystals to the Eu^{3+} ions that act as emitting centers.

Zheng et al. [10] produced Tb^{3+}-doped $BaGdF_5$ glass–ceramics for X-ray detection, showing a very high scintillation efficiency. The Tb^{3+}-doped glass–ceramics with composition $17SiO_2$–$11H_3BO_3$–$10Al_2O_3$–$6Cs_2CO_3$–$32BaF_2$–$17GdF_3$–$7TbF_3$ (mol%) was produced by melt-quenching technique, and the study compared the

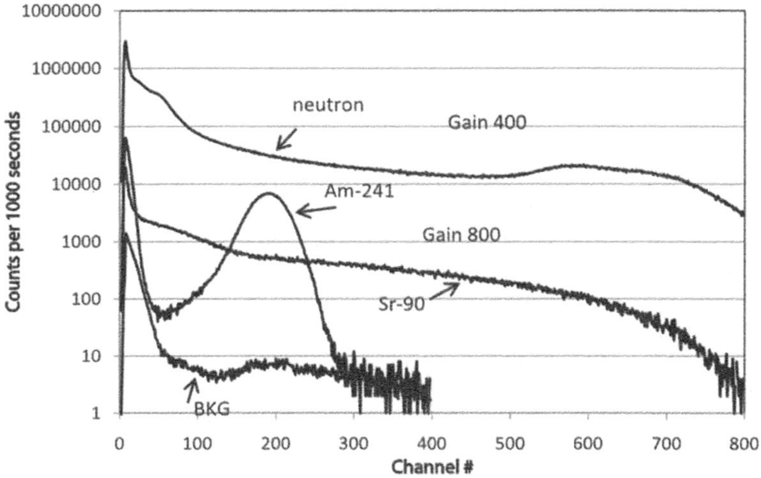

Fig. 2 Spectra of the glass-matrix detector for ^{241}Am alpha particles, $^{90}Sr/^{90}Y$ beta particles, and radiation background at gain 800 and for a $^{239}Pu/Be$ neutron source at gain 400. Reproduced from [11]. Copyright 2013, with permission from Elsevier

performance of this glass system before and after the thermal treatment. They reported emissions of the glass–ceramics at 274 and 377 nm excitation, respectively, 1.4 and 3.1 times stronger than that of the glass sample.

Khang et al. [11] added $GdBr_3$ and $CeBr_3$ to a crystalline phase in a silica-alumina glass matrix to obtain a detector for application in nuclear spectroscopy and gamma-rays detection. They observed an emission peak at 410 nm, which is a higher wavelength than the emission peak of traditional Ce^{3+}-doped silica glasses, and this wavelength shift is likely to be caused by the presence of Gd-rich nanoclusters. As reported in Fig. 2, this device also responds to α particles from ^{241}Am, β particles from $^{90}Sr/^{90}Y$, and neutrons from $^{239}Pu/Be$ source.

2.2 *Waveguides in Glass*

When a light ray strikes at the interface between two media, with an incidence angle θ_i, refraction occurs, and the ray continues its propagation with a refraction angle θ_r given by the Snell's law:

$$n_1 \sin\theta_i = n_2 \sin\theta_r, \qquad (1)$$

where n_1 and n_2 are the refractive indices, respectively, of the first and second medium.

When $\sin\theta_r$ is equal to 1, i.e., the refraction angle assumes the critical value $\frac{\pi}{2}$, there is no longer a refraction wave, and the ray is completely reflected. This situation, represented in Fig. 3 and called ***total internal reflection (TIR)***, always occurs when the incidence angle is equal or greater than a critical value θ_i^{cr} given as:

$$\theta_i^{cr} = arcsin\left(\frac{n_2}{n_1}\right) \tag{2}$$

From Eq. (2), it is clear that the TIR phenomenon only takes place when a light ray impinges the boundary of a medium with a lower refractive index because, according to the codomain of sin function, the ratio $\frac{n_2}{n_1}$ must always be lower than or equal to 1 and that means that n_1 must be greater than or equal to n_2. The case $n_1 = n_2$ is also excluded since it would entail that θ_i is equal to θ_r, and no refraction would be observed.

When a light ray is "forced" to propagate in a medium surrounded by another one with a lower refractive index, a so-called waveguide is obtained by exploiting the total internal reflection phenomenon.

Many configurations may be used to obtain these waveguides, although this paragraph mainly focuses on ***planar waveguides***, whereas Sect. 3 deals more in detail with ***optical fibers***.

The propagation of the light into the waveguide can be affected by the external environment, and this mechanism can be exploited for sensing purposes. Indeed, the sensors so obtained can be classified under four main categories according to which feature of the light is affected by the external environment. These four categories of sensing mechanisms are: intensity-modulation, phase-modulation, polarization-modulation, and wavelength-modulation.

To obtain a waveguide, one can start from a planar sample and must produce a variation of the refractive index on its surface.

The refractive index variation is obtainable either by creating a change in the chemical composition of the material or by introducing some stresses into the material [12].

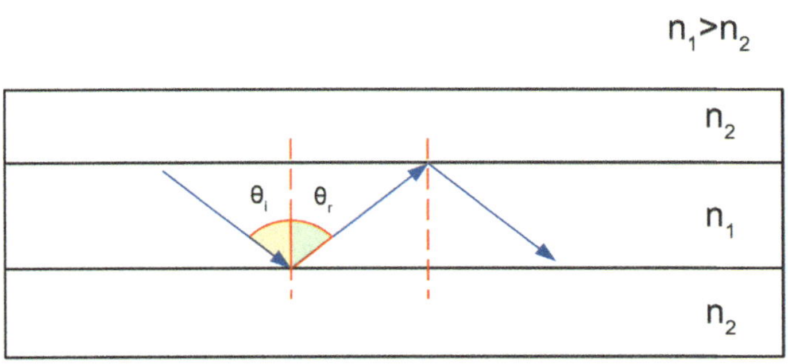

Fig. 3 Illustration of total internal reflection phenomenon in a planar waveguide

As far as glasses are concerned, one of the possible ***glass waveguide (GWG)*** fabrication methods is the ***ion-exchange technique***. The latter consists of immersing the glass in a molten solution of a salt, usually a nitrate.

Walker et al. [13] performed an Ag^+ exchange by immersing a masked soda-lime glass into a molten $AgNO_3$ bath, obtaining a refractive index variation on the surface of 0.09. Moreover, K^+ turned out to be an excellent alternative to Ag^+ for waveguides fabrication by ion-exchange [14]. It is worth underlining that by this technique pretty thick modified layers can be obtained: Giallorenzi et al. [15] treated a soda-lime glass in a KNO_3 bath at 365 °C for 24 h and obtained a K^+ substituted layer with a thickness of around 8 μm, as shown in Fig. 4.

On the other side, GWGs can also be produced by depositing a thin film of a suitable material having a different refractive index on a glass substrate. The thin-film deposition can be realized in a plethora of different ways, including radio-frequency (RF) sputtering, pulsed-laser deposition (PLD), chemical vapor deposition (CVD), or even sol–gel processes. Despite being a much slower process and showing limits on the maximum achievable thickness, sputtering and CVD allow obtaining more uniform and higher quality thin films and the deposition of complex composition layers.

Chiasera et al. [16] proposed SiO_2-P_2O_5-HfO_2-Al_2O_3-Na_2O glasses activated by Er^{3+} ions obtained by multitarget RF sputtering. In particular, they produced a GWG

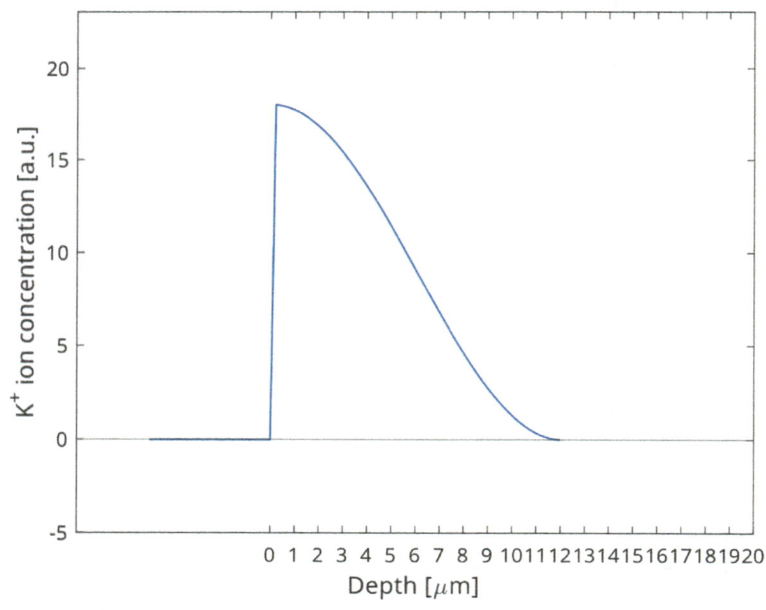

Fig. 4 Ion concentration profile, as a function of the depth, in a soda-lime glass treated for 24 h in a molten KNO_3 bath at 365 °C. Adapted from [15]

by sputtering a SiO_2 target onto a disk of a $69P_2O_5$-$15SiO_2$-$10Al_2O_3$-$5Na_2O$-$1Er_2O$ glass, and one of HfO_2 was positioned.

On the other hand, the sol–gel process is a quick and inexpensive way of depositing thin optical films on flat substrates by spin-coating.

Almeida et al. [17] realized, by a two-step sol–gel method, a co-doped Er^{3+}/Yb^{3+} sulfide glass. This glass was obtained by first depositing oxide gel films on a silica glass or borosilicate glass and then sulfidizing the oxide gel into a sulfide by exposure to H_2S gas in temperature.

In waveguide applications, also glass–ceramic materials have recently attracted noticeable attention. Indeed, ***glass–ceramic waveguides (GCWGs)*** can be obtained by appropriate thermal treatment to produce a phase separation. They typically show a sandwich structure where a glass layer is surrounded by another layer consisting of a crystalline phase dispersed into an amorphous matrix. Moreover, also rare-earth dopants are usually embedded in the form of nanocrystals, and an example of this approach is the GCWG by Jestin et al. [18], who proposed a SiO_2-HfO_2 activated by Er^{3+} GWG obtained via sol–gel route and subsequent dip-coating deposition onto a SiO_2 substrate and annealing treatment at 900 °C. A high-resolution transmission electron microscopy (HRTEM) image of the synthesized silica-based glass–ceramic nanocomposite material is reported in Fig. 5.

In this case, the refractive index of the SiO_2-HfO_2 film can be finely tuned by controlling the HfO_2 concentration and the annealing process parameters, which influence the composition and the phase separation of the glass–ceramic system.

Er^{3+} doping is often considered in GWGs and GCWGs since it displays a very strong emission at around 1500 nm [19], which is the wavelength where the silica glass shows the lowest loss (about 0.2 dB/km) and is, therefore, the most common for telecommunications purposes.

Optical waveguides are suitable for various sensing purposes in analytical chemistry, ranging from the detection of gases [20, 21] to the pH measurement of water solution [22, 23].

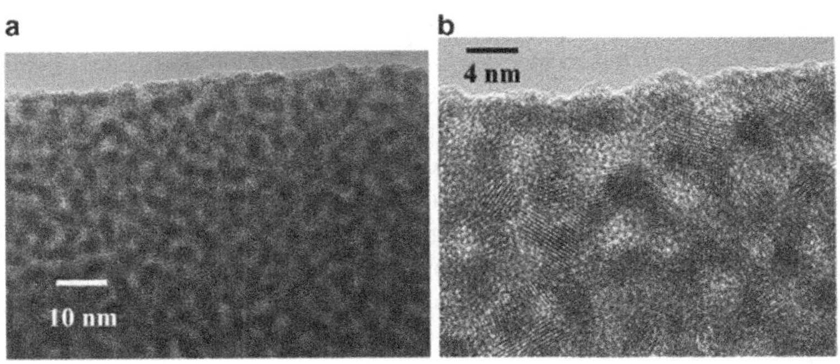

Fig. 5 HRTEM image of SiO_2-HfO_2 thin film at different magnifications. Reproduced from [18]. Copyright 2007, with permission from Elsevier

Han et al. [21] realized a spiral chalcogenide glass waveguide for methane detection to operate in the middle infrared region (MIR) and observed its transmittance.

T follows the Beer-Lambert's law:

$$T = \exp(-\alpha LC), \tag{3}$$

where α is the absorption coefficient, L is the waveguide length, and C is the methane concentration.

The pH sensing, in particular, is usually accomplished by realizing a porous layer on the outer surface of the waveguide where a pH-sensitive dye can be incorporated.

Moreover, waveguides have also been successfully employed in biology to monitor the bacteria growth process [24]. In this case, the outer surface of the waveguide is often functionalized with an antibody that can bind certain bacteria so that a change in the refractive index is observed close to the surface and can be probed employing an evanescent wave.

2.3 Thin Films for Sensing

A wide range of oxide thin film (OTF) sensors has been prepared on glass substrates by deposition with different techniques.

The first OTF sensor for gas detection was proposed in 1962 by Seiyama et al. [25] and was based on the steep electrical conductivity variations that occur when a gas adsorbs or desorbs from a porous semiconductor layer at high temperature. This thin film was deposited onto a borosilicate glass by evaporating a thin layer of metallic Zn which was subsequently oxidated in the air at 450 °C for 10 h.

Although Seiyama's work focused only on a ZnO layer [25], also many other semiconductor metal oxides, such as TiO_2, WO_2, In_2O_3, and SnO_2, have been employed for the preparation of gas thin film-based sensors due to the considerable variation in their conductivity when exposed to chemical vapor.

The sensor has been tested by exposing it to different CO, benzene, toluene, and propane concentrations while its increase in electrical resistivity has been recorded over time. This increase is due to the absorption of the above-mentioned organic molecules which behave as electron donors to the ZnO (n-type doping).

The thin film-based sensors are mainly composed of a porous metal oxide semiconductor thin layer deposited on a glass substrate through different techniques.

Among the techniques commonly employed, it is worth mentioning physical vapor deposition (PVD), laser ablation, molecular beam epitaxy, sputtering, CVD, sol–gel techniques, ultrasonic spray pyrolysis spin, and dip-coating.

Touidjen et al. [26] deposited a SnO_2 thin film on a glass substrate to realize an organic vapor sensor that successfully detected ethanol vapor in the concentration of a few dozen ppm. The deposition was performed using spray-pyrolysis techniques using an isopropanol-aqueous $SnCl_2$ solution as a precursor.

Salman et al. [27] deposited a TiO_2:Mn thin film utilizing RF sputtering with a pure TiO_2 target on an Indium-Tin Oxide (ITO) glass and subsequent annealing thermal treatment at 600 °C.

Salah et al. [28] exploited the spin-coating of a ZnO sol which was prepared using zinc acetate, 2-methoxyethanol, and ethanolamine on a glass substrate to work as a humidity sensor in the range 15–95 relative humidity (RH%). The sensor is based on the electrical resistance decrease when water molecules chemisorb or physisorb (especially at high RH% values) on the ZnO surface.

Glass thin films are also employed for sensing applications. A remarkable example is represented by chalcogenide glasses-based thin-films, which have been extensively investigated over the years as materials for producing membranes for potentiometric techniques of heavy metal ions (e.g., Fe^{2+}, Cu^{2+}, Pb^{2+}, Ag^{+}) in water or even as pH sensor [29]. The probe obtained was tested as an electrode in an electrochemical cell with a double liquid junction, and cell voltage was plotted against the Pb^{2+} ion concentration as reported in Fig. 6.

In the field of thin films for sensing, even metallic glasses (MGs) find a plethora of different applications.

Metallic glasses are usually produced by extremely rapid cooling of a molten alloy so that the metallic system solidifies without crystallizing but rather forming an amorphous structure similar to traditional glasses.

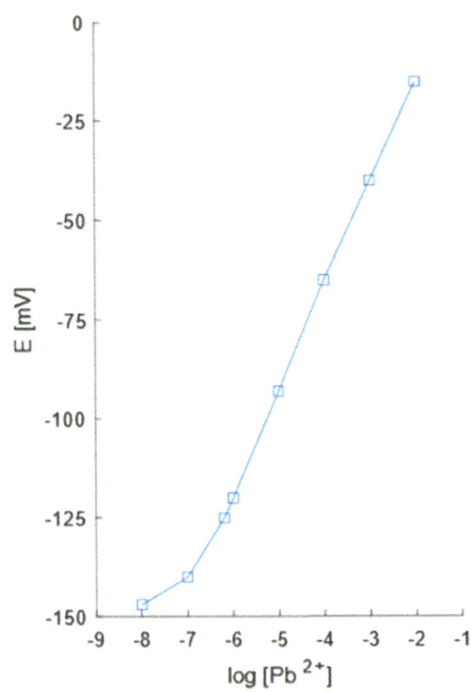

Fig. 6 Calibration curve of a chalcogenide glass thin film sensor for quantification of Pb^{2+} ions in water. Adapted from [29]

This category of materials demonstrates excellent flexibility and elasticity, high chemical resistance due to the lack of grain boundaries, high electrical conductivity, and provides good adhesion to the other layers. Moreover, since diffusion via grain boundary mechanism is not allowed, it is possible to realize a diffusion barrier to prevent interdiffusion at the interface. Also, electron scattering and current leakage are minimized by the MG grain-free structure, limiting the electron scattering [30].

MGTF may show a piezoresistance effect [31] and can thus be used as a strain sensor. Piezoresistance is a change of the electrical resistance of a material when a mechanical deformation is applied, typically due to the change of interatomic distance, which affects the bandgap.

Since the electrical resistance vs. strain dependence in MGTF is usually linear, this behavior is described by introducing a gauge factor (GF) which may be defined as the ratio between the relative variation of resistance and the strain:

$$GF = \frac{R - R_0}{\varepsilon R_0} = \frac{\delta R}{\varepsilon} \tag{4}$$

Xian et al. [31] proposed the metallic glass system $Zr_{55}Cu_{30}Ni_5Al_{10}$ to be deposited as a thin film by ion beam sputtering on a polycarbonate substrate to create an e-skin sensor and observed a GF of 2.86. In Fig. 7, the normalized electrical resistance variation $\Delta R/R_0$ versus the strain for this MG system is reported.

A different approach that uses pyro-resistivity of a graphene layer deposited on top of $LiNbO_3$ has been reported in [32]. In that device, light is shone on the ceramic material that changes its temperature due to light absorption and charge formation on the surface. Those charges dope the graphene layer that changes its resistivity and can be interrogated electrically. In this way, a broadband detector is realized and shows good performance in terms of detectivity, even in the MIR spectral region.

3 Fiber Optics-Based Sensors

Among the approaches that leverage glasses for sensing, fiber optics-based sensors are among the most promising in terms of flexibility, ease of use, and integration. The primary sensing mechanisms in this type of sensor are based on the change in the light propagation in the fiber when the surrounding environment changes. Thus, they share many features in common with the waveguide sensor presented above. However, for their peculiarities and widespread use, the following Section is entirely dedicated to this type of sensors covering the most common configurations for sensing and the glass/glass–ceramic materials used in these applications.

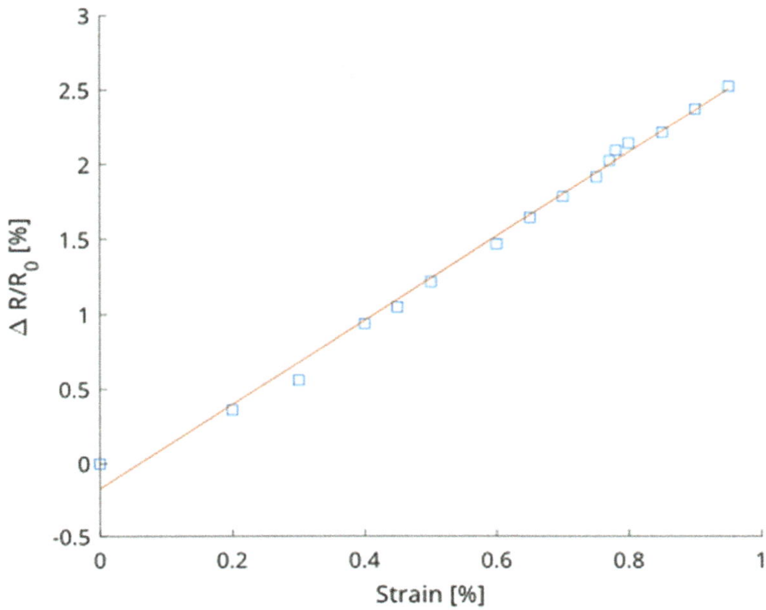

Fig. 7 Dependence of relative resistivity versus applied strain of a $Zr_{55}Cu_{30}Ni_5Al_{10}$ thin film metallic glass. Adapted from [31]

3.1 Gratings-Based Sensors

Among the different optical fiber sensors, Fiber Bragg Gratings (FBGs) are the most widespread. Indeed, they show many interesting features like high sensitivity, and they work at a particular wavelength allowing for multiplexing a high number of sensors along with the same fiber [33]. Moreover, their fabrication is relatively cheap, and their sensitivity can be significantly increased by adding a layer of a polymeric or glass/ceramic material around the fiber [3, 33].

An FBG is a distributed Bragg reflector in a fiber reflecting specific wavelengths of light, typically around 1550 nm. This mechanism is obtained by introducing a periodic variation of the refractive index inside the fiber optics with a periodicity proportional to the Bragg wavelength at which the reflection occurs, as indicated schematically in Fig. 8.

The refractive index variation is typically of the order of 10^{-4}–10^{-2}, and the area affected by such modification ranges from 3 to 10 mm [33]. The index variation is obtained in the fiber core by the formation of color centers or physical damage according to the technique used but always relying on laser irradiation (either ultra-violet or femtosecond). The Bragg wavelength λ_B is related to the refractive index n_f of the fiber and the periodicity Λ of the modification as follows:

$$\lambda_B = 2n_f\Lambda \tag{5}$$

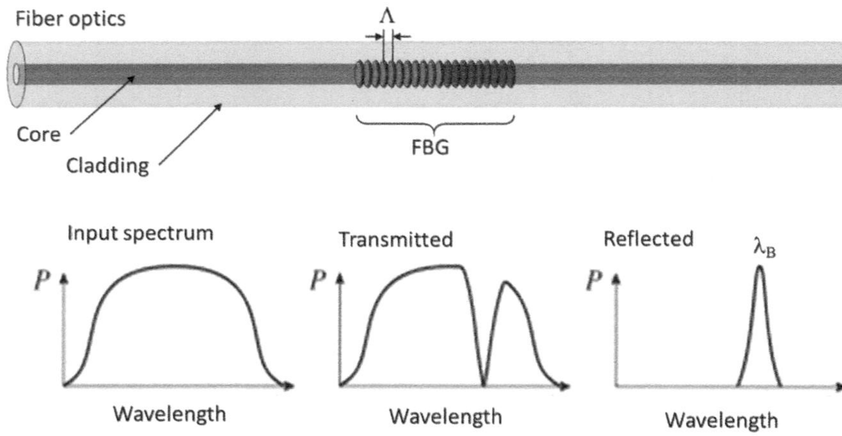

Fig. 8 Periodic index variation of period Λ in the core of a fiber optics and its spectral response in transmission and reflection showing the relationship between Λ and the peak wavelength λ_B

The primary sensing mechanisms for FBG can rely on three refractive index modification phenomena: temperature, strain, and refractive index change around the fiber. While the first two are present in all types of FBG, the last one is only affecting special types of FBG like etched cladding or tilted ones [4, 33]. An example of sensing application with FBG can be found in harsh environments like nuclear sensing [34]. In that case, FBGs have been used to monitor the temperature in a radioactive environment, and the fiber was encapsulated with a special ceramic coating ORMOCER® demonstrating the ability to measure up to 450 °C keeping a tightly sealed environment. The advantage of fiber optics sensors is made evident in harsh environments, given their immunity to electromagnetic interference, the electric power-free operation of the sensor, and the possibility of interrogating it remotely with light propagating in the fiber.

To exploit the sensing mechanism based on refractometry in a fiber, tilted FBGs can be used [4]. As illustrated in Fig. 9a, a tilted FBG is a periodic variation of the index of refraction similar to FBG but tilted at an angle with respect to the propagation direction. The associated transmission spectrum (Fig. 9b) shows many dips related to modes of light that travel in the cladding of the optical fiber. Such cladding modes are sensitive to the environment refractive index and present a shift in wavelength upon its variation. This mechanism can be exploited for refractometric sensing by monitoring the wavelength and transmitted power of one of the dips, as exemplified in Fig. 9c.

Coating the exterior of the fiber with glass or glass–ceramic materials can significantly enhance the sensitivity of the sensor and make it more specifically sensitive to, e.g., humidity, pH, giving the possibility to target many different parameters to be measured. For instance, depositing a 200 nm coating of ITO greatly increases the sensing performance in terms of sensitivity [36], making this sensor a good candidate for biochemical sensing after surface functionalization.

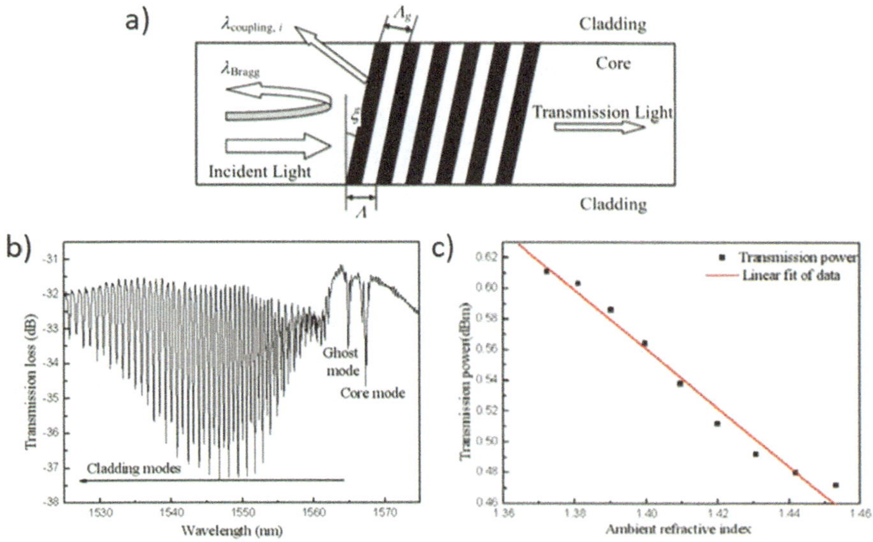

Fig. 9 **a** Light scattering and transmission in a fiber with a tilted FBG. **b** Optical spectral response of a tilted FBG. **c** Dip power variation versus external refractive index showing the sensing principle. Adapted from [35] with permission from the Institute of Electrical and Electronics Engineers

3.2 Lossy Mode Resonance

The integration of fiber optics with micro/nanotechnology has allowed the development of very sensitive photonic devices and fostered novel (bio-)sensing schemes [37]. Coating fibers with nanoparticles or nanofilms allows tailoring the optical features with an excellent spectral resolution, precision, and accuracy, enabling the realization of high-performance optical platforms [38, 39]. Indeed, the interaction of light traveling in a fiber optics with its surrounding environment can generate surface waves of different types, such as evanescent, Bloch, surface plasmon polariton, lossy/leaky mode, or derived from guided modes, etc. Many schemes can be used to excite those waves [40] and are sensitive to every change occurring in the medium surrounding the fiber [41], making them suitable for sensing applications. Among surface wave sensors, the deposition of metal-oxide nanofilms on fiber optics has allowed effective exploitation of lossy mode resonance (LMR) for sensing [42], attracting strong interest from many fields of applications. The basic principle of LMR sensors is illustrated in Fig. 10.

In a multimode fiber optics (MMF), the cladding is removed partially or completely, and the light propagating in the core is allowed to interact with the surrounding environment. As a result, refractive index variations are recorded as a change in the transmitted power at specific wavelengths. In addition, by coating the exposed core with glass/glass–ceramic material, the sensitivity can be significantly enhanced.

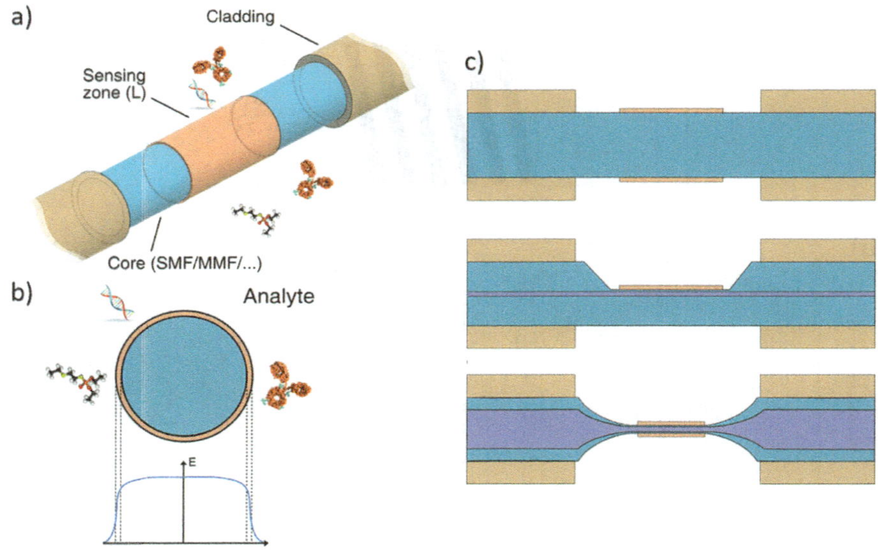

Fig. 10 Working principles of LMR. **a** Sensing zone constituted by a thin film coating the fiber and interacting with **b** the guided modes and the external environment (analyte). **c** Possible configurations of LMR sensors showing the fiber optics and the sensing film. Adapted from [35] with permission from the Institute of Electrical and Electronics Engineers

The complex (relative) permittivity of each material can be expressed as shown in Eq. (6):

$$\varepsilon = \varepsilon' + j\varepsilon'' = \mathrm{Re}\{\varepsilon\} + j\mathrm{Im}\{\varepsilon\} = (n + jk)^2, \quad (6)$$

where ε' describes the lossless permittivity, whereas ε'' the lossy permittivity. The previous equation can also be rewritten in terms of n representing the material refractive index and k describing the extinction coefficient that refers to how strongly a substance absorbs light at a given wavelength. The selection of materials for sensitivity enhancement can be made using this parameter, and, in particular, dielectric materials with high refractive index can be used in LMR. Indeed, many have been proposed [43], such as aluminum-doped zinc oxide (AZO), zinc oxide (ZnO), hafnium dioxide (HfO_2), zirconium dioxide (ZrO_2), tantalum oxide (Ta_xO_y), titanium dioxide (TiO_2), indium oxide (In_2O_3), indium gallium zinc oxide (IGZO), ITO, tin oxide (SnO_{2-x}), amorphous silicon, etc. Some complex high refractive indices (n, k) are shown in Fig. 11. At a first glance, we see that no single material allows attaining the best performance, but the choice is application dependent, considering factors like the average refractive index of the environment and the operating wavelength of the device.

Moreover, n tends to decrease with the wavelength implying less LMR sensitivity, but conversely, the refractive index sensitivity increases at the same time. Therefore, a careful study and modeling of the device are compelling to design reliable sensors.

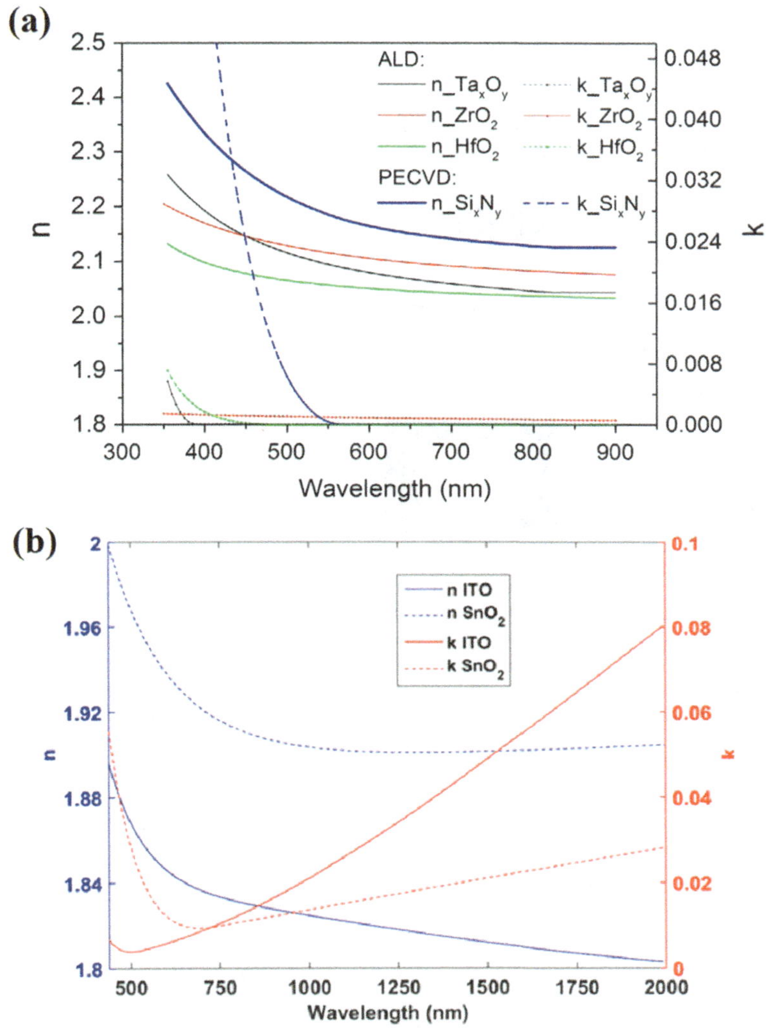

Fig. 11 **a** Dispersion characteristics in terms of *n* and *k* for HfO_2, ZrO_2, Ta_xO_y deposited by atomic layer deposition technique and for Si_xN_y films deposited by plasma-enhanced CVD. Adapted from [44]. Copyright 2018, with permission from Elsevier. **b** Dispersion characteristics in terms of *n* and *k* for ITO and SnO_2 deposited by direct current (DC) sputtering. Adapted from [43]

So far, SnO_2 seems to guarantee the greatest performance in terms of refractive index sensitivity of LMR-based devices for refractive index ranges close to a water environment which is typical for, e.g., biosensing applications.

An example of a sensor obtained by LMR si reported in Fig. 12. The performance has been evaluated by considering two metal oxides, i.e., ITO and SnO_2,

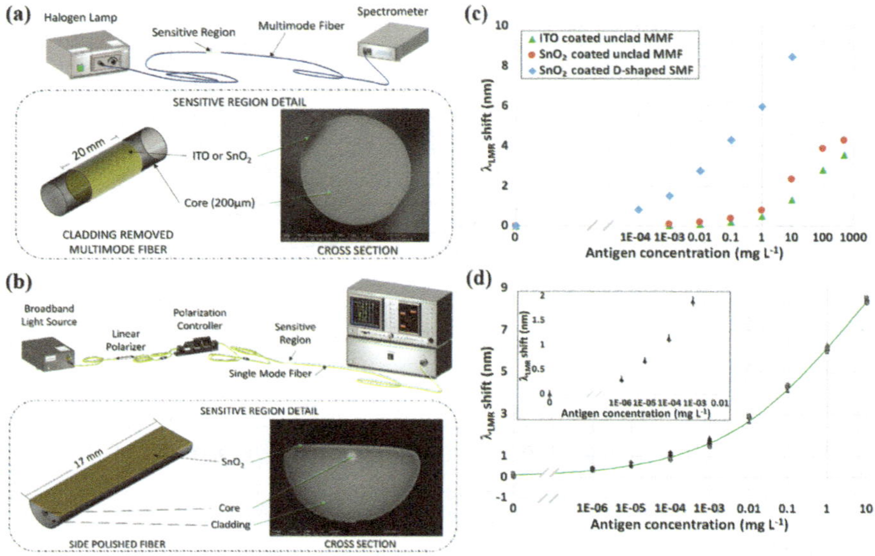

Fig. 12 **a** Scanning Electron Microscopy (SEM) image and a sketch of the sensing device consisting of CR-MMF coated with ITO or SnO_2, together with the experimental set-up. **b** SEM image and sketch of the sensing device consisting of D-shaped single-mode fiber coated with SnO_2, together with the experimental set-up. **c** Comparison of the calibration curves achieved with three LMR sensors: the ITO- (green triangles) and SnO_2-coated (red circles) CR-MMF biosensors, and the SnO_2-coated D-shaped single-mode fiber biosensor (sky blue rhombuses). **d** Calibration curve of the SnO_2-coated D-shaped biosensor repeated several times and displayed together with the sigmoidal fit of the experimental points. Adapted from [43]

and two types of fibers, i.e., cladding-removed (CR)-MMF and side-polished (D-shaped) fiber. Figure 12a, b illustrates the two types of sensors coated with thin films of ITO or SnO_2 and the experimental set-ups. Comparing the calibration curves (Fig. 12c), one can observe that, given the better optical features of SnO_2 (higher *n*, real part of permittivity; lower *k*, the imaginary part of permittivity), the best performance is achieved with D-shaped fibers. The optimized sensor features a SnO_2-coated ($\approx$ 160 nm) D-shaped fiber LMR biosensor, and its calibration curve based on Mouse IgG and goat anti-mouse-IgG is reported in Fig. 12d. Such device shows a limit of detection of 150 pg/L (1 fM), constituting, in terms of sensitivity, the best-performance biosensing platform in the literature so far.

4 Nanocrystals and Quantum Dots Embedded Glass–Ceramics

The incorporation of nanocrystals and nanoparticles in a glass matrix allows enhancing its sensing properties.

On one side, nanocrystal glass–ceramics allows the realization of sensors based on the photon upconversion (UC) phenomenon, i.e., the process in which two or more photons, with a specific wavelength, are absorbed, causing the emission of photons with a shorter wavelength.

This UC-based sensors demonstrated a combination of unique properties, including photochemical and thermal stability, which make them the perfect candidate for the preparation of non-contact temperature probes.

On the other side, the embedding of quantum dots (QDs) opens the door to realize a wide range of gas and chemical sensors due to their environment-sensitive fluorescence properties.

4.1 Nanocrystals

Trivalent rare-earth (RE) ions-doped UC materials have attracted a great deal of attention in the last decades due to their wide variety of applications in biological luminescent labels and drug delivery carriers, solar cells, all-solid-state UC lasers, and display devices [45–49].

Only more recently, luminescent UC materials, whose characteristic parameters including the absolute emission intensity, the fluorescence intensity ratio (FIR), the lifetime, and the peak wavelength could be strongly temperature-sensitive, have been profitably applied in the field of non-contact optical thermometry [50]. Notably, optical thermometry based on the FIR technique (see Fig. 13a) is recognized as one of the most valuable optical thermometry strategies due to its unique merits of contactless feature, fast response, simple data processing, and excellent resolution [51]. Moreover, the temperature-measurement FIR approach also displays some advantages over other optical-based techniques, such as fluorescence lifetime and amplified spontaneous emissions, in that much higher measurement accuracy and reliability could be achieved due to FIR being independent of external interferences, spectrum losses, and fluctuations in excitation density [52]. In light of all the above-mentioned promising features, this type of non-contact optical sensors can be used in many special environments, such as electrical transformer temperature in power stations, building fire detections, biological systems, and corrosive circumstances [53], where conventional contact-mode temperature sensors are not exploitable.

The research of high-performance FIR-based optical temperature sensors mainly focuses on optimizing the active ions and the host matrix. Many RE ions, such as Pr^{3+}, Nd^{3+}, Sm^{3+}, Eu^{3+}, Dy^{3+}, Ho^{3+}, Tm^{3+}, Yb^{3+}, and Er^{3+} have been studied for temperature sensing based on the change of FIR of two thermally-coupled energy

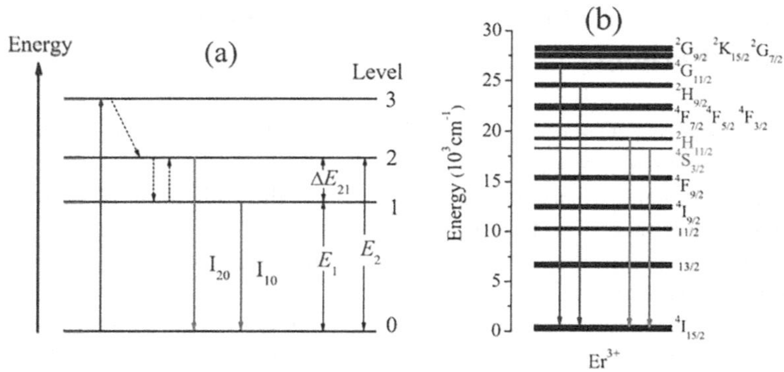

Fig. 13 **a** Schematic energy level diagram showing the energy levels and transitions of interest in the FIR-based optical thermometry technique; **b** energy levels of the Er^{3+} ions. Reproduced from [54]. Copyright 2014, with permission from Elsevier

levels (TCELs) with temperature [54]. Er^{3+} ion is the most extensively investigated because of its intense UC luminescence and appropriate bandgap (around 800 cm^{-1}) between the $^2H_{11/2}$ and $^4S_{3/2}$ energy levels (see Fig. 13b), which exactly falls in the thermally coupled energy range (200–2000 cm^{-1}) [55]. To obtain stronger UC luminescence, Yb^{3+} is usually adopted as the preferred sensitizer for Er^{3+} due to its high absorption cross-section at 980 nm and efficient $Yb^{3+} \rightarrow Er^{3+}$ energy transfer efficiency [56]. Conversely, Yb^{3+} and Pr^{3+} are of less advantage due to slight energy differences, while Eu^{3+}, Ho^{3+}, and Tm^{3+} show energy differences above 1500 cm^{-1}, leading to thermal decoupling at low temperature [54].

Besides optimizing the RE active ions, the selection of a proper host material is also crucial for the application of UC materials in FIR-based temperature sensors. Host materials should exhibit low phonon energy for RE doping and good stability for long-term use, especially under the measured temperature range. In general, oxide materials show good physical and chemical stability. However, they have the shortcoming of high phonon energies, while fluoride materials can provide a low phonon energy environment for dopants and lead to efficient UC luminescence but are easy to undergo phase transformations at high temperatures [57]. Thus, novel materials combining the advantages of both oxide and fluoride materials should be a favorable host for RE doping to achieve high-performance FIR-based optical temperature sensors with a good signal-to-noise ratio.

In this scenario, transparent oxyfluoride glass–ceramic luminescent materials, featured by the precipitation of fluoride nanocrystals among the glassy oxide matrix with suitable chemical compositions, have attracted an increasing interest owing to their facile preparation and superb optical performance [52]. It is worth highlighting that incorporating the RE activators into the precipitated fluoride crystals with low phonon energy is of paramount importance to enhance luminescence in glass–ceramic samples [58]. Moreover, compared to the powder and slice samples, where only the surface portions can be efficiently excited and emit due to the strong scattering of excitation/emission light, the optical properties of glass–ceramics may be markedly improved due to the "volume effect" of transparent material, i.e., the incident and output light can pass through the whole sample [59, 60]. Conventionally, glass–ceramics are formed through a finely controlled post-heat treatment of the glass at two different temperatures, firstly to induce nucleation and allow crystal growth of the nuclei [61]. Only more recently, femtosecond laser irradiation at a high repetition rate has been recognized as a powerful alternative technique to induce the precipitation of nanocrystals only in spatially selected regions of the glass [62].

Interesting studies on the optical temperature sensing behavior of RE ions-doped transparent oxyfluoride glass–ceramics have been reported in the literature. Transparent glass–ceramics containing 25 nm-sized $NaYF_4$:Yb^{3+}/Er^{3+} crystals were successfully fabricated by Jiang et al. [54] and showed highly enhanced green UC emissions. The FIR of the two green UC emission bands corresponding to the $^2H_{11/2}$ and $^4S_{3/2}$ multiplets to the $^4I_{15/2}$ ground multiplet transitions was characterized for optical thermometry ranging from 298 to 693 K, and relative temperature sensitivity of 1.24% K^{-1} at around 300 K was obtained.

Highly transparent bulk NaY_2F_7:Yb^{3+}/Tm^{3+} glass–ceramics with excellent temperature sensing performances, benefiting from the large energy gap of Tm^{3+} between TCELs of $^3F_{2,3}$ and $^1G_4$, were reported by Chen et al. [57]. In particular, the maximal relative and absolute sensitivity reached 1.63 and 10.01% K^{-1} at 415 and 567 K, respectively.

Transparent oxyfluoride glass–ceramics containing Yb^{3+}/Tm^{3+}/Er^{3+}:YF_3 nanocrystals were successfully synthesized by Chen et al. [52], and their potential application in FIR-based optical thermometry was systematically investigated. As a result, dual-modal temperature sensing was achieved with the assistance of FIRs of two emission bands between Er^{3+}:$^2H_{11/2} \rightarrow {}^4I_{15/2}$ and Er^{3+}:$^4S_{3/2} \rightarrow {}^4I_{15/2}$ as well as between Tm^{3+}:$^3F_{2,3} \rightarrow {}^3H_6$ and the combined Tm^{3+}:$^1G_4 \rightarrow {}^3F_4$/$Er^{3+}$:$^4F_{9/2} \rightarrow {}^4I_{15/2}$, showing maximal temperature sensitivities of 1.01 and 1.89%K^{-1} at 293 and 393 K, respectively.

Transparent oxyfluoride glass–ceramics containing hexagonal $NaGdF_4$:Yb^{3+}/Ho^{3+} nanocrystals were fabricated by Li et al. [50] through the melt-quenching method followed by self-crystallization. The FIR of two blue UC emission bands peaked at 446 and 484 nm, originating from the transitions from two TCELs of $^5F_1$/$^5G_6$ and $^5F_{2,3}$/$^3K_8$ states to the $^5I_8$ state, was recorded for optical thermometry in the temperature range of 390–773 K and a relative sensitivity of 1.43% K^{-1} at 390 K, corresponding to an effective energy gap of 1515 cm^{-1}, was achieved.

4.2 *Quantum Dots*

QDs are semiconductor particles whose dimensions are in the range of few nanometers (less than 10 nm) along all the three spatial dimensions and are therefore usually referred to as zero-dimensional (0D) materials.

The typical band structure of QDs, compared to that of bulk semiconductors and single molecules, is reported in Fig. 14.

Like bulk semiconductor materials, a QD can absorb a photon when an electron is promoted from the valence band to the conduction band, leaving a hole in the former, as shown in Fig. 14. Consequently, it can emit radiation when the electron–hole pair gives rise to recombination through a radiative pathway.

When the sizes of the particles are in the order of magnitude of the size of the electron–hole pair, known as Bohr's radius, a quantum effect named quantum confinement is observed.

A very naive, but at the same time effective, way to explain the quantum confinement is to choose the particle in an one-dimensional (1D) box model.

Imagining that the electron is confined in a potential well, represented by the physical boundary of the QDs and solving the Schrodinger's equation under these conditions, the energy spectra of the particle itself can be expressed as:

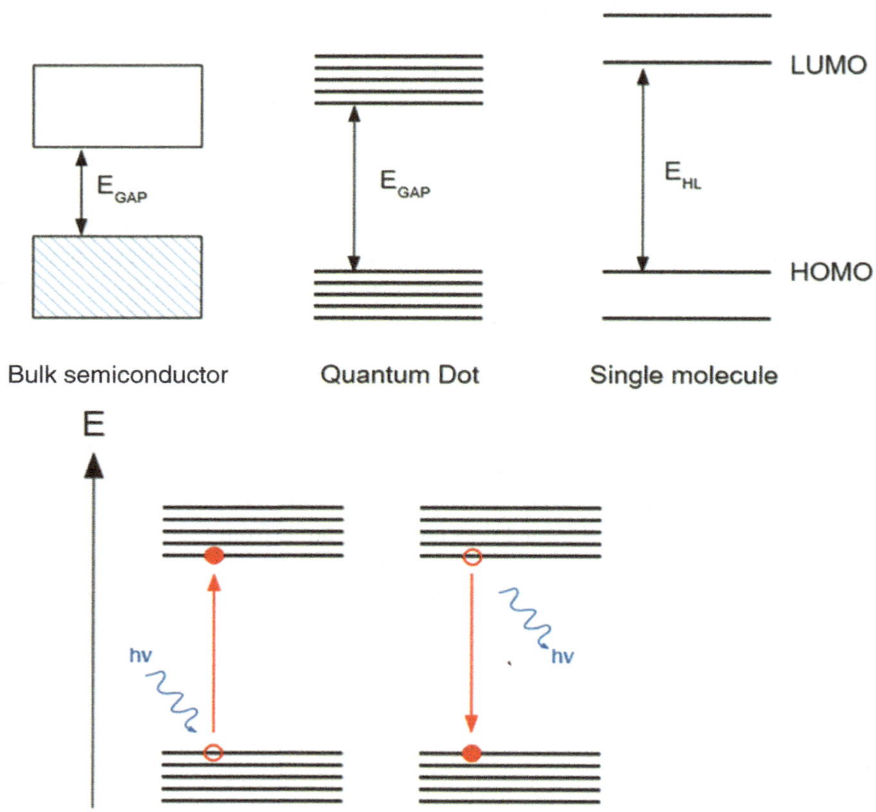

Fig. 14 Top panel: different band structures for bulk semiconductors, quantum dots, and single molecules. Bottom panel: photon absorption and emission mechanisms in QDs

$$\varepsilon_i = \frac{\hbar\pi^2}{2m_e a^2} i, \quad i = 1, 2, 3, \ldots, \tag{7}$$

where $\hbar$ is the reduced Planck's constant, m_e is the effective mass of the electron, and a is the size of the quantum well.

From Eq. (7), it can be observed that, as the particles get smaller, due to this quantum confinement phenomenon, the valence band and the conduction band lose their continuity, and band structure becomes similar to the one typical of the organic molecules, as shown in Fig. 14. The spacing between two adjacent energy levels goes with the size of the QDs and is directly proportional to a^2.

Moreover, the width of the bandgap increases as the size decreases, and then the absorption wavelength can be tuned by changing the particle size.

QDs are widely employed for sensing purposes as fluorescent probes since it has been demonstrated that their fluorescence properties are sensitive to the presence of adsorbates of different nature.

QDs can be employed as an inorganic replacement to organic dyes [63] with two major advantages: first, they usually exhibit much narrower emission peaks [64] whose position is tunable by changing their size, and, secondly, they are also less sensitive to photobleaching phenomena [63].

Different synthesis methods are available for the different types of QDs [22, 65, 66]. For instance, among the methods available for graphene QDs production, there are both some top-down approaches, such as oxidative cutting of the carbon fiber or acidic oxidation of coal, and bottom-up approaches like the glucose/fructose microwave-assisted hydrothermal synthesis or the hydrothermal treatment of nitrated pyrene in NaOH.

As far as Si QDs are concerned, the most common methods include laser ablation, non-thermal plasma synthesis, electrochemical etching, and hydrothermal decomposition of organic precursors containing Si.

CdSe and CdTe QDs are instead usually synthesized by rapid injection of dymethilcadmium and chalcogenide sources into heated tri-n-octilphosphine oxide (TOPO), which result in the formation of a massive amount of CdE nuclei which consume almost totally the reactant and then prevent the subsequent growth.

Usually, after synthesis, QDs are embedded in a solid matrix and are therefore insulated from the rest of the environment: the most significant challenge when realizing QD-based sensors is to obtain this encapsulation without losing the sensing properties of the QDs, and different approaches may be employed.

One of the most common ways is the deposition of QDs by dip coating. For example, Zhang et al. [67] realized an ethanol and water vapor sensor based on Si-QDs deposited on the end of an optical fiber (see Fig. 15b) by simply dipping into a solution of the QDs itself and drying under environmental conditions. The sensor was interrogated through a blue diode laser, and the fluorescence spectra of the Si-QDs were then collected, showing an increase in its intensity when the sensor is exposed to water or ethanol vapor (see Fig. 15a).

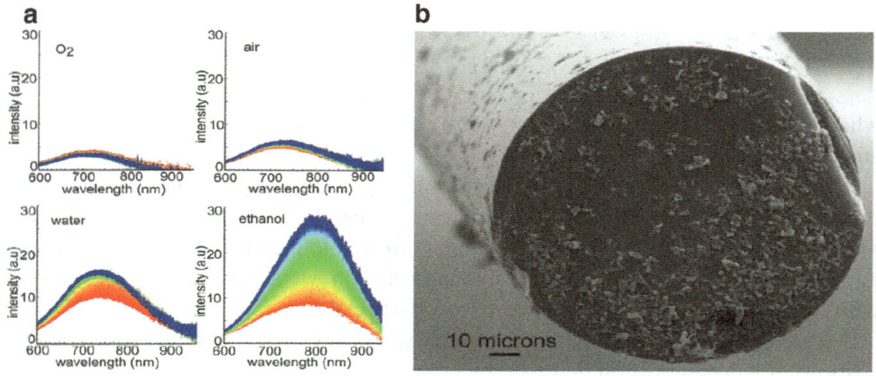

Fig. 15 **a** Fluorescence spectra of the Si-QDs exposed to oxygen, air, water vapor, and ethanol vapor over 1 h. **b** SEM image showing Si-QDs at the end of an optical fiber. Reproduced from [67]. Copyright 2013, with permission from Elsevier

Wu et al. [68] realized a flexible glass surface acoustic wave (SAW) humidity sensor based on ZnO nanowires and graphene QDs (GQDs). First of all, a film of ZnO was deposited on the flexible glass employing DC sputtering using Zn as a target under a mixed atmosphere of Ar/O_2. The sensor was then processed with ultraviolet photolithography and liftoff process to create the SAW resonator on the ZnO coated glass and eventually immersed in a solution of the ZnO nanowires and GQDs.

Yu et al. [69] realized a glucose sensor based on carbon quantum dots-glucose oxidase/cellulose acetate (CQDs-GOD/CA) complex immobilized by dip coating on the surface of an optical fiber. A bare portion of the optical fiber was immersed into a CQDs-GOD/CA precursor casting solution, then dried at 4 °C for slow removal of the solvent and leading to CQDs-GOD/CA complex sensitive film formation. The device exposure to a glucose solution activates some enzymatic reactions that lead to the formation of hydrogen peroxide and a decrease in the QDs fluorescence intensity.

Sensors may also be produced encapsulating the QDs via a sol–gel process like in the work by Jorge et al. [70], where CdTe was dispersed into a Ru-doped glass matrix to realize an oxygen concentration sensor based on the dynamic quenching of fluorescence of a ruthenium complex, such as Ru(bpy) and Ru(dpp).

In this case, the QDs were realized with a core–shell structure covering the CdTe core with a ZnS layer to ensure better stability of the optical properties and avoid some possible oxidation reactions. Moreover, these sensors also demonstrate that the intensity, wavelength, and width of the QDs luminescent peak are affected by the temperature so that they could be used for monitoring both oxygen concentration and temperature simultaneously.

Zhao et al. [71], instead, exploited a 1-ethyl-3-(3-dimethylaminopropyl) carbodiimide hydrochloride/sulfo-N-hydroxy sulfosuccinimide (EDC/sulfo-NHS) coupling reaction to immobilize some CdSe/ZnS QDs with different sizes (and therefore different emission colors) functionalized with carboxylate groups on the surface of an optical fiber and so to obtain a pH sensor.

The silica of the optical fiber was first functionalized with 3-aminopropyltriethoxysilane (APTES) to obtain $-NH_2$ surface groups and then incubated in a solution of EDC/Sulfo-NHS and QDs overnight to obtain a condensation reaction between the amino and the carboxylic groups, as illustrated in Fig. 16.

The QD-COOH functionalized optical fiber was successfully employed as a pH probe as it was observed that the photoluminescence intensity of QDs increased linearly with the increase of pH. Also, a slight red-shift of the emission peak was observed as the pH was increased. Both these phenomena are likely to ionization of the QDs carboxyl ligand.

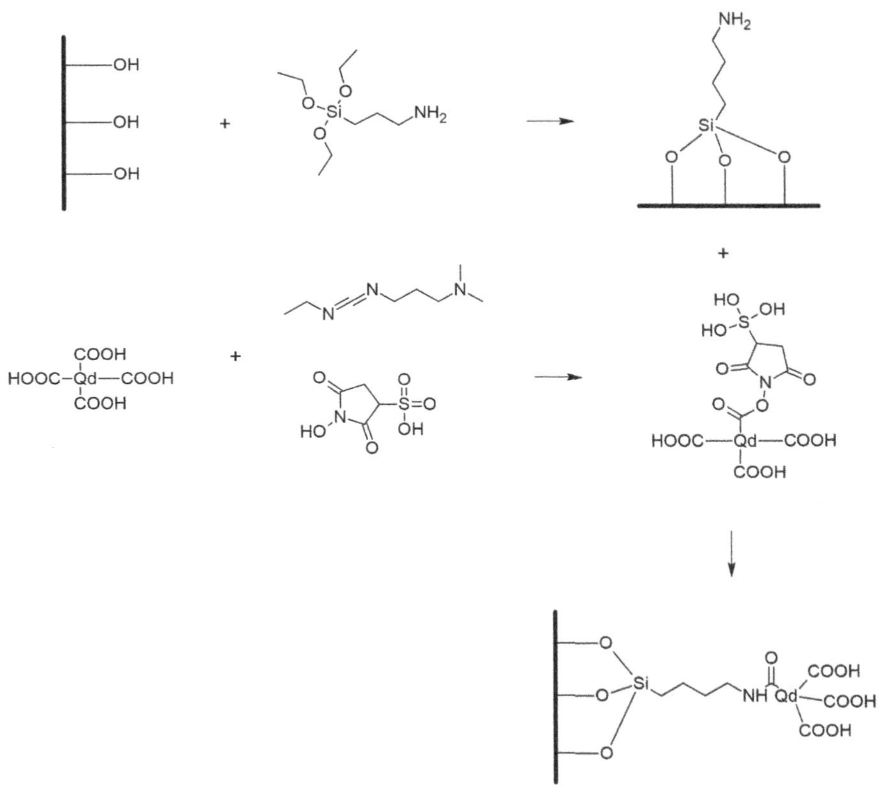

Fig. 16 Carboxylated QDs encapsulation on an optical fiber surface via silanization and subsequent EDC/sulfo-NHS coupling reaction. Adapted from [71]

5 Conclusion and Outlook

The selection of the most suitable material and the development of new ones pushing the limits of the actual performance is of paramount importance in the progress of sensors. While more traditional approaches are commonly used, a great activity in this field of research is ongoing and has much room ahead for development. To confirm this view, state-of-the-art glass/glass–ceramic materials for sensors were covered in this review alongside innovative approaches that leverage, e.g., glass/ceramic nanomaterials, quantum dots, thin films. Although the field is vast, a strong effort in applying the most recent advancement in photonic materials emerges in the literature, indicating a clear path for development to reach more performing and innovative sensors. Indeed, from an economic point of view, the photonic sensors market is expected to garner $18 billion by 2021, registering a compound annual growth rate (CAGR) of 17.7% during the forecast period 2016–2021. Current data also confirm that the winning strategy in this field is based on new product launches (80%) [72]. Those facts testify to the tremendous economic drivers behind the need for new and more performing sensors to pave the way for better control in industrial processes even in harsh environments, better environmental protection, and more sensitive and effective biosensors. To this aim, the development of new or more performing materials, and glass–ceramics in particular, could represent a key enabling technology for the advancement of the photonic sensors field. For this reason, more research and scientific efforts are required calling for a more interdisciplinary approach increasing the fruitful mutual exchanges between the materials science and the photonic communities.

References

1. Stronczek, M.: Glass-ceramics revolutionize exhaust gas sensors. https://www.schott.com/d/epackaging/8af058af-c66a-4399-83bc-92dcaffa40ad. Accessed 18 June 2021
2. Advanced Ceramic Materials for Sensor Applications. https://www.frontiersin.org/research-topics/14727/advanced-ceramic-materials-for-sensor-applications. Accessed 22 Feb 2021
3. Alemohammad, H.: Opto-mechanical fiber optic sensors, 1st edn. Butterworth-Heinemann Elsevier Ltd., Oxford (2018)
4. Dong, X., Zhang, H., Liu, B., Miao, Y.: Tilted fiber Bragg gratings: principle and sensing applications. Photonic Sens. **1**(1), 6–30 (2011)
5. Ozcariz, A., Ruiz-Zamarreño, C., Arregui, F.J.: A comprehensive review: materials for the fabrication of optical fiber refractometers based on lossy mode resonance. Sensors **20**(7), 1972 (2020)
6. Nikl, M.: Scintillation detectors for X-rays. Meas. Sci. Technol. **17**(4), R37–R54 (2006)
7. Leonard, R.L., Johnson, J.A.: Scintillator Glasses. In: Musgraves, J.D., Hu, J., Calvez, F. (eds.) Springer Handbook of Glass, pp. 1555–1584. Springer, Cham (2019)
8. De Faoite, D., Hanlon, L., Roberts, O., Ulyanov, A., McBreen, S., Tobin, I., Stanton, K.T.: Development of glass-ceramic scintillators for gamma-ray astronomy. J. Phys.: Conf. Ser. **620**, 012002 (2015)

9. Struebing, C., Chong, J.Y., Lee, G., Zavala, M., Erickson, A., Ding, Y., Wang, C.-L., Diawara, Y., Engels, R., Wagner, B., Kang, Z.: A neutron scintillator based on transparent nanocrystalline CaF_2:Eu glass ceramic. Appl. Phys. Lett. **108**(15), 153106 (2016)
10. Zheng, Z., Tong, Y., Wei, R., Hu, F., Sun, X., Guo, H.: Tb^{3+}-doped transparent $BaGdF_5$ glass-ceramics scintillator for X-ray detector. J. Am. Ceram. Soc. **103**(4), 2548–2554 (2020)
11. Kang, Z.T., Rosson, R., Barta, B., Han, C., Nadler, J.H., Dorn, M., Wagner, B., Kahn, B.: $GdBr_3$:Ce in glass matrix as nuclear spectroscopy detector. Radiat. Meas. **48**, 7–11 (2013)
12. Glebov, L.B., Nikonorov, N.V., Petrovsky, G.T.: Planar optical waveguides on glasses and glass-ceramic materials. In: SPIE ECO4 (The Hague '91), Glasses for Optoelectronics II, vol. 1513, pp. 56–70. Society of Photo-Optical Instrumentation Engineers, Bellingham, USA (1991)
13. Walker, R.G., Wilkinson, C.D.W., Wilkinson, J.A.H.: Integrated optical waveguiding structures made by silver ion-exchange in glass. 1: The propagation characteristics of stripe ion-exchanged waveguides; a theoretical and experimental investigation. Appl. Opt. **22**(12), 1923–1928 (1983)
14. Gortych, J.E., Hall, D.G.: Fabrication of planar optical waveguides by K^+ ion exchange in BK7 glass. Opt. Lett. **11**(2), 100–102 (1986)
15. Giallorenzi, T.G., West, E.J., Kirk, R., Ginther, R., Andrews, R.A.: Optical waveguides formed by thermal migration of ions in glass. Appl. Opt. **12**(6), 1240–1245 (1973)
16. Chiasera, A., Vasilchenko, I., Dorosz, D., Cotti, M., Varas, S., Iacob, E., Speranza, G., Vaccari, A., Valligatla, S., Zur, L., Lukowiak, A., Righini, G.C., Ferrari, M.: SiO_2-P_2O_5-HfO_2-Al_2O_3-Na_2O glasses activated by Er^{3+} ions: from bulk sample to planar waveguide fabricated by RF-sputtering. Opt. Mater. **63**, 153–157 (2017)
17. Almeida, R.M., Xu, J.: Sulfide glass optical waveguides prepared by sol-gel processing. In: Symposium on Integrated Optoelectronics, 2000, Sol-Gel Optics V, vol. 3943, pp. 58–66. Society of Photo-Optical Instrumentation Engineers, Bellingham, USA (2000)
18. Jestin, Y., Armellini, C., Chiappini, A., Chiasera, A., Ferrari, M., Goyes, C., Montagna, M., Moser, E., Nunzi Conti, G., Pelli, S., Retoux, R., Righini, G.C., Speranza, G.: Erbium activated HfO_2 based glass–ceramics waveguides for photonics. J. Non-Cryst. Solids **353**(5–7), 494–497 (2007)
19. Miniscalco, W.J.: Erbium-doped glasses for fiber amplifiers at 1500 nm. J. Lightwave Technol. **9**(2), 234–250 (1991)
20. Zhang, Y., Nizamidin, P., Abudukeremu, H., Yimit, A.: Optical waveguide xylene gas sensor based on sodium dodecylbenzene sulfonate (SDBS)–TiO_2 film for detection at room temperature. Opt. Mater. Express **10**(9), 2212–2226 (2020)
21. Han, Z., Lin, P., Singh, V., Kimerling, L., Hu, J., Richardson, K., Agarwal, A., Tan, D.T.H.: On-chip mid-infrared gas detection using chalcogenide glass waveguide. Appl. Phys. Lett. **108**(14), 141106 (2016)
22. Gaur, S.S., Singh, P.K., Gupta, A., Singh, R., Kumar, Y.: Synthesis and analysis of planar optical waveguides as pH sensors. Recent Innov. Chem. Eng. **11**(1), 40–44 (2018)
23. MacCraith, B.D., Ruddy, V., Potter, C., O'Kelly, B., McGilp, J.F.: Optical waveguide sensor using evanescent wave excitation of fluorescent dye in sol-gel glass. Electron. Lett. **27**(14), 1247–1248 (1991)
24. Horváth, R., Pedersen, H.C., Skivesen, N., Selmeczi, D., Larsen, N.B.: Optical waveguide sensor for on-line monitoring of bacteria. Opt. Lett. **28**(14), 1233–1235 (2003)
25. Seiyama, T., Kato, A., Fujiishi, K., Nagatani, M.: A new detector for gaseous components using semiconductive thin films. Anal. Chem. **34**(11), 1502–1503 (1962)
26. Touidjen, N.H., Bendahmane, B., Lamri Zeggar, M., Mansour, F., Aida, M.S.: SnO_2 thin film synthesis for organic vapors sensing at ambient temperature. Sens. Bio-Sens. Res. **11**(1), 52–57 (2016)
27. Salman, S.H., Shihab, A.A., Kh. Elttayef, A.-H.: Design and construction of nanostructure TiO_2 thin film gas sensor prepared by R.F magnetron sputtering technique. Energy Procedia **157**, 283–289 (2019)
28. Salah, G., Telia, A., Boukaous, C.: Fabrication of ZnO thin film based humidity sensor with fast response by sol-gel associated to spin coating method. In: 2015 3rd International Conference on Control, Engineering & Information Technology (CEIT), pp. 1–4. Institute of Electrical and Electronics Engineers, Tlemcen, Algeria (2015)

29. Mourzina, Y., Schöning, M.J., Schubert, J., Zander, W., Legin, A.V., Vlasov, Y.G., Kordos, P., Lüth, H.: A new thin-film Pb microsensor based on chalcogenide glasses. Sens. Actuators B Chem. **71**(1–2), 13–18 (2000)
30. Diyatmika, W., Chu, J.P., Kacha, B.T., Yu, C.-C., Lee, C.-M.: Thin film metallic glasses in optoelectronic, magnetic, and electronic applications: a recent update. Curr. Opin. Solid State Mater. Sci. **19**(2), 95–106 (2015)
31. Xian, H.J., Cao, C.R., Shi, J.A., Zhu, X.S., Hu, Y.C., Huang, Y.F., Meng, S., Gu, L., Liu, Y.H., Bai, H.Y., Wang, W.H.: Flexible strain sensors with high performance based on metallic glass thin film. Appl. Phys. Lett. **111**(12), 121906 (2017)
32. Gopalan, K.K., Janner, D., Nanot, S., Parret, R., Lundeberg, M.B., Koppens, F.H.L., Pruneri, V.: Mid-infrared pyroresistive graphene detector on $LiNbO_3$. Adv. Opt. Mater. **5**(4), 1600723 (2017)
33. Kashyap, R.: Fiber Bragg Gratings, 2nd edn. Academic Press Elsevier Ltd., Cambridge, MA, USA (2010)
34. Fan, Z., Diao, X., Hu, K., Zhang, Y., Huang, Z., Kang, Y., Yan, H.: Structural health monitoring of metal-to-glass–ceramics penetration during thermal cycling aging using femto-laser inscribed FBG sensors. Sci. Rep. **10**(1), 12330 (2020)
35. Chiavaioli, F., Janner, D.: Fiber optic sensing with lossy mode resonances: applications and perspectives. J. Lightwave Technol. **39**(12), 3855–3870 (2021)
36. Li, Z., Bao, Q., Zhu, J., Ruan, X., Dai, Y.: Generation of leaky mode resonance by metallic oxide nanocoating in tilted fiber-optic gratings. Opt. Express **28**(7), 9123–9135 (2020)
37. Consales, M., Quero, G., Spaziani, S., Principe, M., Micco, A., Galdi, V., Cutolo, A., Cusano, A.: Metasurface-enhanced lab-on-fiber biosensors. Laser Photonics Rev. **14**(12), 2000180 (2020)
38. Lao, J., Sun, P., Liu, F., Zhang, X., Zhao, C., Mai, W., Guo, T., Xiao, G., Albert, J.: In situ plasmonic optical fiber detection of the state of charge of supercapacitors for renewable energy storage. Light: Sci. Appl. **7**(1), 34 (2018)
39. Liu, Y., Liang, B., Zhang, X., Hu, N., Li, K., Chiavaioli, F., Gui, X., Guo, T.: Plasmonic fiber-optic photothermal anemometers with carbon nanotube coatings. J. Lightwave Technol. **37**(13), 3373–3380 (2019)
40. Chiavaioli, F., Giannetti, A., Baldini, F.: From refractometry to biosensing with optical fibres. In: Del Villar, I., Matias, I.R. (eds.) Optical Fibre Sensors: Fundamentals for Development of Optimized Devices, pp. 331–366. Wiley, Hoboken (2021)
41. Gonzalez-Valencia, E., Del Villar, I., Torres, P.: Bloch waves at the surface of a single-layer coating D-shaped photonic crystal fiber. Opt. Lett. **45**(9), 2547–2550
42. Socorro-Leránoz, A.B., Santano, D., Del Villar, I., Matias, I.R.: Trends in the design of wavelength-based optical fibre biosensors (2008–2018). Biosens. Bioelectron.: X **1**, 100015 (2019)
43. Chiavaioli, F., Zubiate, P., Del Villar, I., Zamarreño, C.R., Giannetti, A., Tombelli, S., Trono, C., Arregui, F.J., Matias, I.R., Baldini, F.: Femtomolar detection by nanocoated fiber label-free biosensors. ACS Sens. **3**(5), 936–943 (2018)
44. Kosiel, K., Koba, M., Masiewicz, M., Śmietana, M.: Tailoring properties of lossy-mode resonance optical fiber sensors with atomic layer deposition technique. Opt. Laser Technol. **102**, 213–221 (2018)
45. Wei, Z., Sun, L., Liu, J., Zhang, J.Z., Yang, H., Yang, Y., Shi, L.: Cysteine modified rare-earth up-converting nanoparticles for *in vitro* and *in vivo* bioimaging. Biomaterials **35**(1), 387–392 (2014)
46. Zhao, P., Zhu, Y., Yang, X., Shen, J., Jiang, X., Zong, J., Li, C.: Multifunctional MnO_2 nanosheet-modified Fe_3O_4@SiO_2/$NaYF_4$:Yb, Er nanocomposites as novel drug carriers. Dalton Trans. **43**(2), 451–457 (2014)
47. Guo, K., Li, M., Fang, X., Luoshan, M., Bai, L., Zhao, X.: Performance enhancement in dye-sensitized solar cells by utilization of a bifunctional layer consisting of core-shell β-$NaYF_4$:Er^{3+}/Yb^{3+}@SiO_2 submicron hexagonal prisms. J. Power Sources **249**, 72–78 (2014)

48. Jiang, S., Guo, H., Wei, X., Duan, C., Yin, M.: Enhanced upconversion in Ho^{3+}-doped transparent glass ceramics containing $BaYbF_5$ nanocrystals. J. Lumin. **152**, 195–198 (2014)
49. Dey, R., Rai, V.K.: Yb^{3+} sensitized Er^{3+} doped La_2O_3 phosphor in temperature sensors and display devices. Dalton Trans. **43**(1), 111–118 (2014)
50. Li, X., Yuan, S., Hu, F., Lu, S., Chen, D., Yin, M.: Near-infrared to short-wavelength upconversion temperature sensing in transparent bulk glass ceramics containing hexagonal $NaGdF_4{:}Yb^{3+}/Ho^{3+}$ nanocrystals. Opt. Mater. Express **7**(8), 3023–3033 (2017)
51. Zhong, J.S., Chen, D.Q., Peng, Y.Z., Lu, Y.D., Chen, X., Li, X.Y., Ji, Z.G.: A review on nanostructured glass ceramics for promising application in optical thermometry. J. Alloy Compd. **763**, 34–48 (2018)
52. Chen, D., Liu, S., Li, X., Yuan, S., Huang, P.: Upconverting luminescence based dual-modal temperature sensing for $Yb^{3+}/Er^{3+}/Tm^{3+}{:}YF_3$ nanocrystals embedded glass ceramic. J. Eur. Ceram. Soc. **37**(15), 4939–4945 (2017)
53. Rami Reddy, B., Kamma, I., Kommidi, P.: Optical sensing techniques for temperature measurement. Appl. Opt. **52**(4), B33–B39 (2013)
54. Jiang, S., Zeng, P., Liao, L., Tian, S., Guo, H., Chen, Y., Duan, C., Yin, M.: Optical thermometry based on upconverted luminescence in transparent glass ceramics containing $NaYF_4{:}Yb^{3+}/Er^{3+}$ nanocrystals. J. Alloy. Compd. **617**, 538–541 (2014)
55. Mukhopadhyay, L., Rai, V.K.: Upconversion based near white light emission, intrinsic optical bistability and temperature sensing in $Er^{3+}/Tm^{3+}/Yb^{3+}/Li^{+}{:}NaZnPO_4$ phosphors. New J. Chem. **41**(15), 7650–7661 (2017)
56. León-Luis, S.F., Rodríguez-Mendoza, U.R., Lalla, E., Lavín, V.: Temperature sensor based on the Er^{3+} green upconverted emission in a fluorotellurite glass. Sens. Actuators B Chem. **158**(1), 208–213 (2011)
57. Chen, S.Y.Z., Song, W.H., Cao, J.K., Hu, F.F., Guo, H.: Highly sensitive optical thermometer based on FIR technique of transparent $NaY_2F_7{:}Tm^{3+}/Yb^{3+}$ glass ceramic. J. Alloy. Compd. **825**, 154011 (2020)
58. Chen, D., Xiang, W., Liang, X., Zhong, J., Yu, H., Ding, M., Lu, H., Ji, Z.: Advances in transparent glass–ceramic phosphors for white light-emitting diodes—a review. J. Eur. Ceram. Soc. **35**(3), 859–869 (2015)
59. Chen, D., Zhou, Y., Xu, W., Zhong, J., Huang, P.: Persistent and photo-stimulated luminescence in Ce^{3+}/Cr^{3+} activated $Y_3Al_2Ga_3O_{12}$ phosphors and transparent phosphor-in-glass. J. Mater. Chem. C **4**(48), 11457–11464 (2016)
60. Xu, J., Ueda, J., Kuroishi, K., Tanabe, S.: Fabrication of Ce^{3+}–Cr^{3+} co-doped yttrium aluminium gallium garnet transparent ceramic phosphors with super long persistent luminescence. Scripta Mater. **102**, 47–50 (2015)
61. Fedorov, P.P., Luginina, A.A., Popov, A.I.: Transparent oxyfluoride glass ceramics. J. Fluorine Chem. **172**, 22–50 (2015)
62. Lin, G., Pan, H., Dai, Y., He, F., Chen, D., Cheng, Y., Jiang, X., Zhang, L., Qiu, J., Zhao, Q.: Formation of Si nanocrystals in glass by femtosecond laser micromachining. Mater. Lett. **65**(23–24), 3544–3547 (2011)
63. Murphy, C.J.: Peer reviewed: optical sensing with quantum dots. Anal. Chem. **74**(19), 520 A–526 A (2002)
64. Resch-Genger, U., Grabolle, M., Cavaliere-Jaricot, S., Nitschke, R., Nann, T.: Quantum dots versus organic dyes as fluorescent labels. Nat. Methods **5**(9), 763–775 (2008)
65. Klimov, V.I.: Semiconductor and Metal Nanocrystals: Synthesis and Electronic and Optical Properties, 1st edn. CRC Press, Boca Raton (2003)
66. Kalluri, A., Debnath, D., Dharmadhikari, B., Patra, P.: Chapter twelve—graphene quantum dots: synthesis and applications. Methods Enzymol. **609**, 335–354 (2018)
67. Zhang, Z.H., Lockwood, R., Veinot, J.G.C., Meldrum, A.: Detection of ethanol and water vapor with silicon quantum dots coupled to an optical fiber. Sens. Actuators B Chem. **181**, 523–528 (2013)

68. Wu, J., Yin, C., Zhou, J., Li, H., Liu, Y., Shen, Y., Garner, S., Fu, Y., Duan, H.: Ultrathin glass-based flexible, transparent, and ultrasensitive surface acoustic wave humidity sensor with ZnO nanowires and graphene quantum dots. ACS Appl. Mater. Interfaces **12**(35), 39817–39825 (2020)
69. Yu, S., Ding, L., Lin, H., Wu, W., Huang, J.: A novel optical fiber glucose biosensor based on carbon quantum dots-glucose oxidase/cellulose acetate complex sensitive film. Biosens. Bioelectron. **146**, 111760 (2019)
70. Jorge, P.A.S., Mayeh, M., Benrashid, R., Caldas, P., Santos, J.L., Farahi, F.: Applications of quantum dots in optical fiber luminescent oxygen sensors. Appl. Opt. **45**(16), 3760–3767 (2006)
71. Zhao, F., Kim, I., Kim, J.: Fabrication of the optical fiber pH sensor based on CdSe/ZnS quantum dot. J. Nanosci. Nanotechnol. **14**(8), 5650–5653 (2014)
72. https://www.alliedmarketresearch.com/photonics-sensor-market. Accessed 30 June 2021

Polymer-Derived Ultra-High Temperature Ceramics (UHTCs) and Related Materials

Emanuel Ionescu, Samuel Bernard, Romain Lucas, Peter Kroll, Sergey Ushakov, Alexandra Navrotsky, and Ralf Riedel

Abstract Ultra-high temperature ceramics (UHTCs) represent an emerging class of materials capable of providing mechanical stability and heat dissipation upon operation in extreme environments, e.g. extreme heat fluxes, chemically reactive plasma conditions. In the last decades, remarkable research efforts and progress were done concerning the physical properties of UHTCs as well as their processing. Moreover, there are vivid research activities related to developing synthetic access pathways to UHTCs and related materials with high purity, tunable composition, nano-scaled morphology or improved sinterability. Among them, synthesis methods considering preceramic polymers as suitable precursors to UHTCs have received increased attention in the last years. As these synthesis techniques allow the processing of UHTCs from the liquid phase, they are highly interesting e.g. for the fabrication of ultra-high temperature ceramic composites (UHT CMCs), additive manufacturing of UHTCs, etc. In the present Chapter, UHTCs are in particular discussed within the context of their physical properties as well as energetics. Moreover, various synthesis methods using preceramic polymers to access UHTCs and related materials (i.e., (nano)composites thereof with silica former phases) are summarized and critically evaluated.

Keywords Ultra-high temperature ceramics (UHTCs) · UHTC (nano)composites · Preceramic polymers · Physical properties · Synthesis · Thermodynamics · Refractoriness

---

E. Ionescu (✉) · R. Riedel
Technische Universität Darmstadt, Fachbereich Material- und Geowissenschaften, Fachgebiet Disperse Feststoffe, Otto-Berndt-Str. 3, 64287 Darmstadt, Germany
e-mail: emanuel.ionescu@tu-darmstadt.de

S. Bernard · R. Lucas
CNRS, IRCER, UMR 7315, Univ. Limoges, 87000 Limoges, France

P. Kroll
Department of Chemistry and Biochemistry, University of Texas at Arlington, 700 Planetarium Place, Arlington, TX 76019-0065, USA

S. Ushakov · A. Navrotsky
School of Molecular Sciences and School for Engineering of Matter, Center for Materials of the Universe, Arizona State University, 871604, Tempe, AZ 85287-1604, USA

F. Baino et al. (eds.), *Ceramics, Glass and Glass-Ceramics*, PoliTO Springer Series,
https://doi.org/10.1007/978-3-030-85776-9_9

1 Introduction

1.1 Definitions, Terms and Limitations

There are two ways to define ultra-high temperature ceramics (UHTCs). Firstly, a strict definition relates to their melting points and requires that UHTCs melt at temperatures beyond 3000 °C (sometimes 3000 K) [1, 2]. These materials are relevant mainly for applications which operate at extremely high temperatures (e.g. $T \gg 2000$ °C) and for short times (typically in the order of minutes, sometimes few hours). Within this context (and due to reasons discussed in some detail below), borides, carbides and nitrides of group 4 transition metals are considered to be typical UHTCs, as they offer the fundamental advantage not to melt at these tremendous high temperatures. However, they suffer from poor environmental compatibility (e.g., poor oxidation resistance), so they usually resist UHTs only for a very short exposure time [3]. Secondly, an alternative definition for UHTCs relates to materials which can be used at temperatures higher than those addressed by typical high-temperature ceramics (such as silicon nitride or silicon carbide) [1, 2]. Thus, materials capable of long-term operation at temperatures beyond 1600 °C (this temperature is generally accepted as the long-term operation limit for Si_3N_4 or SiC) are considered. Their melting points do not necessarily have to exceed 3000 °C (however, the higher the temperature of melting (T_m), the better); whereas their operation time should be significantly longer than those of the materials mentioned within the context of the first definition. Thus, UHTCs are either (i) materials with T_m higher than 3000 °C and which can resist $T \gg 2000$ °C and *short exposure times* to hostile environments or (ii) materials which can withstand $T > 1600$ °C and *long-term exposure* to environmental loading.

When operated at high temperatures (HT) or ultra-high temperatures (UHT) and in extreme environments, structural parts and components are exposed to additional and more severe design constraints as compared to those used in service at or near ambient temperature [1, 4, 5]. A fundamental property which is required in order to be able to operate materials at (ultra)high temperatures relates to their melting point. Naturally, (ultra)high melting points of the chosen materials are necessary, as usually the maximum operation temperature of a material is round half of its melting point with some exceptions, such as Ni-base superalloys, which can be operated at temperatures up to 90% of their melting points [6–9]. Moreover, the environmental degradation of the materials becomes critical at high and ultra-high temperatures and consequently materials capable to withstand thermal decomposition as well as oxidation and corrosion should be taken into consideration. Hence, a basic requirement to be fulfilled relates to thermal stability and environmental (chemical) inertness of the materials of choice. Operation of materials at high and ultra-high temperatures is also related to other factors such as thermally induced stresses and time dependent inelastic strain (creep), which may significantly contribute to catastrophic failure of the components during operation. Depending on the failure consequences (i.e., economic vs. safety-related aspects), those requirements are very strict and materials/components design might suffer here from an extremely high level of conservatism.

Generally, high tolerance to thermal and mechanical stress is desirable in order to avoid catastrophic failure of the materials during operation. Thus, high compliance to thermal stress can be provided by using highly thermal shock resistant materials. Additionally, suitably optimized ductility of the materials or, in the case of brittle (e.g. ceramic based) components, improved fracture toughness is desirable in order to provide high tolerance to mechanical stresses. Finally, no phase transformation of the materials in the range from ambient to operation temperature should occur, as such processes usually are accompanied by significant volume changes which consequently are likely to generate tremendous stresses and induce catastrophic failure. In the case of amorphous materials, increased crystallization resistance may be beneficial for avoiding the generation of mechanical stress upon operation at (ultra)high temperatures.

Intense efforts have been made in the last decades to accommodate those requirements, which are often strongly conflicting. Thus, several types and classes of (ultra)high-temperature materials have been proposed and developed over the years, such as highly refractory metals, intermetallics/alloys, metal matrix composites (MMCs) as well as UHTCs or ceramic matrix composites (CMCs).

When trying to identify suitable materials for operation at UHTs and in extreme environments, one major criterion to be taken into account is their melting point. There are more than 300 materials which exhibit a melting temperature above 2000 °C, including highly refractory metals (such as Hf, Nb, Ta, W, Re, Ir) and numerous ceramics (oxides, carbides, nitrides, borides). However, only few materials are known to exhibit melting points beyond 3000 °C, among them elements (W, Re, C), as well as binaries with high covalent bonding character (e.g., BN, ThO_2, TiC, ZrC, ZrB_2, HfC, HfN, HfB_2, NbC, TaC, TaN, TaB2). [1, 10] They are denoted as ultra-high temperature ceramics (UHTCs). Beside their ultra-high melting points, UHTC materials should withstand non-steady-state interactions with combustion environments and friction with the atmosphere (as for re-entry vehicles). These extreme conditions strongly limit the number of suitable materials among those presented above.

Thus, materials which are left after considering the mentioned criteria are some group 4 carbides, nitrides and diborides (especially that of zirconium and hafnium). [11] This class of materials was first investigated between the 1950s and 1970s, mainly in former USSR and USA and has received increased attention within the last 10–15 years, due to the efforts in developing hypersonic flight vehicles. [4, 12] It has been shown, however, that bulk single-phase UHTCs exhibit rather poor oxidation resistance and damage tolerance under extreme environments. Within this context, the development of UHTCs moved toward composite formulations with improved performance in harsh oxidative environments. Thus, oxidation resistant (mostly particulate) phases have been added to UHTCs, such as SiC or silicides ($TaSi_2$, $MoSi_2$ etc.) [1, 2]. Especially ZrB_2/SiC and HfB_2/SiC (with ca. 20 vol% SiC) composites have been reported to exhibit promising environmental behavior at

UHTs, as they form a protective borosilicate glass coating. [13, 14]. Moreover, intense efforts were made to further improve the environmental stability of SiC-modified diboride-based UHTCs by increasing the viscosity of the liquid silica-based layer, improving the immiscibility of multi-phasic liquid layers formed during oxidation process [15] .

1.2 Thermodynamic Considerations and Solid-State Synthetic Approaches for UHTCs

1.2.1 Borides

The excellent thermal stability as well as high hardness of group 4 transition metal diborides rely on their crystalline structure. They exist in a primitive hexagonal crystal structure (AlB_2-type, space group *P6/mmm*) with two-dimensional graphene-like boron layers alternating with hexagonally close-packed metal layers. Thus, the extremely robust B–B and M–B bonds in MB_2 (M = Zr, Hf) are naturally responsible for their ultra-high refractoriness and hardness (Table 1). Accordingly, ZrB_2 and HfB_2 exhibit extremely high melting temperatures (3245 and 3380 °C, respectively) and the highest hardness values among their periods (23 and 28 GPa, respectively) [13, 16, 17].

In contrast to rocksalt nitrides and carbides, diborides have little variation in stoichiometry and are essentially line compounds (see Fig. 1a as for the Ti-B system [18]). Diborides of Ti, Zr, and Hf are the highest melting borides; however in the Ta-B system (Fig. 1b) [19, 20], the monoboride TaB has a marginally higher melting temperature than that of the corresponding diboride TaB2 (3363 ± 15 K for TaB vs. 3310 ± 20 for TaB2). TaB is not found in NaCl-type structure, but possesses an orthorombic CrB-type (*Cmcm*) structure [21]. The monoboride of Zr was reported [22] in NaCl structure, but it is only stable in a narrow temperature range below 1500 K.

Over the last decade, multiple groups studied the thermodynamic properties of ZrB_2 and HfB_2 [23, 24]. Lawson et al. analyzed in detail the bonding situation in ZrB_2 and HfB_2 [23] and concluded that covalent bonding provides cohesion between boron atoms, while metallic bonding persists within the metal layer. A substantial charge transfer from the metal layer to the boron layer yields to strong ionic bonding

Table 1 Some physical and structural properties of ZrB_2 and HfB_2 (adapted from [13])

	Density [g/cm^3]	Melting point [°C]	CTE [ppm/K]	Thermal conductivity [W/(m K)]	Hardness [GPa]	Young's modulus [GPa]
ZrB_2	6.2	3245	5.9	60	23	489
HfB_2	11.2	3380	6.3	104	28	480

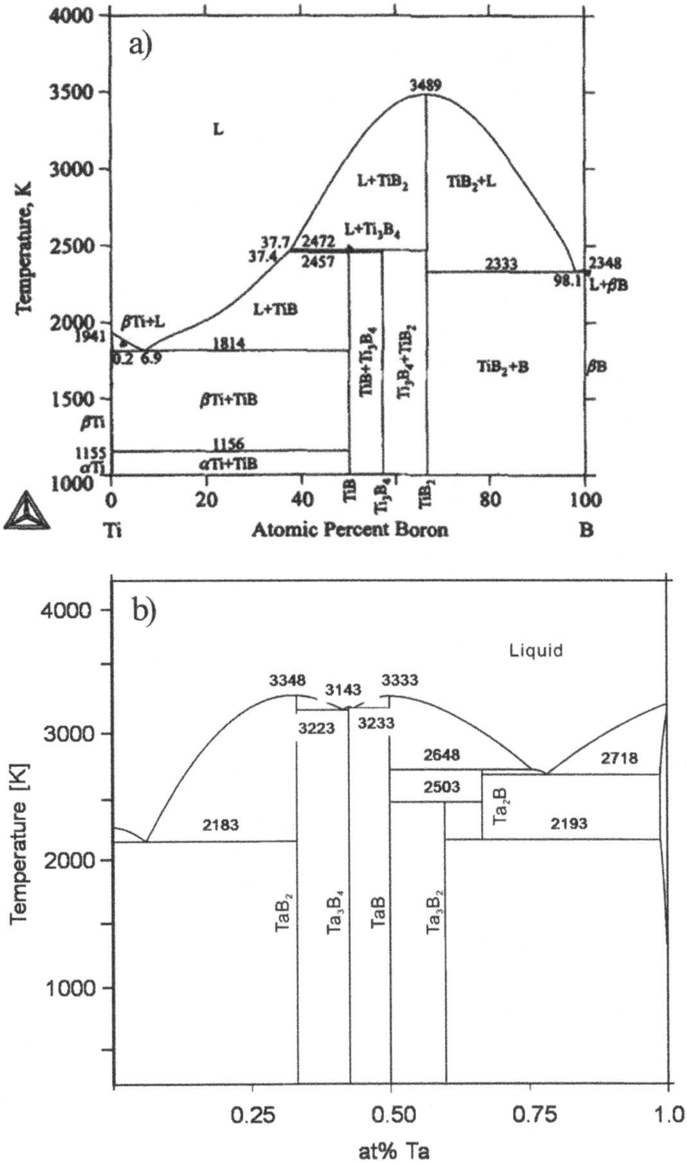

Fig. 1 Phase diagrams of **a** Ti-B (http://www.calphad.com/titanium-boron.html) and **b** Ta-B systems (reprinted from [20] with permission from Elsevier)

between layers and is responsible for the substantial refractory character. Zhang et al. recently reported the remarkable structural stability of ZrB_2 [25]. Even when subjected to extreme pressures, the hexagonal structure remains the only competitive modification of the diboride. Further Zr-B compounds have been investigated as well. Without doubt, ZrB in the rocksalt structure is mechanically stable [26]. However, computed lattice parameters do not agree with experimental data within the usual tolerances of first-principles calculations. Further mechanically and dynamically stable structures are Zr_3B_2, Zr_3B_4, Zr_2B_3, ZrB_2, and ZrB_{12} [27]. Only ZrB_2 appears to be thermochemically stable, and results for ZrB_{12} are still undecided [28]. However, a mononitride ZrB appears as the least competitive phase, and the rocksalt type of ZrB is less favorable than an orthorhombic modification. This, furthermore, suggests that the observed rocksalt-type may either be defect-rich or even stabilized by carbon. Noteworthy in this context is the work of Lee et al., who identified insertion of boron into octahedral sites of the hcp Zr structure as energetically most favorable [29].

The thermochemistry of Hf-B compounds is very similar to those in the Zr-B system. HfB_2 dominates the phase diagram, with orthorhombic HfB and HfB_{12} competitive but not favored by enthalpy [28, 30].

The Ta-B system shows more diversity than borides of Zr and Hf (see Fig. 1b). With exception of the well-known AlB_2-type of TaB2, all appearing structures are quite complex thanks to prevalent Ta-Ta bonding. Boron-rich phases exhibit fragments of the TaB2 structure, e.g. a one-dimensional strip of condensed B_6-rings, interrupted by additional Ta atom. Tantalum-rich structures display motifs of dense packings with B atoms filling interstitial sites, however with significant distortions. Calculations indicate that Ta_3B_2, TaB, Ta_5B_6, Ta_3B_4, Ta_2B_3, and TaB2 are thermochemically stable [28].

Diboride-based UHTCs can be synthesized by different approaches. Most of them rely on reduction processes of the oxides (zirconia, hafnia, boria). Carbon and boron are the most used reducing agents. The carbothermal process has been the mostly used technique to produce diborides and involves the reaction of a mixture consisting of metal oxide (i.e., ZrO_2 or HfO_2) and B_2O_3 with carbon. As these processes are strongly endothermic, temperatures of synthesis typically around 2000 °C are needed [11]. Usually, boria is added in excess in order to promote the formation of the diborides and to inhibit carbide formation. Common impurities in carbothermally synthesized diboride powders are carbon, boria and metal carbides [11, 13]. The reaction of the group 4 transition metal oxides with boron (borothermic reaction) as well as aluminothermic processes (MO_2 and boria reacting with Al) are also strongly endothermic and are performed at temperatures exceeding 1400–1500 °C [11]. The reduction of the metal oxides with boron carbide (B_4C) was shown to become thermodynamically favorable at temperatures as low as 1200 °C. An additional advantage of the reduction process with B_4C relies on the reduced amounts of carbon and oxide impurities in the synthesized diboride powders [31, 32].

Group 4 transition metal diborides (as well as other transition metal borides) can also be synthesized from elements. Depending on the purity of the starting materials, diborides of high purity are accessible upon this process. However, due to the poor oxidation resistance of the group 4 transition metals, these processes have to be performed in inert atmosphere in order to avoid the formation of the corresponding metal oxides. As the reactions of Zr and Hf with B are highly exothermic [11], self-propagating high-temperature synthesis (SHS) techniques have been developed to prepare diboride powders from elements [33, 34].

Table 2 summarizes best available values for melting temperatures, standard entropies and enthalpies of formation at 298 K for selected borides, carbides and nitrides of Ti, Zr, Hf, and Ta. The data for corresponding oxides are included for comparison. The data referenced to NIST-JANAF and Glushko's thermochemical tables are largely based on the evaluation of measurements performed before 1970. Formation enthalpies for TiC, TaB2, and TiB_2 are given after more recent calorimetry measurements performed in Kleppa's group [35, 36]. The uncertainties are given when provided in the original reference.

Synthetic approaches using reactive processes have also been applied to the preparation of diboride-silicon carbide composite materials. Thus, hot-pressing of a mixture of metallic zirconium, carbon and silicon tetraboride, SiB_4, led to monolithic ZrB_2/SiC composites. A small amount of ZrC impurity has been also detected [50]. Reactive hot-pressing of a powder mixture containing zirconium, boron carbide B_4C and silicon was also used to prepare ZrB_2/SiC monoliths with high densities at a temperature as low as 1900 °C [51].

Table 2 Standard entropies, enthalpies of formation and melting temperatures for selected borides, carbides, and nitrides of Ti, Zr, Hf, and Ta

Compound	$\Delta H°f_{298.15}$ [kJ/mol]	$S°_{298.15}$ [J/mol/K]	T_m [°C]
TiC	−186 ± 18 [35]	24.7 ± 0.2 [37]	3067 ± 25 [37]
TiB_2	−328 ± 10 [36]	28.9 ± 0.4 [37]	3225 ± 25 [38, 39]
TiN	−338 ± 4 [40]	30.3 ± 0.2 [40]	2945 ± 30 [41]
TiO_2	−944 ± 1 [37]	50.3 ± 0.2 [37]	1870 ± 15 [37]
ZrC	−207 ± 3 [37]	33.3 ± 0.1 [37]	3572 ± 30 [42]
ZrB_2	−323 ± 7 [40]	35.9 ± 0.1 [40]	3244 [43]
ZrN	−342 [44]	38.9 ± 0.2 [37]	2955 ± 30 [41]
ZrO_2	−1101 ± 1 [37]	50.4 ± 0.4 [37]	2710 [45]
HfC	−208 ± 8 [37]	40.1 ± 0.2 [37]	3982 ± 30 [42]
HfB_2	−329 ± 9 [46]	42.6 [37]	3378 ± 20 [22]
HfN	−374 ± 2 [37]	45 ± 1 [37]	3330 ± 50 [41]
HfO_2	−1118 ± 2 [37]	59.3 ± 0.6 [37]	2800 [45]
TaC	−142 ± 3 [37]	42.3 ± 0.2 [37]	4003 ± 30 [42]
TaB2	−160 ± 6 [47]	47 ± 4 [37]	3037 ± 20 [48]
TaN	−252 ± 2 [49]	41.8 ± 1.2 [49]	†2930 ± 30 [41]
$TaO_{2.5}$	−1024 ± 1 [37]	71 ± 1 [37]	1872 ± 10 [37]

1.2.2 Carbides and Nitrides

Generally, transition metal carbides, nitrides and carbonitrides represent a unique class of materials with outstanding structural properties, such as very high melting points, exceptional high hardness, or high-temperature ablation resistance [52]. Furthermore, they exhibit metallic character with respect to electrical and thermal conductivities, fascinating surface properties and interesting catalytic behavior. In particular, group 4 transition metal carbides and nitrides have been shown to exist over broad ranges of composition with an upper limit to C:M or N:M ratio close to unity. They exhibit tremendous thermal stability (e.g., group 4 transition metal nitrides can melt without suffering decomposition, unlike group 5 or 6 transition metal nitrides), which has been shown to rely on their free energy of formation [52].

Group 4 carbides and nitrides belong to the class of interstitial compounds, having as characteristics simple metallic lattices (face-centered cubic, fcc) and carbon or nitrogen atoms located in their octahedral interstitial sites. Their bonding is characterized by a mixture of metallic, ionic and covalent components. With respect to thermal or electronic transport, carbides and nitrides of group 4 transition metals exhibit metallic behavior. As the carbides show lower density of state (DOS) at the Fermi level than the corresponding nitrides, the electrical and thermal conductivity as well as the T_c values of the group 4 carbides are lower than those of the analogous nitrides [52].

Transition metal carbides are known for their extremely high melting points (Table 3). For instance, ZrC shows a congruent melting point of 3420 °C, whereas HfC melts congruently at 3930 °C. The material predicted to have the highest known melting point (4215 °C) is a solid solution of HfC and TaC in a molar ratio 4:1, with a composition of Ta_4HfC_5 (compare to the melting point of TaC, 3985 °C). Also, transition metal nitrides have extremely high melting points, often above 3000 °C (TiN: 3030 °C; ZrN 3000 °C; HfN 3330 °C).

Table 3 Selected physical and structural properties at room temperature of stoichiometric group 4 transition metal carbides and nitrides (i.e., composition near to 50 at % metal)

	ZrC	HfC	ZrN	HfN
Density [g/cm^3]	6.5	12.3	7.3	13.8
Melting point [°C]	3420	3930	3000	3330
CTE [ppm/K]	7.5	6.1	7.8	8.5
Th. conductivity [W/(m K)]	24.6	25.1	11	11
Hardness [GPa]	25	20	15	18
Young's modulus [GPa]	350	420	460	380
Electrical resistivity [$\mu\Omega$ cm]	75	67	24	27
Curie temperature, T_c [K]	–	–	10.47	6.92

Data taken from Ref. [52]

Group 4 transition metal carbides can be prepared from elements as well. As the reactions are strongly exothermic, SHS techniques can be used to synthesize carbide powders. Transition metal nitrides can be produced similarly [53]. Carbides can also be synthesized by reduction of oxides with carbon at high temperatures. Thus, zirconia or hafnia powders can be reacted with carbon at temperatures as low as 1600 °C to prepare fine and ultrafine carbide powders [54–58]. Due to their high sinter activity (and to some extent to small oxygen impurities), they can be densified at relatively low temperatures. The carbothermal process was combined with the SPS technique to prepare monolithic zirconium and hafnium carbides [59].

The corresponding bulk nitrides can be prepared from group 4 transition metals or metal hydrides via heat treatment in nitrogen or ammonia atmosphere. Carbonitridation of metal oxides at high temperatures also furnishes metal nitrides (which contain oxygen and carbon impurities) [52].

Except for TaN, all compounds listed in Table 2 are reported to melt congruently. TaN decomposes to Ta_2N which has a slightly higher melting temperature (3232 K). It is not uncommon to find discrepancies of a hundred degrees or more in melting temperatures between different reports, especially for borides and nitrides. Melting temperatures of nitrides are reported after measurements by Ettmayer et al. [41], which are the average from experiments performed at 1–80 atm nitrogen pressure.

Ti, Zr, Hf and Ta carbides and nitrides are found in NaCl (rocksalt) structure, in which the metal forms a cubic close-packed lattice and C or N fill some or all octahedral sites comprising from 35 to 50 at%. These compounds possess high electrical conductivities. They are sometimes referred to as δ-phase [60] or interstitial phases [61] referring to intercalation of *fcc* metal lattice with C or N.

Complete miscibility can be found among Ti, Zr, Hf, Ta carbides and nitrides at high temperatures (ca. 2000 °C). The stability of these solid solutions and exact temperatures and shapes of miscibility gaps depend on vacancies concentrations [62]. From the analysis of phase diagrams, the congruently melting compositions for rock salt carbides and nitrides deviate from ideal stoichiometry with 44–48 at % of nonmetal.

Yu et al. investigated several structures in the Ta-C system and showed the elasticity of TaC_{1-x} decreases with loss of carbon in the structure, which is related to a decrease in covalent bonding between Ta and C paralleled by an increase in metallic bonding [63]. This is also the underlying motif of the perfectly ordered ς-like structure of Ta_3N_2 with $[M_6\text{-}C_4/V_2\text{-}M_6\text{-}C_4/V_2{=}{=}M_3C_2]$.

An interesting topic are mixtures of metal carbides forming UHTCs. Not only is the enthalpy of mixing relevant to stability of phases, but also a potential mixing entropy among the metal cations. Vorotilo et al. investigated solid-solutions of TaC and ZrC in the rocksalt structure with full occupation of cation and anion positions [64]. Computed lattice parameters and bulk moduli of solid solutions follow a rule of mixture and agree with experimental data. The enthalpy of mixing ΔH of intermediate compositions $Ta_xZr_{1-x}C$ is negative for all x with a maximum ΔH = −0.030 eV/atom at x = 0.625. Experimental work carried out in parallel did not indicate decomposition of solid solution, suggesting complete solubility without miscibility gap in the $Ta_xZr_{1-x}C$ phase diagram.

The Hf-N system (see phase diagram in Fig. 2a [65]) displays several compounds favored by enthalpy. Two recent comprehensive studies present structures and thermochemistry in this system using DFT calculations [66, 67]. Besides HfN, structures of compositions Hf_4N_3, Hf_3N_2, Hf_2N, Hf_3N, Hf_6N, and, probably, Hf_8N appear thermochemically stable towards the metal-rich side of the phase diagram. All structures can be described as being related to densest packings of metal atoms with N filling octahedral interstitials. Surprisingly, a new polymorph of HfN with hexagonal structure (*P6₃/mmc*; Fig. 2b) was discovered to be the most favorable modification of HfN [67]. Careful work indicated that the rocksalt type of HfN (Fig. 2b), previously assumed to be dominant, only becomes thermodynamically stable above 670 °C. Metal cations in hexagonal $P6_3/mmc$-HfN fall onto the sites of the La structure type, a simple intergrowth of *hcp* and *ccp* with repetition every four layers, also termed double hexagonal closest packing (*dhcp*). Anions fill all octahedral sites, giving the hexagonal HfN structure the flavor of NiAs-type and rocksalt type joined together. Towards the nitrogen-rich side of the phase diagram structures with composition $Hf_{11}N_{12}$, Hf7N8 and Hf4N5 appear, all described as cation-deficient rock-salt structures. Besides the high-pressure phase Hf_3N_4 with Th_3P_4-type structure [60], the most nitrogen-rich compound of Hf thermochemically stable at ambient pressure is Hf_3N_4, for which a new monoclinic structure has been proposed [67]. At pressures beyond 23 GPa, an *Immm*-HfN_{10} phase was predicted via DFT calculations and exhibits Hf atoms eightfold coordinated by nitrogen and consists of infinite polymeric nitrogen chains (N–N bond has a bond order of 1.25) and encapsulated N≡N molecular units (Fig. 2b).

Calculations of the Ta-N system do not yield as many thermochemically stable structures as the Hf-N system. Ta_3N_5, which can be synthesized, appears together with Ta_2N_3 as nitrogen rich phases in the phase diagram at low temperatures. It is well known, however, that upon annealing to 1000 °C, Ta_3N_5 decomposes through a series of phases of intermediate composition into TaN. The phase diagram (Fig. 3a [68, 69]) is then dominated by ε-TaN [70–72] with a hexagonal structure related to the intermetallic CoSb (Fig. 3b). Ta atoms form dense packed layers which are tightly stacked similar to a hexagonal closest packing. With the shorter stacking distance, Ta-Ta bonds persist between next-nearest layers. Anions are then located within the plane of every other Ta layer and do not occupy interstitial sites (Fig. 3b). In comparison to rocksalt-type TaN, ε-TaN has a major energy advantage, but lacks a similar defect chemistry. The next stable phase then is Ta_2N, which structure is best described as a hexagonal closest packing of Ta with N filling octahedral sites in every other layer (Fig. 3b). Although the metal arrangements in both ε-TaN and Ta_2N are described in similar terms, intermediate structures with anion deficiency do not appear with competitive enthalpy of formation [66].

It is worthy to mention that first-principles calculations of thermochemistry for nitride UHTCs face a tough challenge. Reaction enthalpies are computed at zero kelvin. Entropy differences between reactants and products are often neglected, and enthalpy differences are assumed to remain constant over the range of temperatures. Even if those differences are taken into consideration—and there are some fine works that go the extra mile [67]—such an approach is impossible for nitrogen.

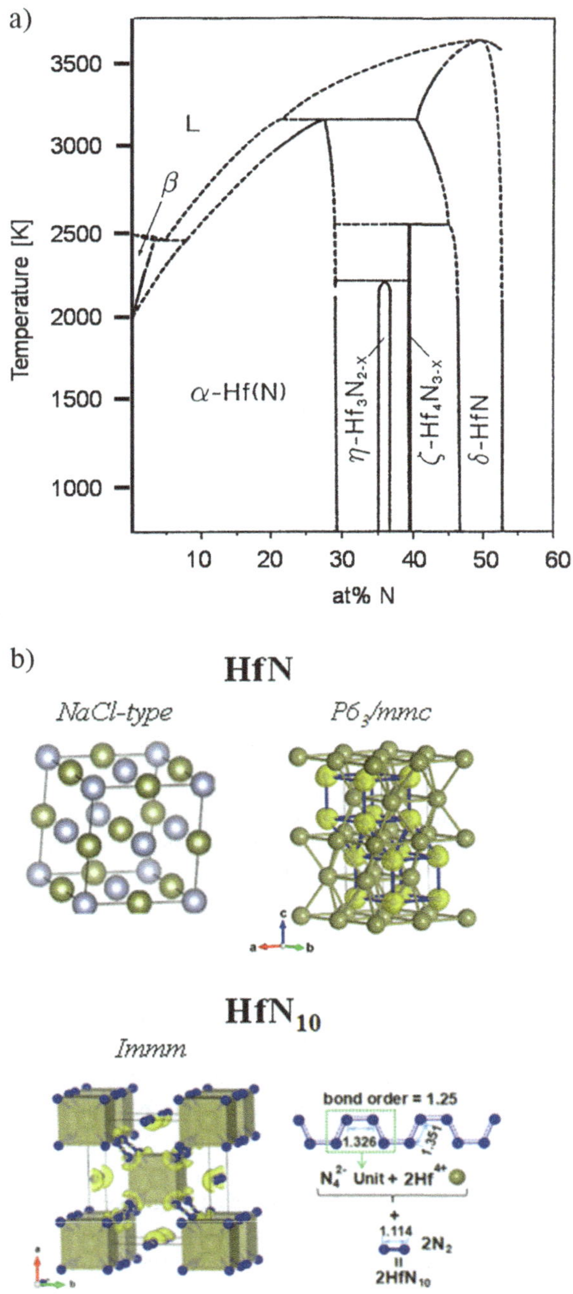

Fig. 2 **a** Phase diagram of the Hf-N system (reprinted from [65] with permission from Springer); **b** crystalline structures of HfN (*NaCl-type* and *P6₃/mmc*) and of HfN_{10} (*Immm*). For *Immm*-HfN_{10}, infinite polymeric nitrogen chains and N≡N bondings are shown (reprinted from [67] with permission of the American Physical Society)

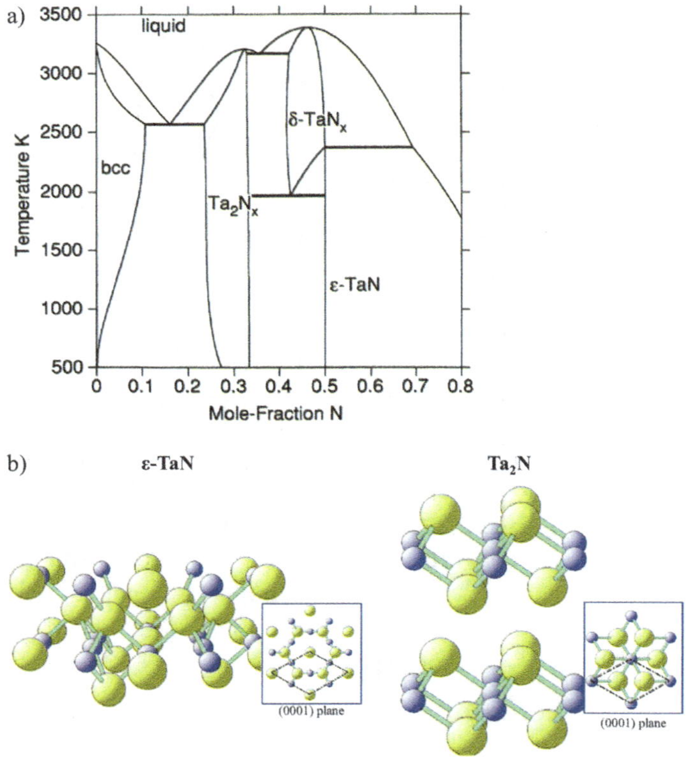

Fig. 3 **a** Phase diagram of the Ta-N system [68, 69] (reprinted from [69] with permission from Springer); **b** crystalline structures of ε-TaN (*P6/mmm*) and Ta_2N ($P\bar{3}m1$): N and Ta: blue and yellow spheres, respectively; dashed lines in the insets indicate the cell unit (reprinted from [72] with permission from the American Physical Society)

N is a gaseous molecule at ambient conditions, quite unlike elemental B and C, which are solid-state compounds. Being a gas, both enthalpy and entropy of nitrogen change significantly larger with increasing temperature than those of the solids, and the computational arsenal to assess heat capacity and vibrational entropy loses its strength. Consequently, one cannot hope for fortuitous error cancellation in reaction enthalpies between products and reactants. This sets N apart from B and C, and makes incorporation of the chemical potential of nitrogen at elevated temperatures indispensable for predicting nitride phases [73–75].

Standard enthalpies of formation are the least negative for carbides and most negative for oxides. This does not reflect the bond strength in compounds, but is mainly related to the difference in the reference states for nonmetal atoms, which are N_2, O_2, and graphite. Formation enthalpies show that carbides, nitrides, and borides are expected to be unstable to oxidation. For Zr, Hf and Ta, melting temperatures decrease in the order of carbides–borides–nitrides–oxides, spanning a range of one to two thousand degrees. Titanium diboride is an exception from this trend, with its melting temperature (3498 K) exceeding that of TiC (3340 K).

No experimental measurements of fusion enthalpies for Ti, Zr, Hf and Ta carbides, borides and nitrides have been reported. First principles computations are increasingly used to evaluate thermodynamic properties of these compounds: thermal expansion and heat capacities were computed for ZrC and ZrN up to 3000 K [76] and the effect of carbon vacancies on miscibility gap in TiC-ZrC carbides was analyzed [62], and melting temperatures were studied in the Hf-C-N system [77].

Since carbides, borides, and nitrides of transition metals are good electrical conductors, most of the measurements of their melting temperatures were performed by direct bulk heating of the samples with electric current using an experimental arrangement known as Pirany furnace. In this method, first used by Pirany and Alterthum in 1923 [78], the melting occurs in the bottom of the blind hole drilled in the test specimen. The hole also serves as black body cavity to aim pyrometer for the temperature measurements, eliminating the uncertainty from sample emissivity. This arrangement was used by Rudy et al. [79] on carbides and borides and by Ettmayer [41] to study nitrides. For metals which can be drawn as a wire, the Joule heating evolved into exploding wire technique, which now also yields enthalpy as a function of temperature in the solid and liquid state [80].

For almost 40 years no new studies on melting temperatures of carbides, borides, and nitrides by Joule heating method have been performed; however, the renewed interest in refractory carbides for UHTCs resulted in a new surge of experimental efforts. Pulsed laser melting technique, originally developed for refractory oxides, was recently applied to study the melting of Ta, Hf and Zr carbides [42, 81]. In this method melting is induced by millisecond laser pulses, the surface temperature is recorded with a fast spectropyrometer and liquidus and solidus temperatures are derived from thermal arrests on cooling. Pulsed laser measurements of HfC melting point performed by Barraza et al. [81] in Germany (3959 ± 84 °C) and by Sheindlin et al. [42] in Russia (3982 ± 30 °C) are in reasonable agreement with each other and with earlier Pirany furnace measurements by Rudy et al. [79] in California (3928 ± 40 °C). However, there is a large discrepancy for reported TaC melting point from pulsed laser measurements: (3768 ± 8 °C) [81] versus (4003 ± 30 °C) [42]. The higher value reported by Sheindling et al. [42] (and included in Table 2) is also consistent with earlier results from Pirany furnace (3983 ± 15 °C) [79]. Recently, a pulsed Joule heating technique was also used on 250 μm thick carbide samples, [82, 83] to analyze melting temperature, heat capacity and fusion enthalpy for Zr carbide and nitride [84, 85].

The prize for highest melting solid was given to $Ta_{0.8}Hf_{0.2}C$ in 1930 by Agte and Alterthum [86] (~60 °C higher than TaC), in 1967 Rudy et al. [79] gave it to TaC (3983 °C). Both teams were using Pirany furnace. Barraza et al. [81] gave it to HfC from laser pulse melting, and recent work using electric pulse method returns it back to $Ta_{0.8}Hf_{0.2}C$ (4027 °C). [83] However, computational work predict that experimentalists are looking in the wrong system and the highest melting solid is a Hf-C-N alloy [77] (Fig. 4).

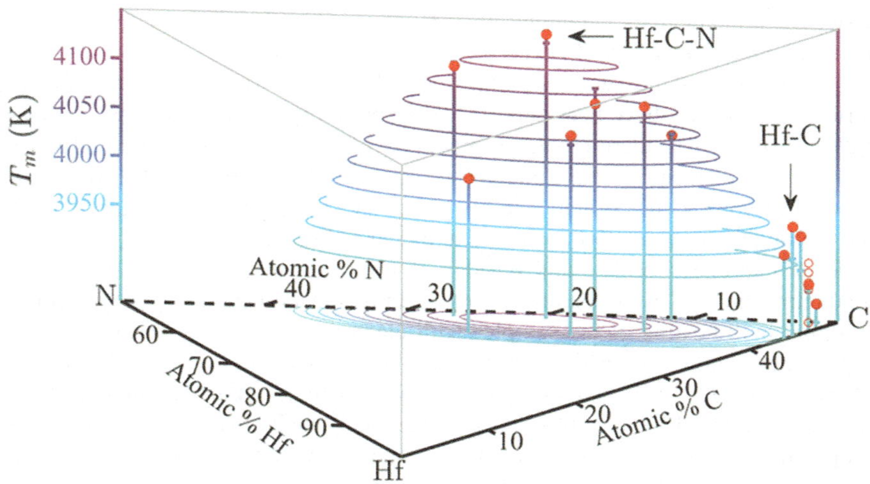

Fig. 4 Melting temperatures of materials compositions in the Hf-C-N and Hf-C (red filled circles) as well as Hf-Ta-C (empty circles) systems (reprinted from [77] with permission from American Physical Society)

There is much to be done in refining experimental measurements and synthesis, and characterization techniques and, for practical applications, oxidation behavior and properties of oxycarbides are more critical than absolute values of melting temperatures for pure compounds.

2 Ceramic Synthesis from Preceramic Polymers (the PDC Route)

Single-source preceramic precursors provide a facile and inexpensive access to various ceramic systems with tailored phase compositions, microstructures and property profiles. Among them, polymeric single-source precursors are of special relevance, as they can be used for the preparation of ceramic bulk powders, dense monoliths, macro, meso and micro porous bodies, thin films and coatings, miniaturized (microstructured) components, fibers, etc. Furthermore, polymeric single-source precursors allow preparation of unique ceramic compositions and microstructures, which are not accessible via other preparative techniques. The molecular structure and chemistry of the single-source precursors greatly influence the composition, microstructure, and phase distribution of the final ceramic product, providing the advantage to control and design ceramic materials with various macroscopic and physical properties [87]. In this context, a crucial role is played by the synthesis of suitable precursors with tailored composition and molecular architecture [87–89].

Polymer-based single-source precursors for the preparation of ceramic nanocomposites were taken into account in the 1980s [90, 91], as researchers started to look for appropriate synthesis techniques for nanoscale non-oxidic ceramics. However, already in the late 1950s and 1960s studies related to the conversion of molecular precursors into non-oxidic ceramics were reported [92–94]; furthermore, in the 1970s, polymeric precursors such as polycarbosilanes (PCS) or polysilazanes (PSZ) were used to prepare small-diameter SiC and SiC/Si_3N_4 ceramic fibers [95, 96]. These preparative efforts culminated in the commercialization of the Yajima production process of continuous SiC ceramic fibers [96, 97]. These SiC fibers were shown to exhibit excellent high temperature stability and promising mechanical properties. However, there are some limitations concerning their use at high temperatures, as they possess limited oxidation resistance and poor mechanical properties above 1300 °C [98, 99]. In this context, numerous studies have been performed with respect to the modification of the SiC-based fibers with metals in order to improve their properties. Mainly, PCS was modified with metal alkoxides or acetyl acetonates (metal = Ti, Zr, Al, etc.) [100–102]. Depending on the metal precursor used for the chemical modification, different reaction pathways of PCS with the metal precursor have been reported. Thus, titanium, zirconium or aluminum alkoxides react with PCS upon alkane evolution and formation of Si-O-M units [101, 102]; whereas metal acetonates were assumed to react upon Si-M bond formation and release of acetyl acetone [101]. However, in both cases the chemical interaction between PCS and the metal precursors has not been clarified unambiguously. For instance, the modification of polycarbosilane with metal alkoxides was supported by spectroscopic data (such as FTIR or NMR spectroscopy) only in some cases [102]; while in other cases no direct evidence for the formation of the Si-O-M units was obtained and the alkoxide was assumed to be physically incorporated within the PCS-based matrix (i.e., the formation of a blend consisting of alkoxide and PCS) [103]. Also, in the case of the reaction of PCS with metal acetylacetonates, no direct evidence of the formation of Si-M units was provided (this was claimed only on the basis of the decrease of the amount of Si–H bonds in PCS; which is, however, not a strong argument) [101]. Recently, a case study on the chemical modification of an allyl-substituted hydrido polycarbosilane with vanadium acetylacetonate indicated that the PCS reacts with metal acetylacetonate complexes upon formation of Si-O-M units, as also reported for the reactions of PCS with metal alkoxides [104].

Thermal treatment of the prepared single-source precursors was shown to lead in a first step to amorphous single-phase ceramics, which subsequently undergo phase separation and crystallization processes to furnish bi- or multi-phase nanocomposites (Fig. 5). The conversion of the single-source precursors into ceramics usually consists of two steps: (i) cross-linking processes at low temperatures (100–400 °C) leading to infusible organic/inorganic networks; and (ii) ceramization at temperatures up to 1000–1400 °C. Whereas mainly amorphous ceramics are obtained upon pyrolysis, subsequent annealing at high temperatures leads to (poly)crystalline materials. During cross-linking, the single-source precursors are converted into organic/inorganic materials. This process prevents the loss of low molecular weight components of the precursor as well as their fragmentation during ceramization, and

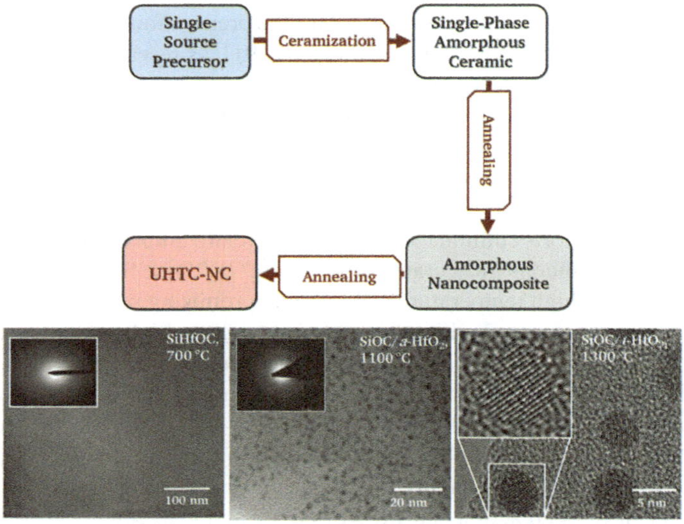

Fig. 5 Typical preparative approach of polymer-derived ceramic nanocomposites starting from single-source precursors [87–89]

thus increases the ceramic yield. Furthermore, the conversion of the single-source precursors into infusible materials prevents their melting during ceramization, while retaining the shape upon pyrolysis. The ceramization process of the cross-linked precursors consists of the thermolysis and evolution of their organic groups at. 600–1000 °C. In this temperature range the conversion of the preceramic materials into amorphous covalent ceramics (usually being single-phasic) occurs. The multifaceted mechanisms involved in the pyrolysis of preceramic polymers are not well understood yet. This is mainly due to the complex structure of the highly cross-linked polymeric materials and the resulting amorphous ceramics. The reactions which occur during pyrolysis can be investigated by solid-state NMR, FTIR and Raman spectroscopy, as well as via thermogravimetric analysis (TGA) coupled with evolved gas analysis (EGA) (i.e., in situ mass spectrometry and FTIR spectroscopy) [89].

Synthesis and phase formation of silicon-based ceramic nanocomposites were studied extensively in the last decade [87]. Pyrolysis of suitable alkoxysilanes, $Si(OR)_4$, or polysiloxanes, $[Si(R)_2$-O]n, chemically modified with metal alkoxides was shown to give MOx/SiOC-based PDC-NCs, as reported for M = Al, Ti, Zr, Hf etc. For M = Zr and Hf a single-phase SiMOC ceramic is obtained upon pyrolysis at rather low temperatures (ca. 700 °C), while at higher temperatures amorphous MO_2 nanoparticles precipitate (800 - 1100 °C). Upon increasing the annealing temperature to 1300 °C, MO_2 nanoparticles crystallize, forming microstructures comprised of tetragonal zirconia/hafnia particles finely dispersed within an amorphous SiOC matrix [87, 105–107]. When using other metals to modify the SiOC ceramic matrix, different and apparently contradictory results have been observed. Thus, in the mentioned case study, the phase composition of different SiMOC systems (M =

Sn, Fe, Mn, V, and Lu) was investigated [108]. Depending on the metal, different ceramic phases formed. For M = Mn and Lu, MO_x/SiOC ceramic nanocomposites were formed, whereas other compositions revealed the formation of M/SiOC (M = Sn), MSi_x/SiOC (M = Fe) or MC_x/SiOC (M = V) upon pyrolysis [87, 108] (Fig. 6). The strong effect of the precursor composition on the phase evolution upon

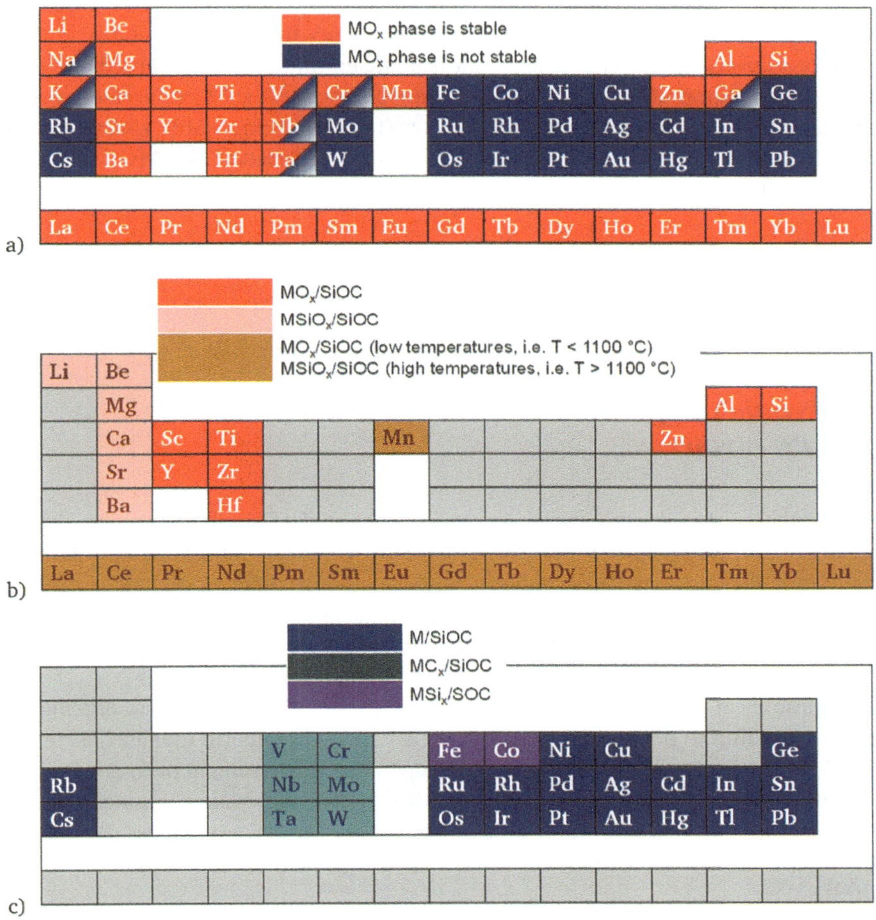

Fig. 6 Thermodynamic control in Si-M-O-C-based ceramic nanocomposites: a oxides of the red marked elements are stable against carbothermal reduction; whereas those of the blue marked elements get reduced in carburizing conditions; **b** metals which upon incorporation into SiOC lead directly to silicate formation (pink); the orange marked elements induce firstly the precipitation of MO_x (at T < 1100 °C), followed by the formation of silicates at higher temperatures; modification of SiOC with the red marked elements leads to MO_x/SiOC nanocomposites; **c** the blue marked metals lead to metallic nanoparticles within the SiOC matrix; the modification of SiOC with the purple marked metals (e.g., Fe, Co) generates metal silicide phases; group 5 and 6 transition metals (green marked) lead to carbide crystallization within SiOC matrix [108]

ceramization reflects the reducing conditions during pyrolysis and annealing. Thus, the thermodynamic stability of the metal oxides generated during pyrolysis, particularly with respect to the system C-O, plays a crucial role. Based on thermodynamic data of the respective oxides, the phase composition of SiMOC/SiMCNO ceramics upon annealing at high temperatures can be predicted for different metals (Fig. 2). The prediction agrees with the experimental results from this study and those reported in the literature for both SiMOC and SiMCNO ceramic nanocomposites. However, in addition to the stability of the oxides with respect to their reduction in carburizing conditions, some other aspects must be taken into account for predicting the phase composition of SiMOC/SiMCNO nanocomposites, such as thermodynamic stabilization through conversion into silicates (for MO_x being stable with respect to carbothermal conversion into M) or into silicides or carbides (for MO_x not being stable against carbothermal reduction).

3 Synthesis of UHTCs and (Nano)composites Thereof from Preceramic Polymers

3.1 Synthesis of Pure UHTC Phases from Polymeric Precursors

The synthesis of polymeric precursors used to access UHTC phases can be divided into two distinct categories: (i) methods using chemical modifications of metal alkoxides with e.g., (poly)alcohols, (poly)acetates or other compounds such as acetylacetone, salicylic acid, etc.; and (ii) methods using chemical modification of organometallic compounds such as metallocene-type compounds or other metal complexes with alkyl, alkenyl, alkynyl or aryl ligands.

Thorne et al. studied the reaction of Ti isopropoxide with various organic diacetates leading to polymeric compounds, so-called "polymeric titanates" (Fig. 7), which were converted into titanium carbide upon thermal treatment in Ar atmosphere at temperatures of 1000–1600 °C [109]. All prepared materials contained specific amounts of excess carbon. Interestingly, despite the presence of excess carbon, the obtained TiC_x titanium carbide materials were sub-stoichiometric, with $x = 0.4–0.6$ [109].

Using the same principle, metal alkoxides may be used to prepare polymeric metallates upon reactions with di-alcohols [110–113], acetylacetone or combinations thereof [114–120] (Fig. 8). Whereas in some cases the prepared polymers were structurally characterized to some extent, many other publications do not consider the molecular structure of the synthesized polymeric metallates and report rather on their conversion into metal carbide compounds upon pyrolysis at elevated temperatures (1000–1600 °C) in inert gas atmosphere.

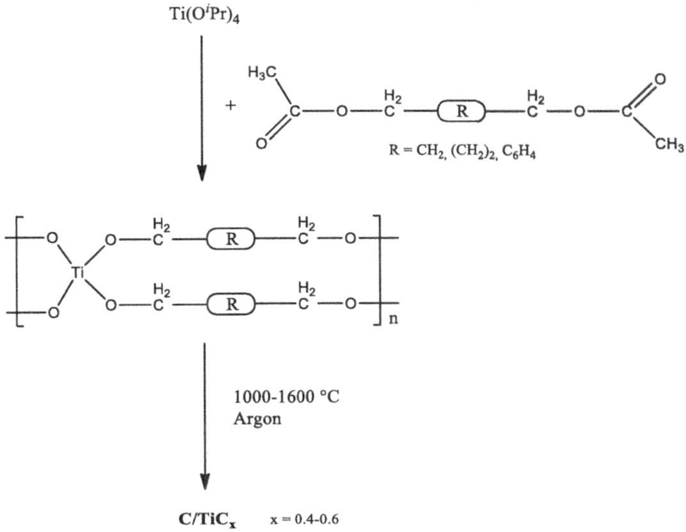

Fig. 7 Schematic description of the synthesis of TiC from polymeric titanates [109]

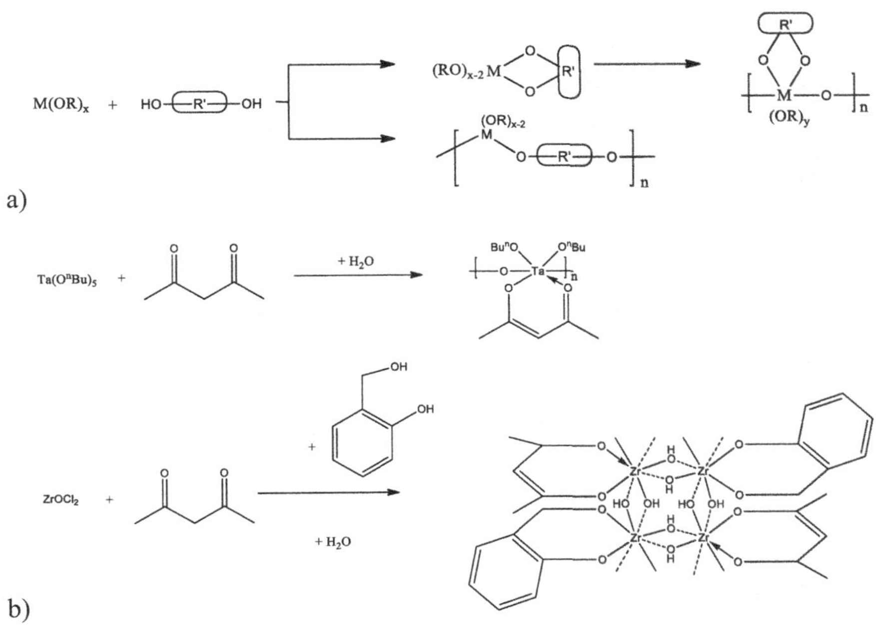

Fig. 8 Synthetic access to polymeric metallates; **a** based on metal alkoxides and di-alcohols (R = alkyl; R′ = C_xH_{2x}) [113, 121]; **b** based on metal alkoxides and acetylacetone [118] as well as on metal salts and a combination of acetylacetone and a di-alcohol [117]

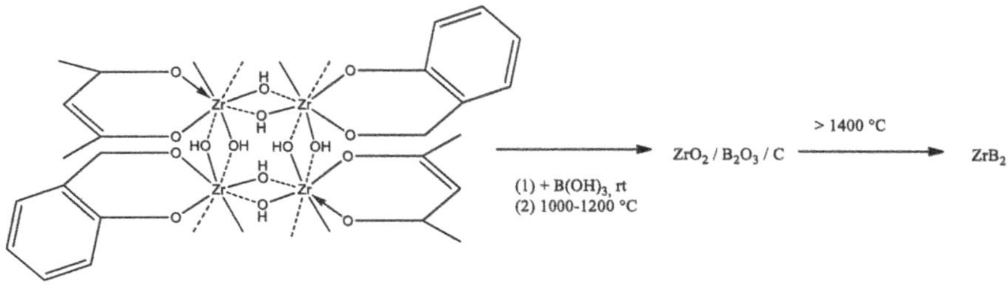

Fig. 9 Synthesis of ZrB_2 upon high thermal conversion of $B(OH)_3$-modified polyzirconoxanesal in argon atmosphere [123]

The conversion of the polymeric metallates into metal carbides is considered to occur via the intermediate formation of metal oxide and excess carbon at temperatures ranging from 800 to 1100–1300 °C; at higher temperatures, the metal oxide phase reacts with carbon and converts into metal carbide. Depending on the molecular structure of the polymeric metallates (i.e., nature and amount of organic substituents), various amounts of excess carbon can be generated in the resulting C/MC_x nanocomposites.

If the thermal conversion of the polymeric metallates is performed in reactive atmospheres, other compounds are accessible. For instance, Kurokawa et al. converted polymeric metallate precursors in ammonia atmosphere into metal nitrides MN_x (M = Ti, Nb, Ta) [122].

In a case study, the thermal conversion of a polymeric precursor consisting of polymeric zirconate (polyzirconoxanesal [117]) modified with boric acid was shown to lead to a mixture of zirconia, boria and segregated carbon at 1000–1200 °C, which finally converted into phase-pure ZrB_2 at temperatures above 1400 °C [123] (Fig. 9).

A second preparative tool to produce UHTC phases from polymeric precursors is based on organometallic precursors [124–126]. Already in the 80s, titanocene derivatives $CpTiCl_3$, Cp_2TiCl_2, $Cp_2Ti(R)Cl$ and Cp_2TiR_2 (Cp = cyclopentadienyl; R = alkyl, aryl) were used to prepare titanium carbide by gas-phase approaches (MO CVD) or thermal decomposition (see [124] and references therein). Moreover, various sandwich compounds of group 4 transition metals $(R\text{-}Cp)_2MR'_2$ (R = H, CH_3, $Si(CH_3)_3$; R′ = Cl, alkyl, aryl, NH_2, SH, OCN, SCN etc. [124]) were prepared and converted via thermal treatment in inert gas atmosphere into the respective metal carbides. It was shown that the decomposition of the molecular organometallic compounds occurs via formation of polymeric (macromolecular) intermediates, as shown in Fig. 10 [124].

Recently, the synthesis of Hf-containing polymeric precursors based on organometallic compounds and their high-temperature conversion into HfC/C composites was reported [125]. Typically, Cp_2HfCl_2 was reacted with bifunctional dilithio diynes (e.g., 1,4-dilithio-1,3-butadiyne or 1,4-bis(lithioethynyl)-benzene) to generate the corresponding polymeric precursors [125]. Additionally, a dialkenyl-substituted hafnocene was polymerized upon thermal treatment in the presence of dicumyl peroxide as radical polymerization catalysts (Fig. 11) [125].

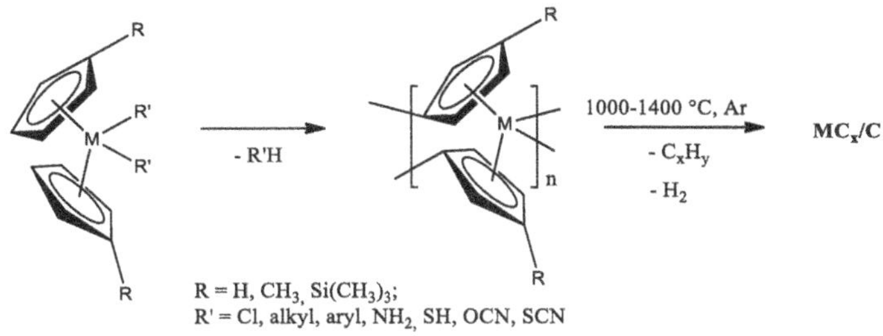

Fig. 10 Use of sandwich metal complexes (M = Ti, Zr, Hf) as precursors for the synthesis of MC_x/C nanocomposites. Their thermal decomposition was shown to occur via formation of polymeric intermediates [124]

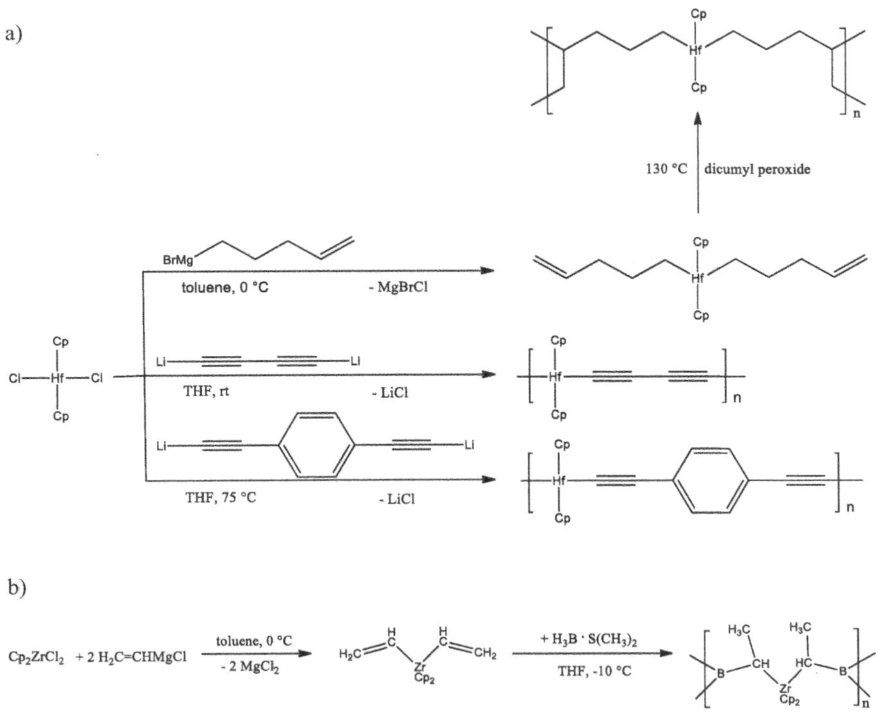

Fig. 11 **a** Synthesis of non-oxidic preceramic polymers for HfC starting from Cp_2HfCl_2 [125]; **b** synthesis of ZrC/ZrB_2 using a preceramic polymer synthesized from a divinyl-substituted zirconocene and borane [126]

Also, di-vinyl substituted zirconocene was reacted with borane to deliver a polymeric precursor which was thermally converted at 1000–1500 °C into a ZrC/ZrB_2 nanocomposite [126].

3.2 Synthesis of (Nano)composites Consisting of UHTCs and Silica Formers

The relatively poor oxidation stability of pure UHTCs phases (especially at moderate temperatures, i.e. 500–600 °C and 1200–1500 °C) pushed forward efforts to find different methods to improve their oxidation resistance. Thus, different approaches such as solid solutions, ternary diboride compositions, or the use of secondary phases were studied. Currently the most promising approach involves the use of additive phases (e.g., silicon carbide [127]), which can alter the composition and behavior of the oxide scale.

The beneficial effect of SiC addition on the oxidation behavior of diborides has been explained to rely on the formation of a borosilicate glass layer on the oxidized ceramics [15]. For instance, extensive studies on the oxidation behavior of SiC-containing diboride materials indicate the formation of a silica layer at the beginning of the oxidation process [127]. Accordingly, the further oxidation is expected to be limited by the inward oxygen transport through the formed silica scale. Additionally, boria is formed as an oxidation product of the diborides and contributes to a significant decrease of the viscosity of the silica scale. Consequently, the oxygen transport through the borosilicate glassy scale is increased and at the same time boria partially vaporizes from the surface of the glass scale. A multilayer structure has been proposed for the oxidized diborides, which show a porous oxide layer (zirconia, hafnia) below the borosilicate glass layer. At high oxidation temperatures (i.e., beyond 1500 °C) an additional layer located between the porous oxide layer and the parent materials was analyzed. It consists of SiC-depleted diboride and was stated to relate to an active oxidation process of silicon carbide. Accordingly, the silicon carbide converts into gaseous silicon monoxide, SiO, which is transported through the porous oxide layer towards the surface, where it reacts with oxygen to form silica, SiO_2 [15, 128–130].

In addition to silicon carbide as a silica former, silicon nitride [131, 132], molybdenum disilicide [133, 134] or other transition metal silicides [135, 136] have been used. These systems exhibit similar oxidation behavior to that of MB_2/SiC composites.

3.2.1 UHTC (Nano)composites with Low Content of Silica Former (i.e., 10–20 vol%)

Similar to the preparative methods presented in Sect. 4.1 for UHTC phases, the access to UHTC-based (nano)composites with typically 10–20 vol % silica former

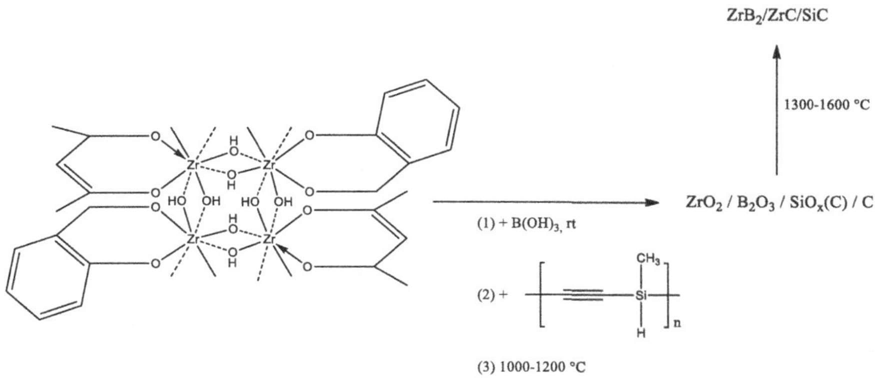

Fig. 12 Synthesis of ZrB_2/ZrC/SiC from a polymer blend consisting of $B(OH)_3$-modified polyzirconoxanesal and poly(methylsilylene-ethynylene) upon high-temperature thermal treatment in argon atmosphere [141]

phase (e.g., SiC) was reported in the last decade using the same synthesis tools. Thus, various publications from the last decade report the synthesis of MC/SiC (M = Zr, Hf, Ta) using polymeric precursors based on polymeric metallates either blended or modified with poly(carbo)silanes [118, 137–139].

For instance, a polymeric zirconate (polyzirconoxanesal [117]) was mixed with boric acid and poly(methylsilylene-ethynylene) [140] and was converted via thermal treatment in argon atmosphere at 1000–1400 °C into ZrB_2/ZrC/SiC/C composites (Fig. 12) [141].

Similarly, ZrC/SiC, HfC/SiC as well as TaC/SiC were prepared from corresponding polymeric precursors based on polymeric metallates blended with e.g. poly(allylhydridocarbosilane) [118, 137–139]. More recently, a case study reported on the synthesis of HfC/SiC from precursors based on polymeric metallates modified with TEOS as a source for SiC [142].

A different approach was reported by Amorós et al., based on poly(dimethylsilane), PDMS, or poly(methylhydridocarbosilane), PCS, which were mixed with Cp_2MCl_2 [143]. The polymeric precursors, resulted from the thermal treatment under reflux condition, were then thermally converted into MC/SiC/C in argon atmosphere at temperatures in the range from 900 to 1400 °C [143]. This synthesis methods allows adjusting the M:Si ratio (i.e., from 1:1 to 1:10) as well as the amount of free carbon (between ca. 15 and 50 wt%) in the prepared nanocomposites. As expected, the use of PCS led to an improved ceramic yield after pyrolysis, which relies on the presence of cross-linking active Si–H bonds in PCS. However, for both PDMS and PCS, rather low ceramic yields (10–25 wt%) after pyrolysis at 900 °C were reported (Fig. 13) [143].

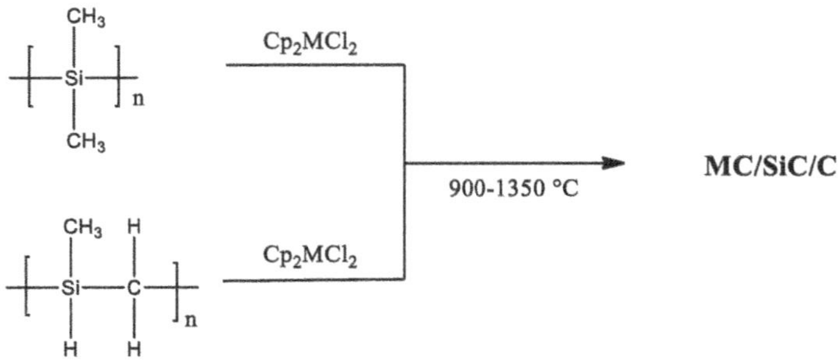

Fig. 13 Synthesis of MC/SiC/C nanocomposites (M = Ti, Zr, Hf) from polydimethylsilane or polymethylhydridocarbosilane modified with metallocene dichlorides (Cp = cyclopentadienyl) [143]

3.2.2 UHTC (Nano)composites with High Content of Silica Former (i.e., 80–90 vol%)

In this part, we consider (nano)composite materials in which the UHTC phase is preferentially distributed in a silica former matrix such as Si_3N_4, SiC and more complex systems like Si-B-C-N. A general strategy is to modify precursors of the silica former phase at molecular scale by metalorganic compounds. This allows the generation of composites or nanocomposites made of (at least) one more phase in addition to the SiC, Si_3N_4 or Si-C-B-N phase. Such materials often exhibit dramatically enhanced mechanical properties and/or structure stability [87, 144, 145]. Additional routes consist of (i) mixing polymers with metal fillers [146], (ii) blending two precursors of the desired composition [147, 148], (iii) copolymerizing two monomeric precursors to form co-polymers [149].

(Nano)composites with Si_3N_4(C,O) Matrix as Silica Former

Riedel et al. described the reaction between $TiCl_4$ and bis(trimethylsilyl)carbodiimide leading to a single source polymeric {[Ti(NCN)$_2$]$_3$NCNSiMe$_3$} compound as precursor suitable to form a $TiC_{0.3}N_{0.7}$/Si-C-N nanocomposite [150]. Accordingly, the main process to design UHTC nanocomposites with Si_3N_4 and Si-N(C,O) matrices as silica former is based on the chemical modification of pre-formed preceramic polymers with low molecular weight species such as metalorganic compounds to form so called polymeric single-source precursors (SSP), in which a uniform chemical composition is established at the molecular scale. Then, the SSPs are converted in a first pyrolysis step under ammonia to reduce or even eliminate the carbon-based organic groups present in the polymer while a single-phase amorphous ceramic is generated at 1000 °C. The latter is subsequently heat-treated at higher temperatures under nitrogen to form the UHTC nanophase providing nanocomposites composed of MN nanocrystals (M = Ti, Hf)

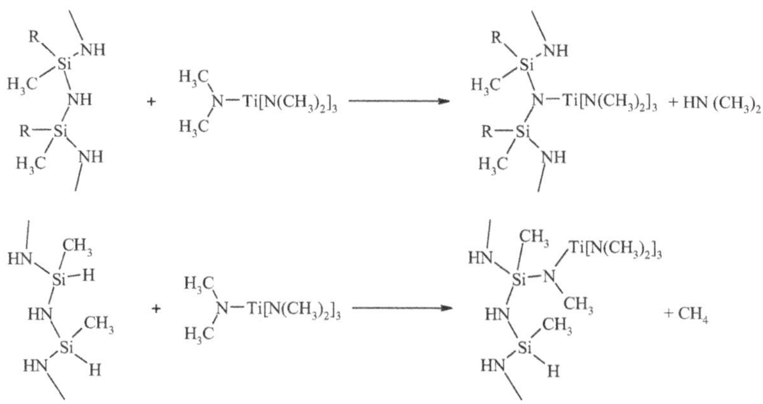

Fig. 14 Reaction between NH and SiH units present in polysilazanes and NCH_3 groups present in TDMATi [154]

homogeneously distributed in an amorphous matrix [151–155]. The first report focused on the design of TiN/Si_3N_4 nanocomposites involved the reaction between a commercially available polysilazane, i.e., perhydropolysilazane (PHPS), and a metallorganic compound namely tetrakisdimethylaminotitanium (TDMATi) in a 2.5:1 molar ratio. The reaction formed a perhydropolytitanosilazane with a chemical formula of $[Si_{1.1}Ti_{0.4}C_{2.2}N_{1.1}H_{6.2}]_n$ and a molecular weight of 4200 g mol^{-1} [152]. After ammonolysis at −40 °C, the polymer could be warm-pressed at 110 °C under 162 MPa uniaxial pressure. Subsequent pyrolysis of the formed body under ammonia (1000 °C) and nitrogen (1300 °C) produced bulk nanocomposites with 70 wt% yield and comprised of TiN nanocrystallites with an average diameter of 3.1 nm homogeneously embedded in an amorphous Si_3N_4 matrix. The in-situ controlled growth of TiN nanocrystals associated with the absence of Si_3N_4 grain growth and the purity of the materials provided Vickers hardness as high as 25.1 GPa. Later, the same authors investigated the chemistry behind the process of preparing such nanocomposites to gain new insight into the reaction mechanisms of these materials and to understand the role of chemistry behind the processability and design of TiN/Si_3N_4 nanocomposites [154]. Based on complementary characterization tools including solid state NMR, elemental analyses and infrared spectroscopy, two mechanisms have been proposed to govern the reaction between polysilazane and TDMATi. Mechanisms involve N–H and Si–H bonds present in polysilazanes as well as $N(CH_3)_2$ groups in TDMATi (Fig. 14).

Detailed solid-state NMR studies in combination with quantitative thermogravimetry provided a comprehensive understanding of the polytitanosilazane-to-nanocomposite transformation process. Accordingly, a predominantly amorphous ceramic network is formed first at 1000 °C under ammonia. Subsequent annealing in N_2 at 1400 °C produces nanocomposites comprised of homogeneously dispersed TiN nanocrystals embedded in an amorphous matrix (Si_3N_4). The structural evolution above 1000 °C is closely related to (i) the chemistry of polysilazanes (precursors of the Si_3N_4 matrix), (ii) the molar Si:Ti ratio fixed during the synthesis of the polytitanosilazane, and (iii) the material shape (powder, mesoporous monoliths or dense materials) [153]. It appears that TiN is the first phase to nucleate at 1400 °C in such materials and the TiN crystallite growth mainly depends on the molar Si:Ti ratio fixed in the polytitanosilazane. Interestingly, the onset of the Si_3N_4 crystallization temperature in MN/Si_3N_4 nanocomposites is shifted upward with increase of the volume fraction of TiN in the ceramics. The Si_3N_4 crystallization was significantly reduced even after annealing at 1800 °C for the nanocomposite with the highest Ti content. Similar effects have also been observed by replacing titanium by hafnium to form HfN/Si_3N_4 nanocomposites from Hf-modified PHPS. In the latter, HfN nanocrystals are embedded in an amorphous SiN_x matrix after annealing at 1300 °C whereas full crystallization of both phases is found after annealing at 1600 °C [155]. Thus, it appears that the crystallization behavior of the Si_3N_4 matrix in MN/Si_3N_4 nanocomposites is governed by the chemistry of the polysilazanes and the Si:M ratio fixed at the polysilazane level.

By changing the chemistry of the precursors and using polyhydridomethylsiloxane (PHMS) as a matrix precursor and titanium tetra-n-butoxide as a precursor of the nanophase, nanocomposites made of TiN nanocrystals of approximately ~ 4 nm size homogeneously distributed in an amorphous SiOCN matrix have been generated at 1200 °C [151]. The photocatalytic properties of the nanocomposites with respect to dye degradation have been evaluated. The visible light photocatalytic activity of the synthesized nanocomposite resulted in a degradation efficiency of 64% and the hydroxyl radicals were found to be the active species.

(Nano)composites with SiC(N) Matrix as Silica Former—MC/SiC(N) and MN/SiC(N) (Nano)composites

The design of (nano)composites with SiC(N) matrix as silica former has been well reported in the literature. The first strategy consists of using passive or active fillers mixed in the SiC(N) precursor. Whereas passive fillers such as ZrC powders do not interact with the SiC precursor upon pyrolysis in order to form additional phases in the SiC(N) material after the pyrolysis and annealing treatments [156, 157], active fillers do interact with preceramic polymers. As an illustration, Seyferth et al. investigated this approach to prepare Si-C-W systems using polysilazanes and W powders [158]. In the same study, Ti and Zr were combined with polysilazane and pyrolyzed under Ar at 1500 °C to form mixtures of TiN, ZrN and SiC. This approach offers easy access to composite materials in a relatively large range of compositions according to the choice of both the metal and the organosilicon precursor. As an illustration, the pyrolysis of polycarbosilane (PCS) mixed with Ti, Zr and Hf to 1500 °C in a stream of argon delivered MC/SiC composites [159]. Using a carbon-rich SiC precursor, i.e., $[Me_2SiC{\equiv}C]_n$, as a reagent and Ti, Zr and Hf as active fillers, different MC/SiC compositions have been prepared [159]. In contrast, the pyrolysis of a Si-rich SiC precursor, i.e., $[(MeSiH)_{\sim 0.4}(MeSi)_{\sim 0.6}]_n$, under argon at 1500 °C led to TiC and Ti_5Si_3

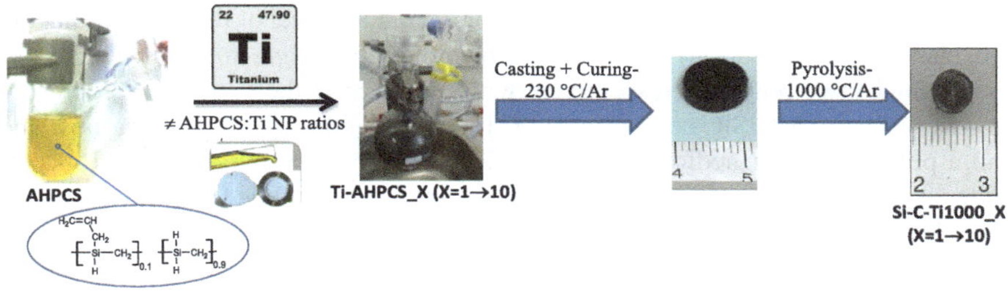

Fig. 15 Overall process for preparing dense monolithic Si-C-Ti components [165]

phases. Riedel et al. mixed titanium powders with polysiloxane to form after pyrolysis to 1100 °C mixtures of titanium carbide and metal silicide [160]. More recently, a series of studies has been reported on the dispersion of titanium powders in polycarbosilanes with the objective to form the well-known MAX phase Ti_3SiC_2 [161–164]. All these works illustrate the simplicity of this polymer approach. However, the use of nanoparticles (diameter $\leq$ 100 nm) as active nanofillers in organosilicon polymers was not reported despite nanostructured materials can provide improved or new properties compared to their microstructured counterparts. In this context, the use of allylhydridopolycarbosilane (AHPCS) as a reagent and titanium (Ti) nanopowders as active fillers (Fig. 15) has been described to generate TiC/SiC nanocomposites [165].

During the pyrolysis under argon, Ti nanopowders reacted with carbon-based gaseous by-products (methane, ethylene, etc.) that evolved from AHPCS during its decomposition and/or with carbon radicals to form TiC as a unique phase after pyrolysis to 800 °C. The dominating crystalline TiC phase further crystallized in the temperature range 800–1000 °C while marking the Ti_3SiC_2 and Ti_5Si_3 nucleation [165]. This procedure suggests that Ti and/or TiC reacted with the AHPCS-derived matrix to form the Ti_5Si_3 and Ti_3SiC_2 phases. Upon further pyrolysis above 1000 °C, monolithic TiC/SiC composites gradually formed at 1400 °C. Most probably, free carbon identified in the samples played a key role in the formation of TiC and SiC from Ti_3SiC_2 and Ti_5Si_3 phases. Based on these studies, it appears that SiC precursors represent appropriate reagents to successfully generate MC/SiC composites via their mixture with metal (M) powders. However, the major disadvantage of this approach is that the processing may result in aggregation of the powders leading to compounds with an inhomogeneous distribution of the phases. Therefore, solution-based methods, in which the SiC and MC (M = Ti, Zr, Hf) precursors are mixed on the molecular scale, as well as single-source precursors (SSP) are preferred although more expensive.

The modification of SiC with Ti, Zr and Hf has been proposed since the 80 s to design (nano)composites with a SiC matrix as a silica former [12, 101–103, 166–176]. Since that time, different synthetic strategies are described in the literature to form MC/SiC(N) (nano)composites, by using either mixed precursors (or blended polymers), or by inserting the metal in the polymer at the level of the macromolecule (i.e. chemical modification) or directly in the monomer (*single-source* precursors).

Mixing of two polymers respectively with M and Si in their structures represent an easy access to MC/SiC composites [177]. Following this route, a polycarbosilane and a polyzirconobutanediol were blended in toluene to be pyrolyzed at 1500 °C under argon [178]. The distribution of the ZrC phase as well as the size of the ZrC crystals correlated with the time of the drying method.

Another route to access MC/SiC composites was based on the addition of a crosslinking agent, namely divinylbenzene (DVB), to the blended polymers consisting of polyhafnoxanesal ($[\text{-Hf-O-Hf-}]_n$), and PCS [139]. HfC was generated at high temperature through a carbothermal reduction of the Hf-O-C phase by the excess carbon in-situ formed from DVB and PCS during pyrolysis. After annealing at 1600 °C under argon, HfC/SiC powder with diameter less than 100 nm and a composition of ca. 13.5 vol% HfC, 64.5 vol% SiC and 22 vol% excess carbon was formed (Fig. 16).

A blend composed of a polyzirconoxanesal (Zr source), a phenyl-terminated polysilane (Si source) and bisphenol-A type benzoxazine was reacted to lead to a multi-element precursor, with ring-opening, condensation and addition polymerizations [179]. This method was used to fabricate ZrC/SiC ceramics, after a suitable pyrolysis of the complex polymeric network up to 1600 °C, under nitrogen atmosphere. The ceramic yields ranged from 40 to 65%, and the microstructures revealed well-distributed Zr, Si and C elements in the final material.

The chemical modification of a selected polysilane or a polycarbosilane containing reactive groups like SiH sites with a metal-containing precursor is probably the most common approach which has been explored to prepare MC/SiC composites and nanocomposites. Starting from polymethylsilane, Wang et al. prepared a series of polyzirconomethylsilanes namely PZMS by adjusting the weight ratio between the Zr-based compound, i.e., $Cp_2Zr(CH=CH_2)_2$ and the polymethylsilane [180]. The pyrolysis of PZMS resulted in a ceramic yield of 78.4 wt%. Crystalline ZrC and SiC phases were well distributed in the composite prepared at 1600 °C with the content of ZrC as high as 54.6% by weight. It was shown that the introduction of ZrC prevented the fast growth of crystalline β-SiC. Based on their work focused on the design of ZrC/SiC composites [174] and on the pioneering report of Latorre et al. [143], Yu et al. reported the preparation of TiC/SiC composites through the reaction of AHPCS and dichlorobis(cyclopentadienyl)titanium (Cp_2TiCl_2) [181]. The reaction was performed with three Cp_2TiCl_2:AHPCS ratios, i.e., 1:3, 1:2 and 1:1, followed by a pyrolysis in an argon atmosphere up to 900 °C and subsequent annealing at 1600 °C in an argon flow. It was shown that Cp_2TiCl_2 played a double role: (1) The cross-linking of the hybrid precursors was significantly catalyzed by Cp_2TiCl_2 and Si–Si dehydrocoupling and hydrosilylation effectively occurred during mixing. (2) Cp_2TiCl_2 reacted with AHPCS through reaction of SiH and TiCl groups forming HCl. These two functions of Cp_2TiCl_2 significantly improved the ceramic yield of the polymer (80.8 wt%) compared to prisitine AHPCS (60.5 wt%). The chemical composition as well as the microstructure could be tailored by the Cp_2TiCl_2:AHPCS ratios. AHPCS was also used to react with borazine via hydroboration before adding

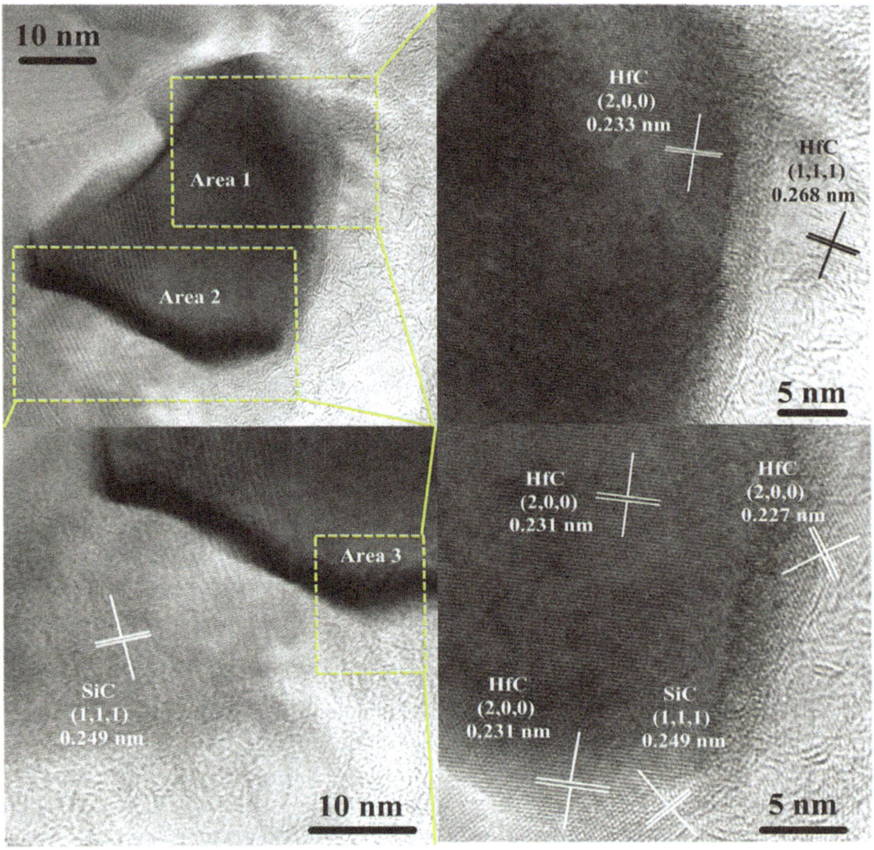

Fig. 16 TEM images of polymer-derived HfC/SiC powders at 1600 °C [139] (reprinted with permission from Wiley, 2018)

titanium butoxide $Ti(OBu)_4$ to be cross-linked at 170 °C [182]. By fixing B and excess C, Ti allowed to form crystalline TiB_2 and TiC phases in addition to the crystalline SiC phase through pyrolysis at 900 °C and annealing at 1800 °C under argon. Polycarbosilanes may also react with alkene functional groups present in metal-containing molecules through hydrosilylation. This approach was developed by Wang et al. using Zr-containing molecules [183, 184]. The authors performed functionalization of zirconocene dichloride ((Cp_2ZrCl_2) by using an organomagnesium reagent (i.e., allyl- ($CH_2 = CHCH_2MgCl$) or vinyl-magnesium chloride ($CH_2 = CHMgCl$)). Then, the as-obtained diallylbis(cyclopentadienyl)zirconium (DACZ) depicted in Fig. 17 could react with a low molecular weight polycarbosilane to give a soluble crosslinked preceramic polymer labelled PCS-DACZ via free radical polymerization [184]. The PCS-DACZ displayed a relatively high ceramic yield of 80 wt%. ZrC/SiC/C nanocomposites were retrieved after a pyrolysis at 1400 °C under

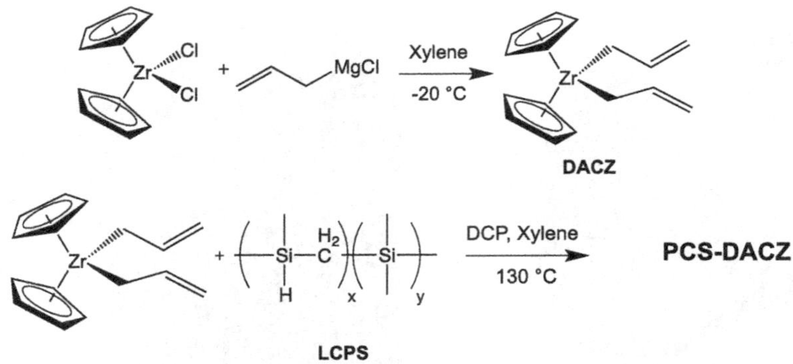

Fig. 17 Synthesis of DACZ and reaction with a low molecular weight PCS [183]

argon atmosphere, displaying an excellent high-temperature resistance up to 2000 °C with well-dispersed ZrC nanoparticles.

This strategy was also investigated with a zirconocene dichloride functionalized with vinyl groups to generate ZrC/SiC ceramics with a high Zr content [184]. The ceramics showed a good resistance to oxidation in air up to 1400 °C. In parallel to the reactivity of alkene groups, other methods were attempted involving N- or O-containing complexes of zirconium. For example, the two-step reaction between allylamine (C_3H_7N), zirconium tetrachloride ($ZrCl_4$), polymethylsilane (PMS), and $CH_2 = CHCH_2MgCl$ led to a soluble ZrC/SiC precursor. The composite was isolated in a 49% yield at 1000 °C, and the material exhibited less than 2 wt% of N in its structure [185]. In the field of polymer-derived ceramics, oxygen containing-zirconium complexes are often used. Reacting with polycarbosilanes, they can lead to ZrC/SiC ceramics after a suitable heat-treatment. In this way, Devasia et al. modified polycarbosilanes with zirconium acetylacetonate ($Zr(Acac)_4$) to synthesize polyzirconocarbosilanes [186]. The authors introduced different Zr:Si ratio and showed that the ceramic yield of the precursors was improved by the introduction of Zr, to reach 76 wt% at 1200 °C, under argon. HRTEM analyses were undertaken on the composites heat-treated at 1650 °C, revealing the presence of an oxide free ZrC/SiC composite.

In the case of hafnium, an earlier work reported on the synthesis of a precursor suitable to form HfC_xN_{1-x}/SiC nanocomposites [187]. The procedure involved the reaction between AHPCS, and tetrakis(dimethylamido)hafnium ($Hf(N(CH_3)_2)_4$, TDMAH) according to Fig. 18.

The functionalization of AHPCS was confirmed by IR and NMR spectroscopy. Amorphous SiHfCN-based ceramic powders were then densified using spark plasma sintering (SPS) at 2200 °C, and led to dense monolithic HfC_xN_{1-x}/SiC nanocomposites, highlighting homogeneously embedded hafnium carbonitride ($HfC_{0.83}N_{0.17}$) grains in a □-SiC matrix, encapsulated by in situ formed carbon layers. These materials were then characterized in detail regarding their dielectric properties [188], oxidation resistance [189–191] and laser ablation behavior [192]. HfC_xN_{1-x}/SiC nanocomposites display improved laser ablation resistance compared

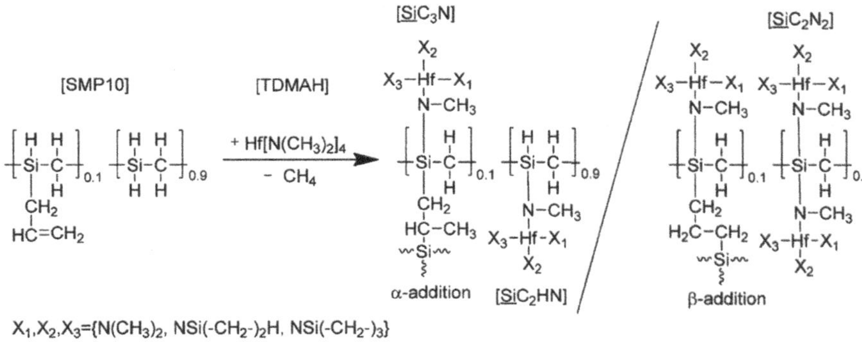

Fig. 18 Modification of AHPCS with $Hf(NEt_2)_4$ leading to two silicon environments in the resulting preceramic polymer (from [187]; we acknowledge the Royal Society of Chemistry for the reproduced material)

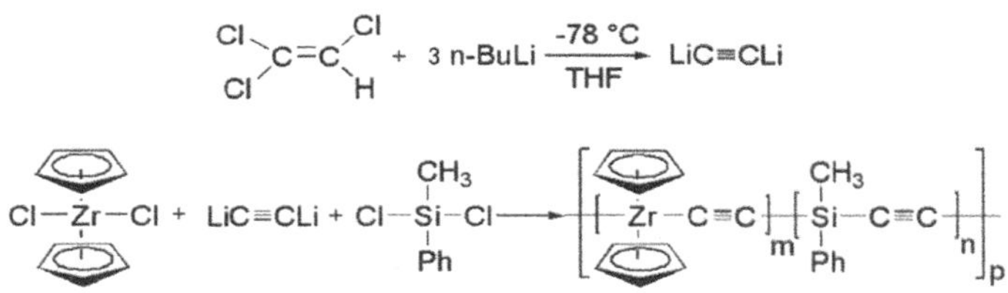

Fig. 19 Synthesis of a Zr and Si containing poly-yne polymer used as a single-source precursor for the synthesis of ZrC/SiC nanocomposites [193]

to pure SiC-based monoliths [192] and are especially suited for electromagnetic shielding in harsh environment [188, 190]. It was reported that the additional incorporation of tantalum (Ta) in the HfC_xN_{1-x} phase leads to an enhanced sintering ability of the oxide scale and to an improved oxidation resistance via a short pre-oxidation step at temperatures sufficiently high to activate the formation of silica [189, 192].

The last—but highly important—method to reach multi-element polymers with metal from the group 4 elements is the molecular synthesis of *single-source* precursors. This route is very useful especially in order to control the chemical composition of the preceramic precursor, and *in fine* of the ceramic, at the molecular scale. Upon salt metathesis reactions of zirconcene dichloride, methylphenyldichlorosilane and dilithioacetylene, Cai et al. synthesized a ZrC/SiC nanocomposite precursor according to Fig. 19 [193].

Thus, the prepared polymer dissolves well in different solvents, such as THF, toluene and chloroform etc.. The obtained polymer could be converted to nanosized ZrC/C/SiC composites at a relatively low temperature, i.e., 1400 °C with a ceramic yield of over 52 wt%. Starting with oxide compounds, Li et al. detailed the synthesis

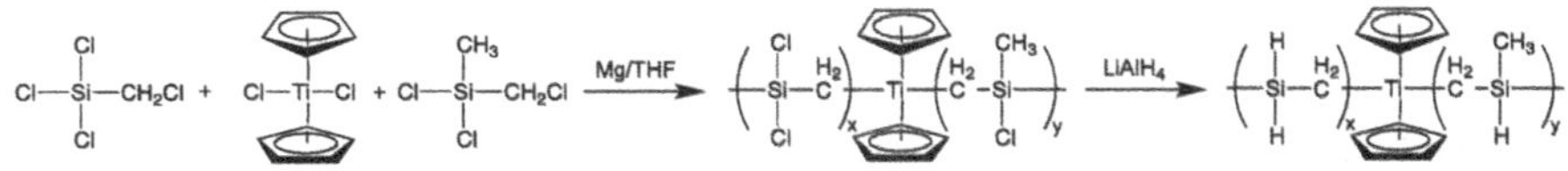

Fig. 20 Two-step synthesis of a polymeric single-source precursor containing Ti and Si mixed at molecular scale [195]

of a polymer with Si and Zr in its structure [194]. They used the reaction of hydrolysis/condensation between tetraethoxysilane ($Si(OC_2H_5)_4$ and zirconium oxychloride octahydrate ($ZrOCl_2 \cdot 8H_2O$) in the presence of phenol, paraformaldehyde and acetylacetone. The as-obtained precursor yielded ZrC/SiC/C ceramics after pyrolysis up to 1800 °C. However, SiO_2 and ZrO_2 phases were still present in the powders. Focusing on the titanium chemistry, Liu et al. used the incorporation of both Ti and Si in the backbone of a hyperbranched polycarbosilane labelled HPCS derived from Grignard coupling of a mixture made of trichloro(chloromethyl)silane (Cl_3SiCH_2Cl), Cp_2TiCl_2 and dichloromethyl(chloromethyl)silane ($Cl_2Si(CH_3)CH_2Cl$) followed by $LiAlH_4$ reduction according to Fig. 20 [195].

The thermal decomposition of the final liquid hyperbranched structure denoted as HPTiCS indicated a ceramic yield of 76 wt% at 1200 °C (higher by around 31% than that of the ceramic yield measured with the pure hydridopolycarbosilane) and led to TiC/SiC composites.

An original approach was recently described, using organolithium reagents instead of organomagnesium compounds. Moreover, the reactivity of the cyclopendatienyl groups from the zirconocene dimethyl ($Cp_2Zr(CH_3)_2$) was investigated this time. Wang et al. synthesized three preceramic polymers from an intermediate dilithiozirconocene complex as a key Zr-Li salt obtained by the reaction of $Cp_2Zr(CH_3)_2$ with tetramethylethylenediamine (TMEDA) and butyl lithium (*n*-BuLi) according to Fig. 21 [196].

Di- and trifunctional chlorosilanes were employed to react with the Zr-Li salt. After removal of LiCl by filtration, the respective polymers were obtained (Fig. 17). The polymer containing a vinyl group attached to the silicon atoms displayed the highest ceramic yield of 77 wt% under argon atmosphere at 800 °C. Annealing experiments from 1000 to 2000 °C showed the crystallization from zirconium silicate ($ZrSiO_4$) and zirconia (ZrO_2) at lower temperature to ZrC at higher temperature. The prepared micrometer-sized spheres were shown to possess interesting electromagnetic wave absorbing capacity [196].

It is also worth mentioning two other studies using the monomer route: Firstly, bis(cyclopentadienyl)hafnium dichloride (Cp_2HfCl_2) and trans-1,4-dibromo-2-butene were reacted in the presence of magnesium powder in THF to yield a macromolecular HfC precursor. A chain termination with chloromethyltrimethylsilane ($(CH_3)_3SiCH_2Cl$) is then added to introduce silicon into the structure of the polymer [197]. This preceramic precursor exhibited a good solubility in common organic solvents, and pyrolysis at 1600 °C resulted in HfC/SiC nanocomposite ceramics. Secondly, a polycarbosilane was reacted with zirconocene at the end of the silicon polymer chain [198]. Zhang et al. used a zirconocene as a catalyst with metallic

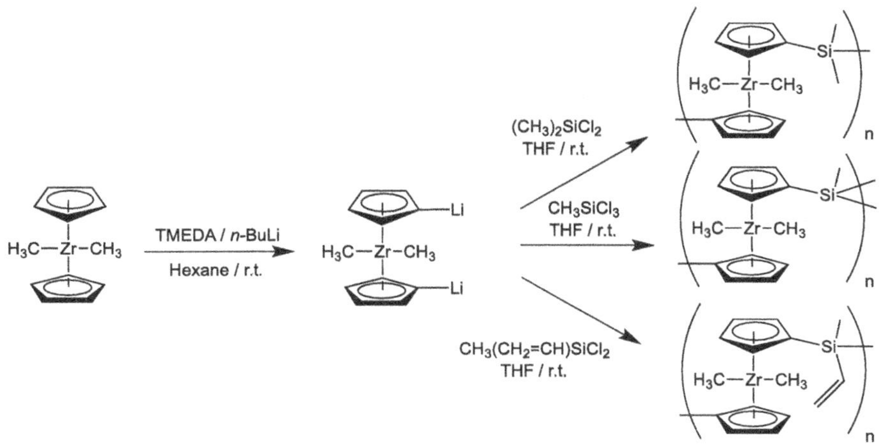

Fig. 21 Synthesis of preceramic polymers via the reaction of dilithiozirconocene complex with various chlorosilanes [196]

sodium (Na) for an insertion of 1-methylsilene ($H(CH_3)Si{=}CH_2$) into a preceramic precursor. The reaction product was denoted as polyzirconocenecarbosilane.

Metallic sodium was used to in-situ obtain 1-methylsylene, which polymerizes spontaneously at the Na-surface to PCS with zirconocene at the chain end groups. The thermal decomposition of the polymer with narrow molar-mass dispersity (M_w/M_n) led to ZrC/SiC ceramic composites with a ZrC concentration of about 15 wt%. This method was extended to titanocene and hafnocene, with apparently attractive rheological properties for fiber melt spinning.

In 2012, Li et al. reported mixing a polyzirconoxanesal with boric acid and a poly(methylsilylene)ethynylene to produce ZrB_2/ZrC/SiC composites by pyrolysis at 1400 °C under argon [141]. Following this study, He et al. reported the design of ZrB_2/SiC nanocomposites via a precursor route based on a similar simple one-pot method [199]. In this approach, a ZrB_2 precursor and a polycarbosilane (SiC precursor) were mixed in the desired ratio to form a molecular hybrid compound as a ZrB_2/SiC precursor with a targeted ZrB_2-to-SiC volume ratio of 25%. After pyrolysis at 1300 °C, a ceramic yield close to 61 wt% has been measured and t-ZrO_2 and β-SiC phases were identified exclusively by X-ray powder diffraction. Above this temperature, reaction of ZrO_2 with B_2O_3 and C occurred and led to the formation of the ZrB_2 phase along with SiC. Thus, after pyrolysis at 1700 °C for 4 h, a nanocomposite made of ZrB_2 nanoparticles embedded within a SiC matrix was generated, although ZrC was identified as an additional phase present in the powders. Later, the same authors performed spark plasma sintering (SPS) of these powders by changing the ZrB_2-to-SiC volume ratio from 6.4 to 61.5 vol% and investigated the mechanical properties of the as-prepared pieces [200]. SPS was performed at 1600 °C for 30 min under an argon atmosphere and a uniaxial pressure of 35 MPa. As expected, the Vicker's hardness of the monolithic samples increased with the

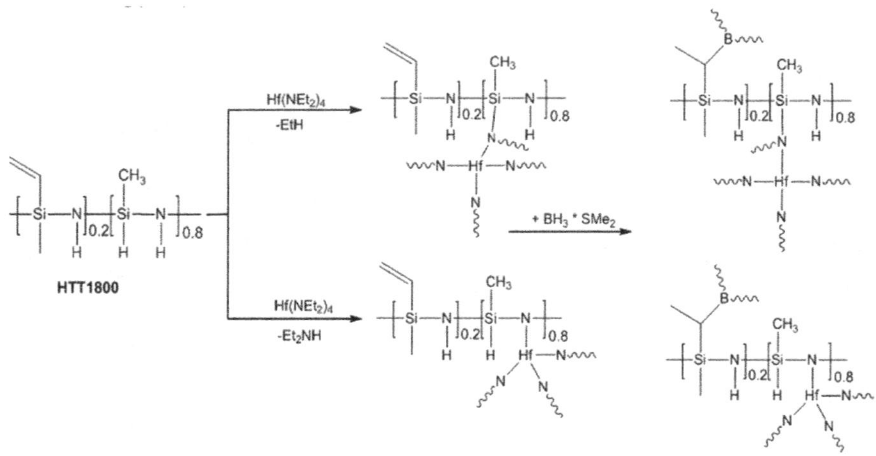

Fig. 22 Modification of HTT1800 with $Hf(NEt_2)_4$ and $BH_3{\cdot}SMe_2$ (from [202]; reproduced with permission from ACS, 2014)

increase of the density of the materials from 7.42 ± 0.3 GPa (relative density of 87.6%) to 17.54 ± 0.9 GPa (relative density of 99.1%).

(Nano)composites with Si-B-C-N Matrix as Silica Former

Adding boron to Si-C-N ceramics, i.e., leading to silicon boron carbonitride (Si-B-C-N) ceramics, improves the thermal stability of the amorphous structure of Si-C-N ceramics in terms of crystallization and decomposition [201]. This phenomenon is related with the strong local covalent bonds associated with the presence of boron in SiBCN. As a consequence, the carbothermal reaction of Si_3N_4 occurring in Si-C-N ceramics is shifted to higher temperature. The use of this amorphous inorganic network appears also to be beneficial for the high temperature creep behavior of these multicomponent ceramics. (Nano)composites with a Si-B-C-N matrix have been prepared following the synthesis process implemented for dense SiC/HfC_xN_{1-x}-based ultra-high temperature ceramic nanocomposites [187]. A commercially available polysilazane (HTT1800 (now Durazane® 1800), Merck)) was modified with $Hf(NEt_2)_4$ and $BH_3{\cdot}SMe_2$ (Fig. 22) and subsequently cross-linked and pyrolyzed [202].

The presence of Hf and B within the molecular structure of the polysilazane leads to low-temperature phase separation processes within the resulting Si-Hf-B-C-N. Depending on the annealing atmosphere, $HfC/HfB_2/SiC$ (annealing in argon) and $HfN/Si_3N_4/SiBCN$ (annealing in nitrogen) nanocomposites were obtained. The Si-Hf-B-C-N precursor could be warm-pressed after cross-linking at 250 °C then pyrolyzed and annealed to 1300 °C to form dense pieces [203]. The oxidation behavior of these bulk samples (and powders) was studied at 1200–1400 °C in air. The measurement of the surface-specific weight gain as a function of the isothermal oxidation time showed that both, powder and monolithic samples, exhibited a parabolic oxidation behavior characterized by low parabolic rates. Observations of the oxide scale that formed on the monoliths at different oxidation temperatures revealed that a continuous oxide scale consisting of borosilicate, silica (cristobalite), m-and t-HfO_2 formed at temperatures below 1300 °C, limiting the capacity for oxygen transport. At temperatures above 1300 °C, oxide evaporation increased and the oxide scale became discontinuous and consisted of silica, $HfSiO_4$ and m-and t-HfO_2, and large pores were formed in the bulk. Alternatively, the molecular Si-Hf-B-C-N precursor was cross-linked then pyrolyzed at 900 °C to deliver powders that could be sintered by SPS at 1850 or 1950 °C into dense monolithic samples [204]. The high hardness and high elastic modulus of the dense pieces highlighted superior mechanical properties as compared to other polymer-derived ceramics, which rely on their phase compositions and their fine microstructure. The investigation of the oxidation of the prepared dense ceramic nanocomposites at high temperature revealed that the parabolic oxidation rates of Si-Hf-C-N were comparable to those of ultra-high temperature ceramics such as metal-diboride/SiC composites, whereas the parabolic oxidation rates of Si-Hf-B-C-N were several orders of magnitude lower than those reported for UHTCs and even lower than values reported for SiC. Additionally, the Si-Hf-B-C-N precursor could be used to infiltrate carbon fiber preforms (C_f) and form after curing and pyrolysis C_f/Si-Hf-B-C-N composites [205, 206]. These ceramic matrix composites (CMCs) were tested regarding hydrothermal corrosion and laser ablation tests. The C_f/Si-Hf-B-C-N composites exhibited enhanced resistance towards hydrothermal corrosion as compared to C_f/Si-C-N due to the improved kinetics upon Hf and B incorporation. Additionally, a tight C_f/matrix interface (which is rather disadvantageous for mechanical behavior) was found to be beneficial for improved corrosion behavior in C_f/Si-Hf-B-C-N composites [205]. Various ablation mechanisms occurred in three distinct ablation regions, as detected by the study of the evolution of microstructure and phase composition of the CMCs by scanning electron microscopy (SEM) and energy dispersive spectroscopy (EDS) [206]: (i) in the central region, the erosion of C_f/Si-Hf-B-C-N is controlled by the sublimation of the components of the CMCs; (ii) in the transition region, the degradation of C_f/Si-Hf-B-C-N is dominated by the passive oxidation of the Hf-containing components and the active oxidation of the fibers and Si-containing components; (iii) in the outer region, the pronounced degradation process of C_f/Si-Hf-B-C-N is the passive oxidation of the Si-containing components. It has been found that the ablation mechanism and the microstructure evolution of CMCs are mainly dependent on the local temperature which directly depends on the distance from the laser spot and the thermal conductivity of the carbon fibers. Similarly, a hafnium chloride compound has been used to modify a liquid polyborosilazane containing borazine rings as bridges between two monomeric silazane units [207]. After pyrolysis at 1100 °C in an argon atmosphere, the final material consisted of HfN nanoparticles embedded in a Si-B-C-N matrix. The HfN nanoparticles distributed uniformly in the Si-B-C-N matrix provided an excellent oxidation resistance: the HfN/Si-B-C-N nanocomposite was shown to be more oxidation-resistive with a weight loss of 2.8 wt% at 1500 °C in air, compared to a weight loss of 5.4 wt% for Si-B-C-N ceramics under the same condition.

The addition of Zr in polymer-derived Si-B-C-N ceramics through the modification of a liquid polyborosilazane with Cp_2ZrCl_2 according to different weight ratios has been also reported [208]. In particular, fibers were prepared from this polymer to be cured with a BCl_3/N_2 mix flow gas, then pyrolyzed and annealed at 1400 °C under Ar. The Zr-modified Si-B-C-N ceramic fibers were amorphous after pyrolysis at 1200 °C. As the temperature was increased to 1400 °C, crystalline phases, namely ZrC_xN_y, ZrB_2 and Si_3N_4 were identified. Zr was also shown to improve the oxidation resistance of the Si-B-C-N ceramics as reported for the Hf counterpart.

4 Conclusions and Outlook

The polymer-derived ceramic (PDC) route represents a unique preparative access to UHTCs and their derived (nano)composites containing silica formers. These materials possess adjustable phase compositions and microstructures according to the fact that the chemical composition of the materials is controlled at a molecular scale. Thus, two strategies may be applied to deliver UHTC: (i) methods using chemical modifications of UHT-metal containing alkoxides with e.g., (poly)alcohols, (poly)acetates or other compounds such as acetylacetone, salicylic acid, etc.; and (ii) methods using chemical modification of organometallic compounds such as metallocene-type compounds or other metal complexes with alkyl, alkenyl, alkynyl or aryl ligands. In order to make accessible UHTC-containing nanocomposites (with either low or high content of silica former phase), a Si-containing precursor is in general added to the UHTC precursor forming either precursor blends or, in the case of chemical reactions, single-source polymeric precursors which are further pyrolyzed at elevated temperatures ($\gg$ 1000 °C). It may be mentioned that typically polymeric blends are used to prepare nanocomposites with low volume fractions of silica former phase (i.e., 10–20 vol%); whereas single-source precursors are highly suitable for preparing nanocomposites with high content of silica former phase (i.e., 80–90 vol%).

There are obvious advantages of the polymeric-precursor-based synthesis methods of UHTCs and their nanocomposites, among them e.g. nanoscopic particle size, high purity materials, relatively low synthesis temperatures. However, specific issues related to the use of polymeric/organometallic precursors should be also mentioned such as relatively high synthetic efforts (especially for the single-source precursors) or, in some cases, the environmental sensitivity of the polymeric precursors (i.e., for hydrolysis, oxidation processes). The industrial potential of UHTCs and derived (nano)composites is nevertheless evident, despite the PDCs process needs to be further developed in details for industrial and technological applications, e.g. raw materials availability, process scalability etc. In particular, following aspects should be intensively addressed in the future in order to provide a rational design of UHTCs and their derived (nano)composites with tailored compositions, microstructures and properties: (i) a straight-forward synthetic access to single-source precursors with a detailed knowledge on their chemical compositions and molecular architectures; (ii) a fundamental understanding of the polymer-to-ceramic conversion process; (iii) polymer chemistry–rheology correlation to develop the processability of these polymers and design complex textures and morphologies of materials, (iv) joint and tailored experimental and computational studies are necessary in order to assess the properties of the UHTCs and their derived (nano)composites and to understand how the microstructural features of the materials affect their properties; and (v) detailed short and long-term analysis of their high-temperature behavior and mechanical properties (hardness, Young Modulus, fracture toughness, etc.). It is anticipated that polymer-derived UHTCs and their derived (nano)composites have the potential to provide specific functionalities as well as improved and/or extended properties to be applied in particular in the societal relevant fields of energy conversion and future aeronautics and re-entry systems.

Acknowledgements The authors team is grateful to Prof. Yanchun Zhou (Aerospace Research Institute of Materials and Processing Technology, Beijing, China) for encouraging to compile the present review on polymer-derived UHTC systems. EI and RR thank the German Science Foundation (DFG, Bonn, Germany) for financial support (project "Micropatterned polymer-derived ceramic catalysts and sensors"). EI furthermore acknowledges the Heisenberg Program of DFG. PK gratefully acknowledges support by the National Science Foundation (CMMI-1634448 and OISE 1743701).

References

1. Wuchina, E.J., Opila, E., Opeka, M.M., Fahrenholtz, W., Talmy, I.G.: Interface **16**, 30–36 (2007)
2. Fahrenholtz, W.G., Hilmas, G.E.: Scripta Mater. **129**, 94–99 (2017)
3. Wuchina, E., Opila, E., Opeka, M., Fahrenholtz, W., Talmy, I.: Electrochem. Soc. Interface **16**(4), 30–36 (2007)
4. Squire, T.H., Marschall, J.: J. Eur. Ceram. Soc. **30**, 2239–2251 (2010)
5. Van Wie, D.M., Drewry, D.G., King, D.E., Hudson, C.M.: J. Mater. Sci. **39**, 5915–5924 (2004)
6. Fairbank, G.B., Humphreys, C.J., Kelly, A., Jones, C.N.: Intermetallics **8**, 1091–1100 (2000)
7. Fuchs, G.E.: High temperature alloys. In: Kirk-Othmer Encyclopedia of Chemical Technology. Wiley (2000)
8. Muktinutalapati, N.R.: Materials for gas turbines—an overview. In: Benini, E. (ed.) Advances in Gas Turbine Technology, Intech, pp. 293–314 (2011)
9. Reed, R.C.: The Superalloys—Fundamentals and Applications. Cambridge University Press (2006)
10. Spear, K.E., Visco, S., Wuchina, E.J., Wachsman, E.D.: Interface **15**, 48–51 (2006)
11. Blum, Y.D., Kleebe, H.J.: J. Mater. Sci. **39**, 6023–6042 (2004)
12. Opeka, M.M., Talmy, I.G., Zaykoski, J.A.: J. Mater. Sci. **39**, 5887–5904 (2004)
13. Fahrenholtz, W.G., Hilmas, G.E., Talmy, I.G., Zaykoski, J.A.: J. Am. Ceram. Soc. **90**, 1347–1364 (2007)
14. Paul, A., Jayaseelan, D.D., Venugopal, S., Zapata-Solvas, E., Binner, J., Vaidhyanathan, B., Heaton, A., Brown, P., Lee, W.E.: Am. Ceram. Soc. Bull. **91**, 22–28 (2012)
15. Fahrenholtz, W.G., Hilmas, G.E.: Int. Mater. Rev. **57**, 61–72 (2012)
16. Chamberlain, A.L., Fahrenholtz, W.G., Hilmas, G.E., Ellerby, D.T.: J. Am. Ceram. Soc. **87**, 1170–1172 (2004)

17. Dub, S.N., Goncharov, A.A., Ponomarev, S.S., Filippov, V.B., Tolmacheva, G.N., Agulov, A.V.: J. Superhard Mater. **33**, 151–158 (2011)
18. Predel, B.: B-Ti (Boron-Titanium). In: B. (ed.) B-Ba ... Cu-Zr: supplement to subvolumes IV/5B, IV/5C and IV/5D Predel, pp. 77–78. Springer, Berlin, Heidelberg (2012)
19. OuYang, X., Yin, F., Hu, J., Liu, Y., Long, Z.: J. Phase Equilib. Diffus. **38**, 874–886 (2017)
20. Frisk, K., Fernández Guillermet, A.: J. Alloy Compd. **238**, 167–179 (1996)
21. da Silva, A.A.A.P., Ferreira, F., de Lima-Kuhn, B.B., Coelho, G.C., Nunes, C.A., Vilasi, P., Fiorani, J.-M., David, N., Vilasi, M.: CALPHAD: Comput. Coupling Phase Diagrams Thermochem. **63**, 107–115 (2018)
22. Cacciamani, G., Riani, P., Valenza, F.: Calphad **35**, 601–619 (2011)
23. Lawson, J.W., Bauschlicher, C.W., Daw, M.S.: J. Am. Ceram. Soc. **94**, 3494–3499 (2011)
24. Xiang, H.M., Feng, Z.H., Li, Z.P., Zhou, Y.C.: J. Am. Ceram. Soc. **100**, 3662–3672 (2017)
25. Zhang, G., Zhao, Y.X., Zhu, J., Hao, Y.J., Zhang, L.: Int. J. Mod. Phys. B **32**, 15 (2018)
26. Li, H., Zhang, L.T., Zeng, Q.F., Wang, J.J., Cheng, L.F., Ren, H.T., Guan, K.: Comput. Mater. Sci. **49**, 814–819 (2010)
27. Li, J., Fan, C.Z.: Phys. Chem. Chem. Phys. **17**, 1180–1188 (2015)
28. Van der Geest, A.G., Kolmogorov, A.N.: Calphad-Comput. Coupling Phase Diagrams Thermochem. **46**, 184–204 (2014)
29. Lee, D., Vassak, J.J., Zhao, K.J.: Nano Lett. **15**, 6553–6558 (2015)
30. Pan, Y., Huang, H.W., Wang, X., Lin, Y.H.: Comput. Mater. Sci. **109**, 1–6 (2015)
31. Goldstein, A., Geffen, Y., Goldenberg, A.: J. Am. Ceram. Soc. **84**, 642–644 (2001)
32. Zhang, S.C., Hilmas, G.E., Fahrenholtz, W.G.: J. Am. Ceram. Soc. **89**, 1544–1550 (2006)
33. Pathak, S.K.M., Das, S., Das, S.K., Ramachandrarao, P.: J. Mater. Res. **15**, 2499–2504 (2000)
34. Feng, H.J., Moore, J.J., Wirth, D.G.: Metall. Trans. A **23**, 2373–2379 (1992)
35. Meschel, S.V., Kleppa, O.J.: J. Alloys Compd. **321**, 183–200 (2001)
36. Topor, L., Kleppa, O.J.: J. Chem. Thermodyn. **17**, 1003–1016 (1985)
37. Medvedev, V.A., Bergman, G.A., Vasil'ev, V.P., et al.: In: Glushko, V.P. (ed.) Thermal Constants of Substances, vol. 7, 343 pp. (1974)
38. Ma, X., Li, C., Du, Z., Zhang, W.: J. Alloy Compd. **370**, 149–158 (2004)
39. Rudy, E., Windisch, S.: Ternary phase equilibriums in transition metal-boron-carbon-silicon systems. Part I. Related Binary System, vol. VII, Ti–B System, p. 234. Aerojet-Gen. Corp. (1966)
40. Chase, M.W.J.: NIST-JANAF Thermochemical Tables. American Inst. of Physics (1998)
41. Ettmayer, P., Kieffer, R., Hattinger, F.: Metall (Berlin) **28**, 1151–1156 (1974)
42. Sheindlin, M., Falyakhov, T., Petukhov, S., Valyano, G., Vasin, A.: Adv. Appl. Ceram. **117**, s48–s55 (2018)
43. Okamoto, H.: J. Phase Equilib. **14**, 261–262 (1993)
44. Sridar, S., Kumar, R., Hari Kumar, K.C.: CALPHAD: Comput. Coupling Phase Diagrams Thermochem. 56, 102–107 (2017)
45. Wang, C., Zinkevich, M., Aldinger, F.: J. Am. Ceram. Soc. **89**, 3751–3758 (2006)
46. Johnson, G.K., Greenberg, E., Margrave, J.L., Hubbard, W.N.: J. Chem. Eng. Data **12**, 137–141 (1967)
47. Meschel, S., Kleppa, O.: J. Chim. Phys. **90**, 349–354 (1993)
48. Rudy, E., Windisch, S.: Ternary phase equilibriums in transition metal-boron-carbon-silicon systems. I. Related Binary systems. 10. Systems Vanadium-Boron, Niobium-Boron, and Tantalum-Boron, p. 119Aerojet-Gen. Corp. (1966)
49. Mah, A.D., Gellert, N.L.: J. Am. Chem. Soc. **78**, 3261–3263 (1956)
50. Talmy, I., Zaykoski, J.: Ceram. Eng. Sci. Proc. **19**, 105–112 (1998)
51. Zhang, G.J., Deng, Z.Y., Kondo, N., Yang, J.F., Ohji, T.: J. Am. Ceram. Soc. **83**, 2330–2332 (2000)
52. Lengauer, W.: Transition metal carbides, nitrides, and carbonitrides. In: Riedel, R. (ed.) Handbook of Ceramic Hard Materials, pp. 202–252. Wiley, Weinheim (2000)
53. Pampuch, R.: J. Eur. Ceram. Soc. **19**, 2395–2404 (1999)
54. Perry, A.J.: Powder Metall Int **19**, 32–36 (1987)

55. Perry, A.J.: Powder Metall. Int. **19**, 29–35 (1987)
56. Charbonnier, F.M., Mackie, W.A., Hartman, R.L., Xie, T.B.: J. Vac. Sci. Technol. B **19**, 1064–1072 (2001)
57. Curtis, C.E., Doney, L.M., Johnson, J.R.: J. Am. Ceram. Soc. **37**, 458–465 (1954)
58. Cotter, P.G., Kohn, J.A.: J. Am. Ceram. Soc. **37**, 415–420 (1954)
59. Sun, S.-K., Zhang, G.-J., Wu, W.-W., Liu, J.-X., Suzuki, T., Sakka, Y.: Scripta Mater. **69**, 139–142 (2013)
60. Zerr, A., Miehe, G., Riedel, R.: Nat. Mater. **2**, 185 (2003)
61. Burton, J.J., Garten, R.L.: Advanced Materials in Catalysis. Academic (1977)
62. Razumovskiy, V.I., Ruban, A.V., Odqvist, J., Dilner, D., Korzhavyi, P.A.: Calphad **46**, 87–91 (2014)
63. Yu, X.X., Weinberger, C.R., Thompson, G.B.: Acta Mater. **80**, 341–349 (2014)
64. Vorotilo, S., Sidnov, K., Mosyagin, I.Y., Khvan, A.V., Levashov, E.A., Patsera, E.I., Abrikosov, I.A.: J. Alloy Compd. **778**, 480–486 (2019)
65. Okamoto, H.: Bulletin of Alloy Phase Diagrams **11**, 146–149 (1990)
66. Weinberger, C.R., Yu, X.X., Yu, H., Thompson, G.B.: Comput. Mater. Sci. **138**, 333–345 (2017)
67. Zhang, J., Oganov, A.R., Li, X.F., Niu, H.Y.: Phys. Rev. B 95 (2017)
68. Predel, F.: Phase diagram of N-Ta (nitrogen-tantalum) system. In: Predel, F. (ed.) Phase Equilibria, Crystallographic and Thermodynamic Data of Binary Alloys: K-O … Y-Zr, pp. 89–89. Springer, Berlin, Heidelberg (2016)
69. Frisk, K.: J. Alloy Compd. **278**, 216–226 (1998)
70. Christensen, A.N., Lebech, B.: Acta Crystallogr. B **34**, 261–263 (1978)
71. Terao, N.: Jpn. J. Appl. Phys. **10**, 248–259 (1971)
72. Stampfl, C., Freeman, A.J.: Phys. Rev. B 71, 024111 (2005)
73. Kroll, P.: Phys. Rev. Lett. 90 (2003)
74. Kroll, P., Schroter, T., Peters, M.: Angew. Chem. Int. Ed. **44**, 4249–4254 (2005)
75. Alkhaldi, H., Kroll, P.: J. Phys. Chem. C (2019)
76. Abdollahi, A.: Phys. B (Amsterdam, Neth.) **410**, 57–62 (2013)
77. Hong, Q.-J., van de Walle, A.: Phys. Rev. B 92, 020104 (2015)
78. Pirani, M., Alterthum, H., Elektrochem, Z.: Angew. Phys. Chem. **29**, 5–8 (1923)
79. Rudy, E., Progulski, J.: Planseeber. Pulvermetall. **15**, 13–45 (1967)
80. Huepf, T., Cagran, C., Pottlacher, G.: EPJ Web Conf. **15**, 01018 (2011)
81. Cedillos-Barraza, O., Manara, D., Boboridis, K., Watkins, T., Grasso, S., Jayaseelan, D.D., Konings, R.J.M., Reece, M.J.: W. E. Lee **6**, 37962 (2016)
82. Savvatimskiy, A.I., Onufriev, S.V., Muboyadzhyan, S.A.: J. Mater. Res. **32**, 1287–1294 (2017)
83. Savvatimskiy, A.I., Onufriev, S.V., Muboyadzhyan, S.A.: J. Eur. Ceram. Soc. **39**, 907–914 (2019)
84. Kondratyev, A., Muboyajan, S., Onufriev, S., Savvatimskiy, A.: J. Alloy Compd. **631**, 52–59 (2015)
85. Onufriev, S.V., Kondratiev, A.M., Savvatimskiy, A.I., Val'yano, G.E., Muboyajan, S.A.: High Temp. 53, 455–457 (2015)
86. Agte, C., Alterthum, H.: Z. Tech. Phys. **11**, 182–191 (1930)
87. Ionescu, E., Kleebe, H.J., Riedel, R.: Chem. Soc. Rev. **41**, 5032–5052 (2012)
88. Mera, G., Gallei, M., Bernard, S., Ionescu, E.: Nanomaterials-Basel **5**, 468–540 (2015)
89. Mera, G., Ionescu, E.: Silicon-containing preceramic polymers. In: Encyclopedia of Polymer Science and Technology
90. Interrante, L.V., Carpenter, L.E., Whitmarsh, C., Hackney, M.: J. Electrochem. Soc. **133**, C119–C119 (1986)
91. Narsavage, D.M., Interrante, L.V., Marchetti, P.S., Maciel, G.E.: Chem. Mater. **3**, 721–730 (1991)
92. Fritz, G., Raabe, B.: Z. Anorg. Allg. Chem. **286**, 149–167 (1956)
93. Fritz, G., Raabe, B.: Angew. Chem. Int. Ed. **68**, 381–382 (1956)
94. Fritz, G., Raabe, B.: Z. Naturforsch. Pt B **11**, 57–59 (1956)

95. S. Yajima, J. Hayashi, M. Omori, *Chem Lett* **1975**, 931–934.
96. Yajima, S., Hasegawa, Y., Okamura, K., Matsuzawa, T.: Nature **273**, 525–527 (1978)
97. Yajima, S., Hasegawa, Y., Hayashi, J., Iimura, M.: J. Mater. Sci. **13**, 2569–2576 (1978)
98. Shimoo, T., Toyoda, F., Okamura, K.: J. Mater. Sci. **35**, 3301–3306 (2000)
99. S. Yajima, *Philos T R Soc A* **1980**, *294*, 419-+.
100. Ishikawa, T., Kajii, S., Matsunaga, K., Hogami, T., Kohtoku, Y., Nagasawa, T.: Science **282**, 1295–1297 (1998)
101. Ishikawa, T., Kohtoku, Y., Kumagawa, K.: J. Mater. Sci. **33**, 161–166 (1998)
102. Ishikawa, T., Yamamura, T., Okamura, K.: J. Mater. Sci. **27**, 6627–6634 (1992)
103. Babonneau, F., Sorarù, G.D.: J. Eur. Ceram. Soc. **8**, 29–34 (1991)
104. Kaur, S., Cherkashinin, G., Fasel, C., Kleebe, H.J., Ionescu, E., Riedel, R.: J. Eur. Ceram. Soc. **36**, 3553–3563 (2016)
105. Ionescu, E., Papendorf, B., Kleebe, H.J., Poli, F., Muller, K., Riedel, R.: J. Am. Ceram. Soc. **93**, 1774–1782 (2010)
106. Ionescu, E., Papendorf, B., Kleebe, H.J., Riedel, R.: J. Am. Ceram. Soc. **93**, 1783–1789 (2010)
107. Ionescu, E., Linck, C., Fasel, C., Muller, M., Kleebe, H.J., Riedel, R.: J. Am. Ceram. Soc. **93**, 241–250 (2010)
108. Ionescu, E., Terzioglu, C., Linck, C., Kaspar, J., Navrotsky, A., Riedel, R.: **96**, 1899–1903 (2013)
109. Thorne, K., Ting, S.J., Chu, C.J., Mackenzie, J.D., Getman, T.D., Hawthorne, M.F.: J. Mater. Sci. **27**, 4406–4414 (1992)
110. Sham, E.L., Farfantorres, E.M., Bruquegamez, S., Rodriguezjimenez, J.J.: Solid State Ionics **63–5**, 45–51 (1993)
111. Takahiro, G., Hiroshi, Y., Takaaki, H., Kyoko, B.K., Yoshimoto, A.: Appl. Organomet. Chem. **14**, 119–126 (2000)
112. Sacks, M.D., Wang, C.A., Yang, Z.H., Jain, A.: J. Mater. Sci. **39**, 6057–6066 (2004)
113. Inoue, M., Kominami, H., Inui, T.: J. Chem. Soc. Dalton Trans. 3331–3336 (1991)
114. Pan, M., Liu, J.R., Lu, M.K., Xu, D., Yuan, D.R., Chen, D.R., Yang, P., Yang, Z.H.: Thermochim. Acta **376**, 77–82 (2001)
115. Abe, Y., Kudo, T., Tomioka, H., Gunji, T., Nagao, Y., Misono, T.: J. Mater. Sci. **33**, 1863–1870 (1998)
116. Liu, H.Y., Hou, X.Q., Wang, X.Q., Wang, Y.L., Xu, D., Wang, C., Du, W., Lu, M.K., Yuan, D.R.: J. Am. Ceram. Soc. **87**, 2237–2241 (2004)
117. Tao, X.Y., Qiu, W.F., Li, H., Zhao, T.: Polym. Adv. Technol. **21**, 300–304 (2010)
118. Lu, Y., Chen, F.H., An, P.F., Ye, L., Qiu, W.F., Zhao, T.: RSC Adv. **6**, 88770–88776 (2016)
119. Preiss, H., Berger, L.M., Szulzewsky, K.: Carbon **34**, 109–119 (1996)
120. Preiss, H., Schierhorn, E., Brzezinka, K.W.: J. Mater. Sci. **33**, 4697–4706 (1998)
121. McMahon, C.N., Alemany, L., Callender, R.L., Bott, S.G., Barron, A.R.: Chem. Mater. **11**, 3181–3188 (1999)
122. Kurokawa, Y., Ishizaka, T., Suzuki, M.: J. Mater. Sci. **36**, 301–306 (2001)
123. Tao, X., Xiang, Z., Zhou, S., Zhu, Y., Qiu, W., Zhao, T.: J. Ceram. Sci. Technol. **7**, 107–111 (2016)
124. Lang, H., Seyferth, D.: Appl. Organomet. Chem. **4**, 599–606 (1990)
125. Inzenhofer, K., Schmalz, T., Wrackmeyer, B., Motz, G.: Dalton Trans. **40**, 4741–4745 (2011)
126. Wang, H., Chen, X.B., Gao, B., Wang, J., Wang, Y.D., Chen, S.G., Gou, Y.Z.: Appl. Organomet. Chem. **27**, 79–84 (2013)
127. Tripp, W.C., Davis, H.H., Graham, H.C.: Am. Ceram. Soc. Bull. **52**, 612–616 (1973)
128. Fahrenholtz, W.G.: J. Am. Ceram. Soc. **90**, 143–148 (2007)
129. Rezaie, A., Fahrenholtz, W.G., Hilmas, G.E.: J. Eur. Ceram. Soc. **27**, 2495–2501 (2007)
130. Rezaie, A., Fahrenholtz, W.G., Hilmas, G.E.: J. Am. Ceram. Soc. **89**, 3240–3245 (2006)
131. Tampieri, A., Landi, E., Bellosi, A.: J. Therm. Anal. **38**, 2657–2668 (1992)
132. Salikova, A.P., Ban'kovskaya, I.B., Kolovertnov, D.V., Popov, V.P.: Glass Phys. Chem.+ **36**, 225–231 (2010)
133. Sciti, D., Brach, M., Bellosi, A.: J. Mater. Res. **20**, 922–930 (2005)

134. Sciti, D., Brach, M., Bellosi, A.: Scripta Mater. **53**, 1297–1302 (2005)
135. Talmy, I.G., Zaykoski, J.A., Opeka, M.M., Smith, A.H.: J. Mater. Res. **21**, 2593–2599 (2006)
136. Talmy, I.G., Zaykoski, J.A., Opeka, M.M.: J. Am. Ceram. Soc. **91**, 2250–2257 (2008)
137. Cai, T., Qiu, W.F., Liu, D., Han, W.J., Ye, L., Zhao, A.J., Zhao, T.: J. Am. Ceram. Soc. **96**, 3023–3026 (2013)
138. Liu, D., Qiu, W.F., Cai, T., Sun, Y.N., Zhao, A.J., Zhao, T.: J. Am. Ceram. Soc. **97**, 1242–1247 (2014)
139. Cai, T., Liu, D., Qiu, W.F., Han, W.J., Zhao, T.: J. Am. Ceram. Soc. **101**, 20–24 (2018)
140. Kim, D.S., Suh, M.C., Shim, S.C.: J. Polym. Sci. Polym. Chem. **36**, 2275–2282 (1998)
141. Li, Y.T., Han, W.J., Li, H., Zhao, J.B., Zhao, T.: Mater. Lett. **68**, 101–103 (2012)
142. Patra, N., Lee, W.E.: ACS Appl. Nano Mater. **1**, 4502–4508 (2018)
143. Amoros, P., Beltran, D., Guillem, C., Latorre, J.: Chem. Mater. **14**, 1585–1590 (2002)
144. Fonblanc, D., Lopez-Ferber, D., Wynn, M., Lale, A., Soleilhavoup, A., Leriche, A., Iwamoto, Y., Rossignol, F., Gervais, C., Bernard, S.: Dalton Trans. **47**, 14580–14593 (2018)
145. Paine, R.T., Janik, J.F., Fan, M.: Polyhedron **13**, 1225–1232 (1994)
146. Bernardo, E., Fiocco, L., Parcianello, G., Storti, E., Colombo, P.: Materials **7**, 1927–1956 (2014)
147. Guron, M.M., Wei, X.L., Welna, D., Krogman, N., Kim, M.J., Allcock, H., Sneddon, L.G.: Chem. Mater. **21**, 1708–1715 (2009)
148. Majoulet, O., Salameh, C., Schuster, M.E., Demirci, U.B., Sugahara, Y., Bernard, S., Miele, P.: Chem. Mater. **25**, 3957–3970 (2013)
149. Weinmann, M., Zern, A., Aldinger, F.: Adv. Mater. **13**, 1704–1708 (2001)
150. Hering, N., Schreiber, K., Riedel, R., Lichtenberger, O., Woltersdorf, J.: Appl. Organomet. Chem. **15**, 879–886 (2001)
151. Awin, E.W., Lale, A., Kumar, K.C.H., Demirci, U.B., Bernard, S., Kumar, R.: Mater. Des. **157**, 87–96 (2018)
152. Bechelany, M.C., Proust, V., Gervais, C., Ghisleni, R., Bernard, S., Miele, P.: Adv. Mater. **26**, 6548–6553 (2014)
153. Lale, A., Proust, V., Bechelany, M.C., Viard, A., Malo, S., Bernard, S.: J. Eur. Ceram. Soc. **37**, 5167–5175 (2017)
154. Bechelany, M.C., Proust, V., Lale, A., Miele, P., Malo, S., Gervais, C., Bernard, S.: Chem. Eur. J. **23**, 832–845 (2017)
155. Zhou, C., Gao, X., Xu, Y., Buntkowsky, G., Ikuhara, Y., Riedel, R., Ionescu, E.: J. Eur. Ceram. Soc. **35**, 2007–2015 (2015)
156. Pizon, D., Lucas, R., Foucaud, S., Maitre, A.: Adv. Eng. Mater. **13**, 599–603 (2011)
157. Pizon, D., Lucas, R., Chehaidi, S., Foucaud, S., Maitre, A.: J. Eur. Ceram. Soc. **31**, 2687–2690 (2011)
158. Seyferth, D., Bryson, N., Workman, D.P., Sobon, C.A.: J. Am. Ceram. Soc. **74**, 2687–2689 (1991)
159. Seyferth, D., Lang, H., Tracy, H.J., Sobon, C., Borm, J.: Abstr. Pap. Am. Chem. S **202**, 359-Poly (1991)
160. Honack, F., Riedel, R.: Adv. Eng. Mater. **5**, 122–125 (2003)
161. Luo, Y.M., Zheng, Z.M., Xu, C.H., Mei, X.N.: Mater. Lett. **62**, 3570–3572 (2008)
162. Luo, Y.M., Zheng, Z.M., Mei, X.N., Xu, C.H.: J. Cryst. Growth **310**, 3372–3375 (2008)
163. Luo, Y.M., Zheng, Z.M., Mei, X.N., Xu, C.H.: Mater. Chem. Phys. **113**, 26–30 (2009)
164. Yang, J.S., Dong, S.M., Ding, Y.S., Wang, Z., Zhou, H.J., Lu, B.: J. Am. Ceram. Soc. **93**, 2117–2120 (2010)
165. Proust, V., Bechelany, M.C., Ghisleni, R., Beaufort, M.F., Miele, P., Bernard, S.: J. Eur. Ceram. Soc. **36**, 3671–3679 (2016)
166. Yajima, S., Iwai, T., Yamamura, T., Okamura, K., Hasegawa, Y.: J. Mater. Sci. **16**, 1349–1355 (1981)
167. Ichikawa, H., Machino, F., Mitsuno, S., Ishikawa, T., Okamura, K., Hasegawa, Y.: J. Mater. Sci. **21**, 4352–4358 (1986)
168. Song, Y., Feng, C., Tan, Z., Lu, Y.: J. Mater. Sci. Lett. **9**, 1310–1313 (1990)

169. Babonneau, F., Soraru, G.D., Mackenzie, J.D.: J. Mater. Sci. **25**, 3664–3670 (1990)
170. Soraru, G.D., Glisenti, A., Granozzi, G., Babonneau, F., Mackenzie, J.D.: J. Mater. Res. **5**, 1958–1962 (1990)
171. Maeda, K., Higashi, M., Siritanaratkul, B., Abe, R., Domen, K.: J. Am. Chem. Soc. **133**, 12334–12337 (2011)
172. Chollon, G., Aldacourrou, B., Capes, L., Pailler, R., Naslain, R.: J. Mater. Sci. **33**, 901–911 (1998)
173. Levine, S.R., Opila, E.J., Halbig, M.C., Kiser, J.D., Singh, M., Salem, J.A.: J. Eur. Ceram. Soc. **22**, 2757–2767 (2002)
174. Yu, Z.J., Yang, L., Zhan, J.Y., Zhou, C., Min, H., Zheng, Q., Xia, H.P.: J. Eur. Ceram. Soc. **32**, 1291–1298 (2012)
175. Wang, X.F., Liu, J.C., Hou, F., Hu, J.D., Sun, X., Zhou, Y.C.: J. Am. Ceram. Soc. **98**, 197–204 (2015)
176. Bouzat, F., Darsy, G., Foucaud, S., Lucas, R.: Polym. Rev. **56**, 187–224 (2016)
177. Li, Q.G., Dong, S.M., He, P., Zhou, H.J., Wang, Z., Yang, J.S., Wu, B., Hu, J.B.: Ceram. Int. **38**, 6041–6045 (2012)
178. Li, J.W., Yu, S.Q., Zhou, Y., Ge, M., Qian, Y.B., Zhang, W.G.: Mater. Lett. **156**, 72–75 (2015)
179. Huang, C., Wang, Z., Wang, M.: Adv. Appl. Ceram. **115**, 166–173 (2016)
180. Wang, H., Gao, B., Chen, X.B., Wang, J., Chen, S.G., Gou, Y.Z.: Appl. Organomet. Chem. **27**, 166–173 (2013)
181. Yu, Z.J., Min, H., Zhan, J.Y., Yang, L.: Ceram. Int. **39**, 3999–4007 (2013)
182. Li, S.W., Zhang, L.T., Huang, M.H., Yu, Z.J., Xia, H.P., Feng, Z.D., Cheng, L.F.: Mater. Chem. Phys. **133**, 946–953 (2012)
183. Long, X., Shao, C.W., Wang, J., Gou, Y.Z.: Appl. Organomet. Chem. 32 (2018)
184. Chen, S.G., Wang, J., Wang, H.: Mater. Des. **90**, 84–90 (2016)
185. Chen, S.G., Gou, Y.Z., Wang, H., Wang, J.: J. Eur. Ceram. Soc. **36**, 3843–3850 (2016)
186. Vijay, V.V., Nair, S.G., Sreejith, K.J., Devasia, R.: J. Inorg. Organomet. P **26**, 302–311 (2016)
187. Wen, Q.B., Xu, Y.P., Xu, B.B., Fasel, C., Guillon, O., Buntkowsky, G., Yu, Z.J., Riedel, R., Ionescu, E.: Nanoscale **6**, 13678–13689 (2014)
188. Wen, Q.B., Feng, Y., Yu, Z.J., Peng, D.L., Nicoloso, N., Ionescu, E., Riedel, R.: J. Am. Ceram. Soc. **99**, 2655–2663 (2016)
189. Wen, Q.B., Riedel, R., Ionescu, E.: Corros. Sci. **145**, 191–198 (2018)
190. Wen, Q.B., Yu, Z.J., Xu, Y.P., Lu, Y., Fasel, C., Morita, K., Guillon, O., Buntkowsky, G., Ionescu, E., Riede, R.: J. Mater. Chem. C **6**, 855–864 (2018)
191. Wen, Q., Riedel, R., Ionescu, E.: 1800879
192. Wen, Q., Luan, X., Wang, L., Xu, X., Ionescu, E., Riedel, R.: J. Eur. Ceram. Soc. (2019)
193. Cai, T., Qiu, W.F., Liu, D., Han, W.J., Ye, L., Zhao, A.J., Zhao, T.: Dalton Trans. **42**, 4285–4290 (2013)
194. Liu, C.Q., Li, K.Z., Li, H.J., Zhang, S.Y., Zhang, Y.L., Hou, X.H.: J. Mater. Sci. **50**, 2824–2831 (2015)
195. Yu, Z.J., Yang, L., Min, H., Zhang, P., Liu, A.H., Riedel, R.: J. Eur. Ceram. Soc. **35**, 851–858 (2015)
196. Zhang, Q.C., Gou, Y.Z., Wang, J.L., Wang, H., Jian, K., Wang, Y.F.: J. Eur. Ceram. Soc. **37**, 1909–1916 (2017)
197. Cheng, J., Wang, X.Z., Wang, H., Shao, C.W., Wang, J.: J. Am. Ceram. Soc. **100**, 5044–5055 (2017)
198. Tian, Y.L., Ge, M., Zhang, W.G., Lv, X.X., Yu, S.Q.: Sci. Rep. UK **5** (2015)
199. He, J.B., Gao, Y., Wang, Y.G., Fang, J.Y., An, L.N.: Ceram. Int. **43**, 1602–1607 (2017)
200. He, J.B., Cao, Y.J., Zhang, Y.X., Wang, Y.G.: Ceram. Int. **44**, 6520–6526 (2018)
201. Viard, A., Fonblanc, D., Schmidt, M., Lale, A., Salameh, C., Soleilhavoup, A., Wynn, M., Champagne, P., Cerneaux, S., Babonneau, F., Chollon, G., Rossignol, F., Gervais, C., Bernard, S.: Chem. Eur. J. **23**, 9076–9090 (2017)
202. Yuan, J., Hapis, S., Breitzke, H., Xu, Y., Fasel, C., Kleebe, H.-J., Buntkowsky, G., Riedel, R., Ionescu, E.: Inorg. Chem. **53**, 10443–10455 (2014)

203. Yuan, J., Galetz, M., Luan, X.G., Fasel, C., Riedel, R., Ionescu, E.: J. Eur. Ceram. Soc. **36**, 3021–3028 (2016)
204. Yuan, J., Li, D., Johanns, K.E., Fasel, C., Durst, K., Kleebe, H.J., Shen, Z.J., Riedel, R., Ionescu, E.: J. Eur. Ceram. Soc. **37**, 5157–5165 (2017)
205. Yuan, J., Luan, X.G., Riedel, R., Ionescu, E.: J. Eur. Ceram. Soc. **35**, 3329–3337 (2015)
206. Luan, X., Yuan, J., Wang, J., Tian, M., Cheng, L., Ionescu, E., Riedel, R.: J. Eur. Ceram. Soc. **36**, 3761–3768 (2016)
207. Wang, S.H., Zhang, Y.C., Sun, Y., Xu, Y., Yang, M.: J. Alloy Compd. **685**, 828–835 (2016)
208. Long, X., Shao, C., Wang, H., Wang, J.: Ceram. Int. **42**, 19206–19211 (2016)

Potential Role of Vitrification and Waste Vitrification in the Circular Economy

Elham Sharifikolouei and Monica Ferraris

Abstract Following the general increase of population and industrial development, the management of resources through use of secondary source of raw materials, recycling and energy recovery from waste products are growing in importance. There are many challenges associated with each approach in which waste treatment and handling its by-products are among the most environmentally concerned matters. This chapter is divided into two main parts for two types of waste management: municipal solid waste management, and wastewater management. In the first part, management of municipal solid waste, incineration as a tool for energy recovery and its hazardous by-products fly ash and bottom ash are introduced. It has shown that vitrification of these ashes could open lots of potential for their utilization in various products such as cement and clinker substitution, or preparation of foam glass–ceramics as construction materials. In this chapter, a guide to successfully vitrify fly ash is also introduced and discussed. The second part of this chapter is dedicated to wastewater management, focusing on the potential use of rice-husk, an agricultural by-product to be utilized as an adsorbent for removal of heavy metals from wastewater. The final heavy-metal containing rice husk introduces new environmental concerns prior to landfill, and we have shown that vitrification has the potential to safely encapsulate this material and even give the opportunity to prepare foam glass–ceramics after its vitrification.

Keywords Circular economy · Municipal-solid-waste incinerator · Fly ash · Bottom ash · Rice-husk vitrification · Glass–ceramics · Waste glass

1 Introduction

The term *circular economy* has both a linguistic and a descriptive meaning. Linguistically it is an antonym of a *linear economy*. In linear economy natural resources are exploited to manufacture products which will eventually lead to the waste production.

E. Sharifikolouei (✉) · M. Ferraris
Department of Applied Science and Technology (DISAT), Institute of Materials Physics and Engineering, Politecnico di Torino, 10129 Turin, Italy
e-mail: elham.sharifikolouei@polito.it

F. Baino et al. (eds.), *Ceramics, Glass and Glass-Ceramics*, PoliTO Springer Series,
https://doi.org/10.1007/978-3-030-85776-9_10

The term "circular" suggests closing the loop of product lifecycle. The concept of circular economy was introduced by policy makers from China and European Union as a solution to the growing problem of natural resource depletion, and waste generation [1]. European commission has introduced the first circular economy action plan in 2015. In these action plans, factsheets and strategies for closing the loop are introduced. These factsheets target different phases of production, consumption, waste management and in short, they cover the whole cycle. They have introduced strategies to help consumers select more sustainable products and services, factsheets on the production phase, introduced tools for better waste management, and finally has suggested ways for utilization of waste as a resource.

In this book chapter, we will focus on the solid waste and wastewater management and the role vitrification plays on safe encapsulation of hazardous waste by-products. In the first part, we focus on the municipal solid waste management through incineration. We will introduce the benefits and challenges associated with incineration, and how the incineration hazardous by-products, fly ash and bottom ash, should be treated. Different opportunities for utilization of ashes are introduced in this section and the primary focus is dedicated to their vitrification and utilization of vitrified ashes for fabrication of different products such as porous glass–ceramics, bricks and cements. Since vitrification of fly ash is of primary importance and there are lots of technical challenges associated with it, we have reviewed an approach to get closer to a more scientific and reproducible vitrification batch preparation.

The second part of the chapter is dedicated to wastewater management and in particular the removal of heavy metals through adsorption by biomass waste products such as rice husk. In the end, the possibility of fabrication of glass–ceramic foam from heavy-metal containing rice husk is introduced.

2 Municipal Solid Waste Management: Recycling, Energy Recovery, Treatment and Disposal

Following the general increase of population and industrial development, the production of municipal solid waste in the world is expected to increase by 60% from 1.3 billion tons in 2015 to 2.2 billion tons in 2025 [2]. In Europe, the European commission has set waste legislation, mainly divided into five subcategories and organizes various action plans as guidance for sustainable waste management for its member states. These directives focus mainly on recycling waste. The circular economy is one of the main European policies and action plans that focusses on reducing landfilling to a minimum [3]. However, when recycling is not economically viable, waste-to-energy technologies (WTE-T) are used to turn waste into a useable form of energy. There are different waste-to-energy technologies such as biological treatment technologies, thermal treatment technologies, landfill gas utilization, and biorefineries [4]. In this section, we will focus on the incineration process.

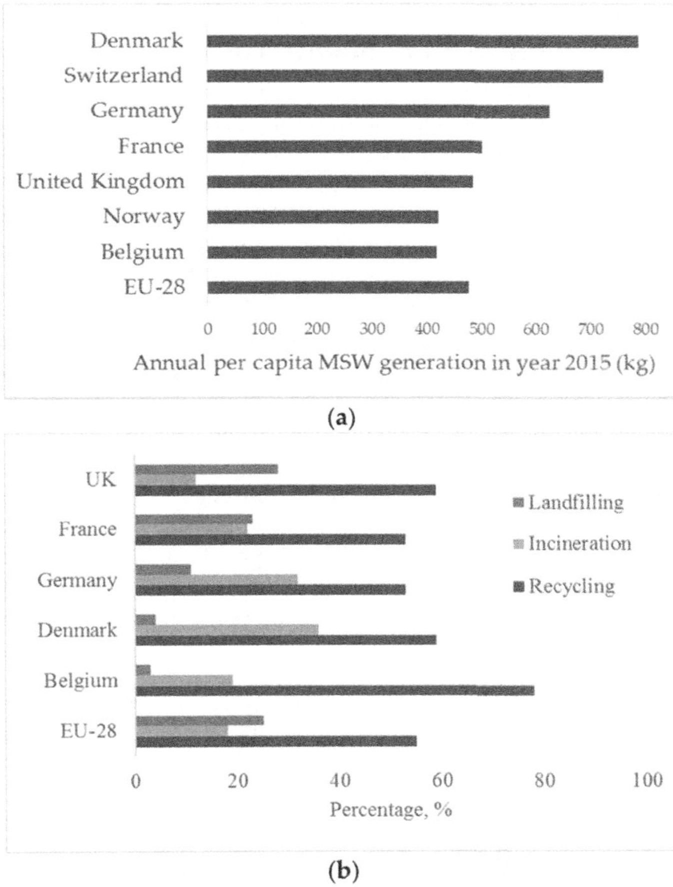

Fig. 1 **a** Annual per capita generation of MSW in Belgium compared to the average value of EU nations and other prominent neighbors; **b** percentage of recycled, incinerated and landfilled waste cf. Reproduced with permission [3]

Figure 1a shows the comparison between the annual municipal solid waste (MSW) generation among major European countries and its average in Europe. The annual generation of MSW in countries such as Denmark reaches as high as 800 kg per capita. Figure 1b shows the different waste management strategies for each country. Landfill is still high in some countries such as UK, but overall the major part of MSW gets recycled. The major driving force for recycling and incineration is the policy measures such as landfill taxes, and the environmental quality requirements to be met.

2.1 Municipal Solid Waste Incineration: Challenges in Handling Fly Ash and Bottom Ash

Incineration is one of the most common waste-to-energy technologies that can help reducing the mass and volume of municipal solid waste [5, 6]. Figure 2 illustrates a classic incineration WTE-T. In the first step, waste is directly burned in the combustion chamber at an adequate temperature (900–950 °C) using flue gas and pre-heated air. The produced superheated steam is then used within a cogeneration system to produce energy and heat. The electric energy is produced by a turbine connected to a generator and the heat by a district heating system [4, 7]. It is worth mentioning that the emerging new technologies are offering the possibility of partially recovering materials such as metals inside the incinerators plant and in some pioneer countries, e.g. Switzerland, recovery of metals such as copper, aluminum, zinc and others from the combustion residues is becoming standard practice [8]. The most common residues of incineration are bottom ash and fly ash.

Bottom ash (BA) is the major residue in waste-to-energy plants which accounts for around 80%wt. of the total residues, and it is composed of heterogeneous solid granules with a broad particle size distribution varying from 0.01 to 100 mm and has different constituents including broken glass cullet, ceramic, mineral, slag/sintered phases and unburnt materials [9, 10]. Figure 3 shows the SEM images of Municipal

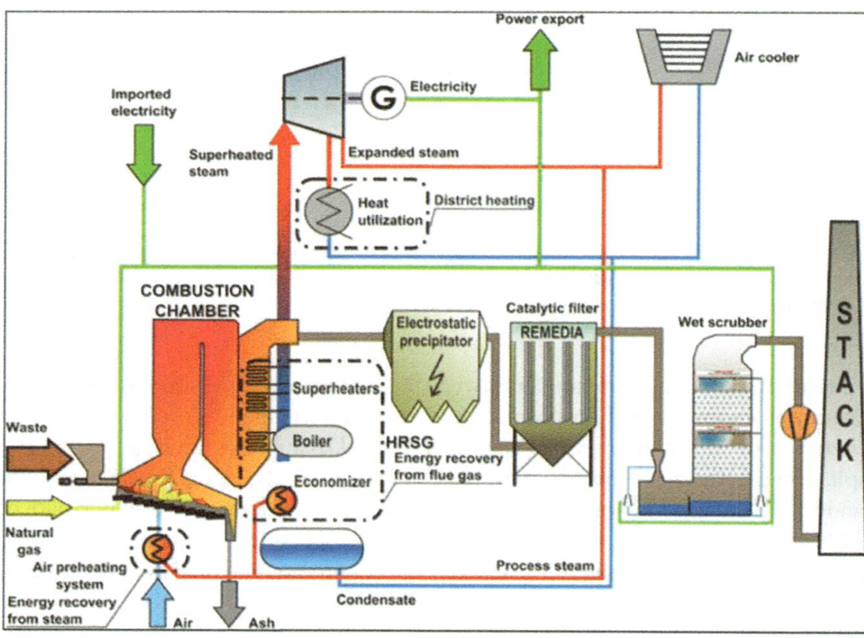

Fig. 2 Incineration plant to produce electricity and heat from MSW. Reproduced with permission [7]

Fig. 3 SEM images of MSWI BA fine particle size fractions. Reproduced with permission [9]

Solid Waste Incinerator (MSWI) BA fine particle size fractions. The major mineral found in BA is quartz. Other minerals frequently found in BA are Feldspar, Melilite, Spinel, Pyroxene, Sylvite, Calcite, Halite, and even traces of Zeolite [9, 11, 12]. Bottom ash also contain heavy metals which are prone to leach and therefore, they should be treated before the final landfill.

On the contrary, fly ash represents a light fine particle. The composition of fly ash varies a lot depending on the type of the waste used in the incinerator. They mostly contain soluble salts, high concentration of heavy metals, organic pollutants such as dioxins. The concentration of heavy metals in fly ash is significantly higher than bottom ash and that is why it is always classified under the hazardous waste materials. Some of the most common heavy metals found in fly ash are Hg, Pb, Zn, Cd, As, Sb, Cu, Sn, Ni, Cr, V [13, 14]. Management of bottom ash and fly ash is a major environmental concern because the heavy metal leaching can find its way to soil and water reservoirs if landfills without a proper treatment. Treatment of bottom ash and fly ash introduces a new cost and sometimes new pollutants after the treatment.

Some countries such as Switzerland has developed an incinerators system which recovers the heavy metals while treating fly ash. This is particularly interesting for the recovery of Zn. The acid-washing process of fly ash is developed under the name FLUWA. In this process, scrub water from the wet flue gas cleaning process is used for extraction of the heavy metals. Most often acidic and neutral scrub water is used in the FLUWA process that is purified from mercury by selective ion exchanger. This leads to the removal of > 90 wt.% of the mercury entering the incineration process.

Figure 4 shows the schematic representation of FLUWA process. In this figure,

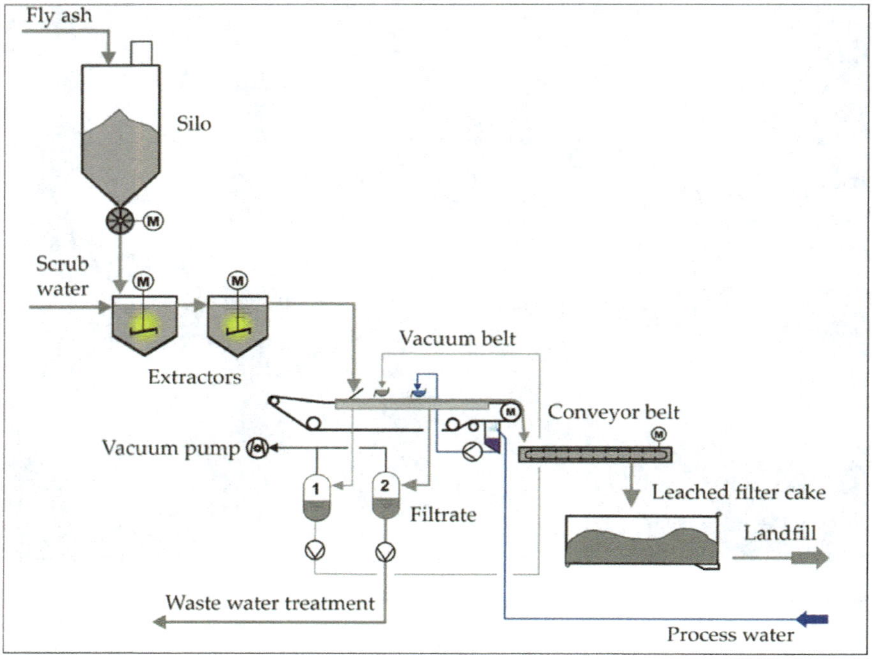

Fig. 4 Scheme of the FLUWA process. Reproduced with permission [15]

fly ash is leached with acidic scrub water in a multistage cascade. There are different parameters affecting the final extraction of heavy metals among which are alkalinity of the fly ash, acidity of the scrub water, liquid to solid ratio (LS), temperature and leaching time. It is reported that up to 80% Zn, and 85% Cd can be extracted through FLUWA. To reach high purity zinc (> 99.99%), and additional process, "FLUREC", is used. The FLUREC process separates Cd, Pb and Cu from the FLUWA filtrate by a cementation process (reductive separation). FLUREC process is depicted in Fig. 5 [15, 16].

The introduced fly ash treatment process can be applied in an industrial scale inside the incinerators plant. However, some have criticized the acid leaching process as it introduces new pollutant to the environment. In the context of resource management, bottom ash and fly ash have the potential to partially substitute raw materials especially in the case of construction materials such as clinker, cement, concrete, and bricks. Concerning the heavy metal leaching, different pre-treatments are suggested and tested such as vitrification and cementation. Even though costly, since the treated fly ash and bottom ash will partially substitute the raw materials, their treatment cost will be still less than the cost of mining and exploitation of raw materials and would help preserve the natural resources. As mentioned before, the chemical composition of fly ash strongly depends on the initial incinerated waste which varies from region to region. Therefore, depending on the type and the properties of fly ash, different

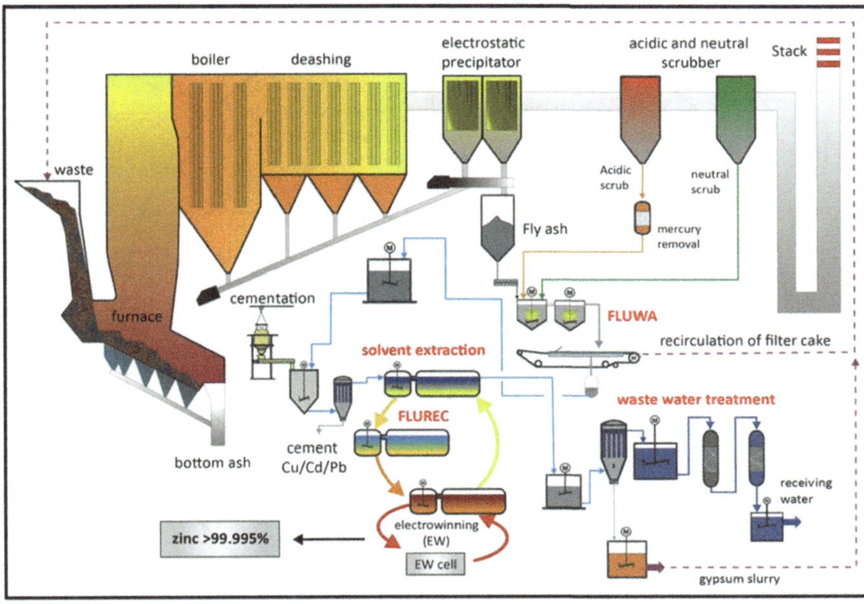

Fig. 5 Scheme of the FLUWA and FLUREC process adapted from Bühler and Schlumberge. Reproduced with permission [15]

routes and applications can be considered. Yao et al. [17], has made a comprehensive review on how to utilize coal fly ash based on its properties (Fig. 6). Even though, coal fly ash often shows different properties compared to the incinerators fly ash, the review provides a good guide on how to approach the utilization of fly ash and bottom ash for different applications.

2.2 *Utilization of Fly Ash and Bottom Ash, and the Benefit of Vitrification*

Among different applications, we focus on the utilization of bottom ash and fly ash for construction materials. Construction materials are used in large scale and are responsible for depletion of many natural resources, as well as a high share in the global CO_2 emission. Therefore, even partial substitution of waste by-products can greatly impact the overall life cycle of materials. There are multiple researches which suggest promising results from utilization of MSWI bottom ash to substitute ordinary Portland cement (OPC) clinker and cement in concrete if they are pretreated. This is an important achievement because production of ordinary Portland cement as the most common cement used in concrete and mortar is accountable for contribution of 5–7% of total global CO_2 emission [18, 19]. Therefore, policies contributing to

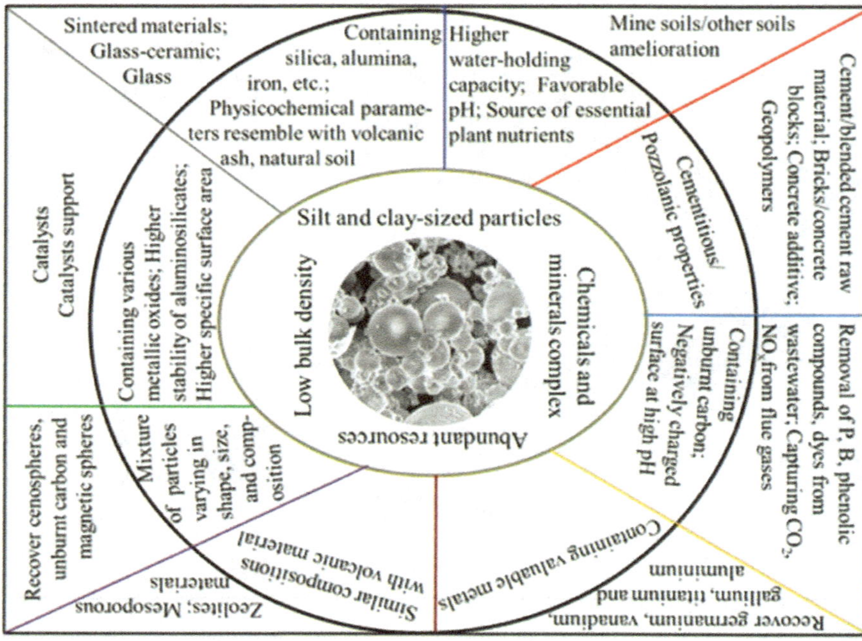

Fig. 6 Various applications of coal fly ash. Reproduced with permission [17]

the reduction or replacement of OPC consumption would have a significant impact on cutting down the CO_2 emission. Some popular low-cost treatments are $Ca(OH)_2$ washing and carbonation. Carbonation is shown to be mostly effective in accelerating strength gain in concrete. $Ca(OH)_2$ washing is performed because incinerator bottom ashes contain considerable amount of chlorine content. Chlorine, in contact with $Ca(OH)_2$ (portlandite) can react and form $CaCl_2$ that is prone to leaching out, making the concrete porous. The porosity will eventually reduce the strength of concrete [20]. The larger bottom ash aggregates have been reported to substitute sand in mortars and concretes as well [21–24]. Furthermore, numerous reports shows that vitrification of bottom ash could help their performance as a substitution material to OPC, as well as substitution of standard aggregates in concrete and mortars [25]. This would bring additional benefits since the global aggregate demand is projected to exceed 50 billion tons per annum by 2019 whereas the available aggregate resources are limited. Similarly, bottom ash could substitute aggregate in road pavements [26].

Utilization of fly ash in cement and concrete production is more limited to "coal fly ash" and the environmental concerns of heavy metal leaching from municipal solid waste incinerators fly ash inhibits its utilization in many countries. Additionally, the Chlorine content in fly ash is significantly higher than bottom ash. However, vitrification is one of the safest techniques for encapsulation of heavy metals and is widely used for the treatment of the most dangerous class of waste materials, radioactive nuclear wastes [27, 28]. The disadvantage of vitrification is mostly its cost. Many incinerators are equipped with direct melting system where fly ash and bottom ash are vitrified in the plant to form slag and vitrified safe products [29, 30]. This approach is popular in countries such as Japan which landfill is not an option since land is scarce. Countries such as France, with various nuclear power plants are equipped with plasma torches which are capable of vitrification of hazardous waste in high temperatures [13, 31, 32]. Therefore, vitrification of fly ash by plasma is an affordable option in France.

Vitrification of bottom ash is feasible due to its high silica content and can be further facilitated by small addition of waste glass (glass cullet). On the other hand, fly ash cannot be easily vitrified due to its low silica and alumina content. A recent comprehensive study suggests a range of composition for the successful design of fly ash vitrification batches by addition of glass cullet or other silica-containing resources. The study calculated the number of non-bridging oxygens per tetrahedron (NBO/T) reported in literature as a degree of depolymerization, as well as the degree of polymerization in the oxide melt, Q ($Q = 4\text{-}NBO/T$). The identified values bring us closer to formulate the design of fly ash vitrification batch. Figure 7 shows an identified region in the SiO_2–Al_2O_3–CaO, and SiO_2–ΣM_2O_3–$\Sigma(MO + M_2O)$ ternary phase diagrams where successfully vitrified fly ash batches are reported. The normalized values for their composition in a SiO_2–Al_2O_3–CaO system show that most of them have the CaO/SiO_2 (C/S) ratio of about 0.44. Considering the effect of all major oxides in the vitrification batches reported in literature, ternary diagram of SiO_2–ΣM_2O_3–$\Sigma(MO + M_2O)$ shows similar trends to the SiO_2–Al_2O_3–CaO ternary phase diagram. The degree of polymerization (Q) in the batches was calculated to be around 3.45 for 75% of reported data [33].

The use for vitrified fly ash and bottom ash is not limited to clinker and concrete and can be used in the production of glass–ceramics foam (thermal insulating bricks). The foamed glass–ceramics can be easily prepared by addition of calcium carbonate to the vitrified ash and its heat-treatment [34]. The reported standard leaching test values confirms the safety of this approach [35]. The know-how of foam glass–ceramic preparation is well developed and well-studied for glass, and prior knowledge of experimental parameters such as the role of heating rate, and the percentage of calcium carbonate addition can be readily used and implemented for vitrified ashes as well. E. Bernardo and F. Albertini have studies those experimental parameters on the characteristics of the final foam glass–ceramics and have concluded that fast heating rates can promote the formation of finer foam microstructures with higher relative density because the foaming process is concentrated in the holding time at the foaming temperature [36]. In a similar study by D. Tulyaganov et al., foam glass–ceramics was prepared from sheet glass. His results show promising opportunities for preparation of high strength glass–ceramics foams in which can be readily used for vitrified fly ash and bottom ash with similar chemical composition [37]. Figure 8 shows the picture of these foam glass–ceramics.

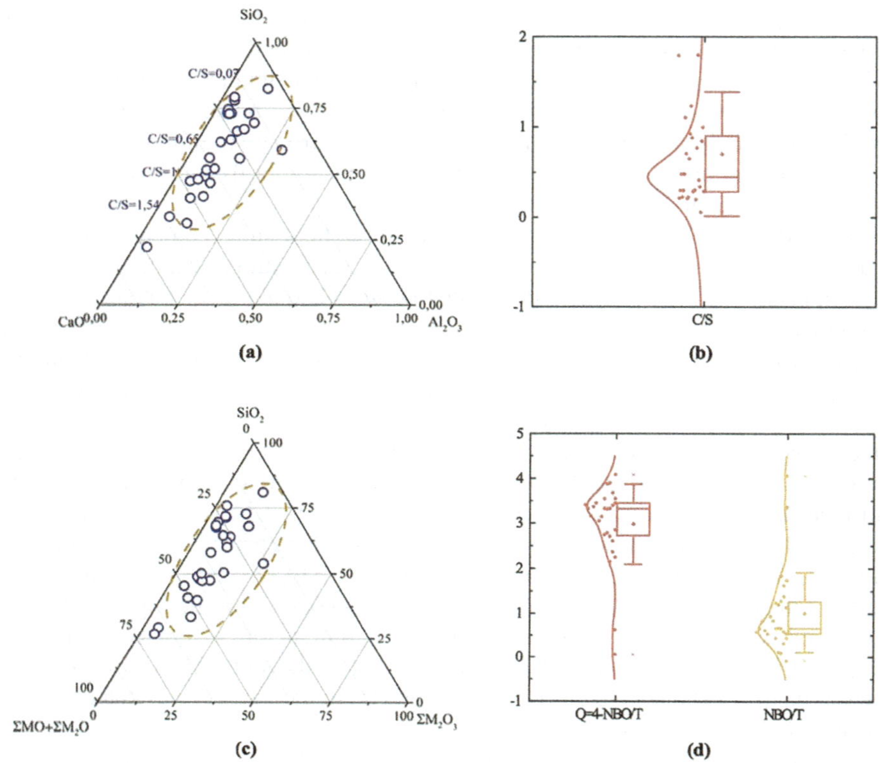

Fig. 7 **a** SiO_2–Al_2O_3–CaO phase diagram for successful vitrification batches containing fly ash. **b** The distribution of CaO/SiO_2 (C/S) values for the vitrification batches. **c** SiO_2–ΣM_2O_3–$\Sigma(MO + M_2O)$ phase diagram for successful vitrification batches containing fly ash. **d** The distribution of NBO/T and Q for vitrification batches to present their degree of polymerization. Reproduced with permission [33]

3 Wastewater Treatment: Focus on the Removal of Heavy Metals

The circular economy can also be applied in the wastewater treatment sector. Reclamation and reuse of wastewater is a vital step to reduce depletion of water resources and to reduce the environmental impact of water treatments, nutrient recovery and energy production. The treated wastewater are mostly used for agricultural, industrial, urban or recreational purposes [38–40]. Additionally, it is used for nutrient recovery and in this regard, the urban wastewater treatment plants (WWTPs) is an important part of circular sustainability due to integration of energy production and resource recovery during clean water production [41–43]. It is claimed that the wastewater treatment plants are soon to become ecologically sustainable technology. WWTPs in the near future are to become "ecologically sustainable" technological systems.

Fig. 8 Produced glass foams. Reproduced with permission [37]

The main driving forces are the global need for nutrients and energy recovery from wastewater [44].

Nutrient recycling from WWTPs has positive impact on environment because it reduced the global demand for fossil-based fertilizers and consequently, reduce the consumption of water and energy. It is possible to recover nutrient from raw wastewater, semi-treated wastewater streams and sewage sludge (biosolids) [44, 45]. For instance, phosphorus can be recycled from sewage sludge or wastewater. Currently, the main technique for phosphorous recycling is struvite crystallization process such as Pearl and AirPrex technologies which have been already implemented at full scale [11].

Another important role of wastewater treatment in circular economy is the energy recovery. The energy recovery can be done through heat pumps in power plant, biogas production or other high temperature streams though heat-exchangers. The biogas produced in a digester via anaerobic digestion (AD) has the energy potential of 6.5 kWh/m^3 (65% methane content) and is the main energy source in WWTP. Rodríguez-Chueca et al. [40] has indicated that sludge digestion in the wastewater treatment plant can reduce the net energy consumption down to 40% less compared to plants without AD. The most adopted technology in the existing self-sufficient WWTPs is Combined Heat and Power (CHP) technologies which generate both electricity and heat from biogas at the same time.

The reuse of wastewater is often challenged by numerous factors which are not limited to its safety and technical parameters. In fact, public acceptance and social perception plays an important role in it [46]. However, in this chapter we will focus on the most technical challenge for wastewater treatment: removal of heavy metals. Heavy metals like mercury (Hg), chromium (Cr), cadmium (Cd), thallium (Tl), lead (Pb), and nickel (Ni) are potentially hazardous elements whether in their elemental or complex forms. Unfortunately, they are highly soluble in aqueous solutions and therefore can be easily absorbed by living organisms. There are many studies that found traces of heavy metals in the liver, and muscle tissues of various species of fish in contaminated marine ecosystems [47]. Besides, the heavy metal containing waters can easily contaminate the soil and all agricultural products. Wuana and Okieimen [48] have indicated the estimation of 20 million hectares of arable land worldwide, are irrigated with wastewater. The conducted studies in several Asian and African cities suggests that that theses contaminated lands are responsible for production of approximately 50% of agricultural products for urban areas [49]. The consumption of contaminated products can have severe effect on humans.

Several methods have been investigated for heavy metal removal from aqueous solutions such as membrane filtration, chemical precipitation, reverse osmosis, ion exchange and adsorption [50–53]. Figure 9 shows the schematic representation of the principles used for metal removal.

3.1 Adsorption of Heavy Metal by Agricultural By-Products

Adsorption is one of the simplest yet very effective techniques for heavy metal removal from wastewater. Adsorption is defined as the adhesion of a chemical species onto the surface of particles. Adsorption can be further divided into physical adsorption, physisorption, or chemical adsorption, chemisorption. Physisorption relies mainly on relatively weak intermolecular forces such as Van der Waals while chemisorption is more permanent and involves the formation of a chemical bond between the sorbate molecule and the surface. The most important adsorbents include zeolites [55–57], activated carbon [58–60], expanded perlite [61], and rice husks [62–65]. In the context of circular economy, rice husk, being an agricultural by-product, has the advantage of lower cost and reducing the environmental impact even further. To reduce its volume and to better use it as an adsorbent, in most cases rice husk is pyrolyzed and its ash is used as an adsorbent [66]. Some of the characteristics of rice husk ash such as high silica contents (87–97 wt.% SiO_2), high porosity, lightweight and very high external surface area makes it an ideal low cost adsorbent material [67]. Figure 10 shows the optical and SEM picture of rice husk, and its possible chemical bonding. The kinetics of heavy metal adsorption of rice husk depends on numbers of parameters including type of the ion, pH, adsorbent concentration, contact time, and temperature [68].

It is worth pointing out that the environmental concern does not end after adsorbing heavy metals. In fact, the adsorbed heavy metals need to be encapsulated within a safe structure to prevent further introduction of these elements into the environment. The most common approaches are encapsulation of heavy metals within a glass, glass–ceramic, or cementitious structures [69–72]. Glass structures are used

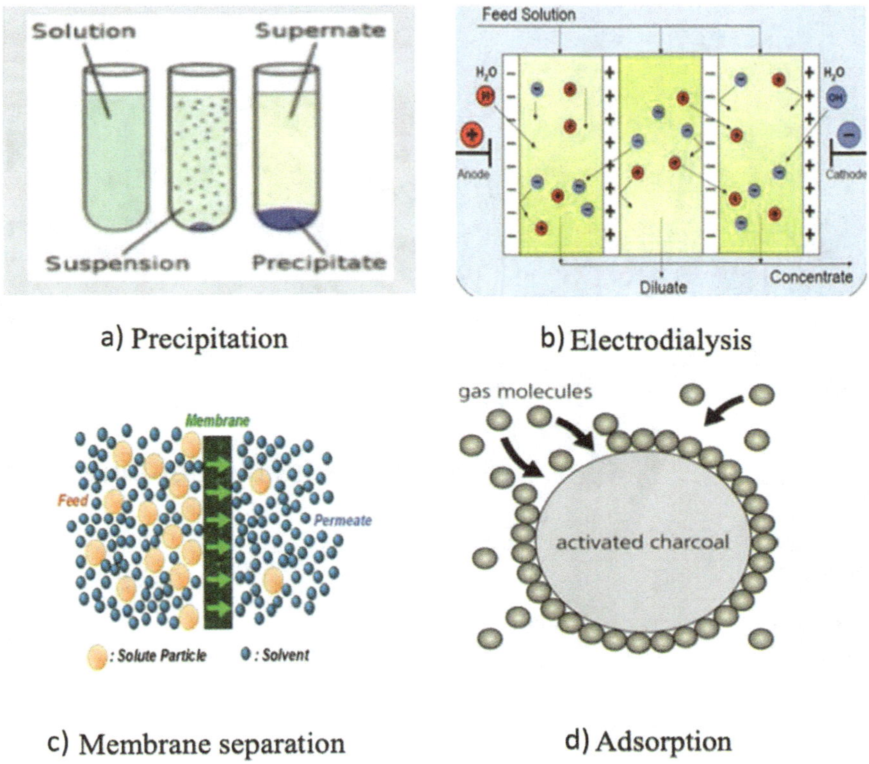

Fig. 9 Review on the physicochemical treatments of rice husk for production of advanced materials. **a** precipitation, **b** electrodialysis, **c** membrane separation, **d** adsorption. Reproduced with permission [54]

for encapsulation of radioactive wastes as well and are promising options because the heavy metal can be entrapped within the glass structure. Furthermore, the produced glass or cementitious structure can be further used for the production of insulation bricks or constructive materials. To be able to evaluate the final safety of the products containing encapsulated heavy metals, leaching test is conducted on the products. The standard leaching test in Europe, EN12457-2, is performed by the final products confirms their safety. Therefore, not only the agricultural by-product, rice husk, acts as an adsorbent in wastewater treatment, but also provides the possibility of bringing back products into the consumption cycle.

Fig. 10 Rice husk origin from agricultural waste. The area marked as A is observed under scanning electron microscopy (SEM) and its possible chemical bonding is presented

4 Conclusion

In this chapter, we have introduced a growingly important research area for material science, circular economy. The by-products of municipal solid waste incineration, fly ash, and bottom ash were introduced together with the methods for their safe treatments. We have shown that fly ash and bottom ash, for their similarity in terms of chemical composition to some minerals, can be implemented as a secondary source of raw material for preparation of clinker, cement, bricks and glass–ceramic foams. It was discussed that the final performance of ash-based products, both in terms of environmental safety and mechanical properties, can be greatly improved through vitrification. This is in particular important for constructive applications such as clinker and cements.

The challenging task of fly ash vitrification was discussed and a new vitrification batch formulation for fly ash was suggested. Based on this approach, the composition of vitrification batches in a SiO_2–Al_2O_3–CaO system has shown that the most successful batches have the CaO/SiO_2 (C/S) ratio of about 0.44. Considering the effect of all major oxides in the vitrification batches reported in literature, the same trend was observed for the ternary diagram of $SiO_2-\Sigma M_2O_3-\Sigma(MO + M_2O)$. Therefore, successful vitrification batches of fly ash could be prepared by the right amount of silica to meet these criteria.

The last part of the chapter was dedicated to wastewater treatment, focusing on the removal of heavy metals though adsorption. We have shown and discussed that the adsorption can successfully take place using agricultural by products such as rice husk, and the final heavy metal containing material can be safely encapsulated in a glass structure through vitrification. The final vitrified product can be used for production of foam glass–ceramics similar to what discussed for vitrified fly ash and bottom ash.

References

1. Prieto-Sandoval, V., Jaca, C., Ormazabal, M.: Towards a consensus on the circular economy. J. Clean. Prod. (2018). https://doi.org/10.1016/j.jclepro.2017.12.224
2. Kawai, K., Tasaki, T.: Revisiting estimates of municipal solid waste generation per capita and their reliability. J. Mater. Cycles Waste Manag. (2016). https://doi.org/10.1007/s10163-015-0355-1
3. Joseph, A.M., Snellings, R., Van den Heede, P., Matthys, S., De Belie, N.: The use of municipal solidwaste incineration ash in various building materials: a Belgian point of view. Materials (Basel) 11 (2018). https://doi.org/10.3390/ma11010141
4. Moya, D., Aldás, C., López, G., Kaparaju, P.: Municipal solid waste as a valuable renewable energy resource: a worldwide opportunity of energy recovery by using waste-to-energy technologies. Energy Procedia **134**, 286–295 (2017). https://doi.org/10.1016/j.egypro.2017.09.618
5. Ji, L., Lu, S., Yang, J., Du, C., Chen, Z., Buekens, A., et al.: Municipal solid waste incineration in China and the issue of acidification: a review. Waste Manag. Res. (2016). https://doi.org/10.1177/0734242X16633776
6. Perrot, J.F., Subiantoro, A.: Municipal waste management strategy review and waste-to-energy potentials in New Zealand. Sustain (2018). https://doi.org/10.3390/su10093114
7. Kilkovsky, B., Stehlik, P., Jegla, Z., Tovazhnyansky, L.L., Arsenyeva, O., Kapustenko, P.O.: Heat exchangers for energy recovery in waste and biomass to energy technologies—I. Energy recovery from flue gas. Appl. Therm. Eng. **64**, 213–223 (2014). https://doi.org/10.1016/j.applthermaleng.2013.11.041
8. Boesch, M.E., Vadenbo, C., Saner, D., Huter, C., Hellweg, S.: An LCA model for waste incineration enhanced with new technologies for metal recovery and application to the case of Switzerland. Waste Manag. **34**, 378–389 (2014). https://doi.org/10.1016/j.wasman.2013.10.019
9. Loginova, E., Volkov, D.S., van de Wouw, P.M.F., Florea, M.V.A., Brouwers, H.J.H.: Detailed characterization of particle size fractions of municipal solid waste incineration bottom ash. J. Clean. Prod. **207**, 866–874 (2019). https://doi.org/10.1016/j.jclepro.2018.10.022

10. Xuan, D., Tang, P., Poon, C.S.: Limitations and quality upgrading techniques for utilization of MSW incineration bottom ash in engineering applications—a review. Constr. Build. Mater. (2018). https://doi.org/10.1016/j.conbuildmat.2018.09.174
11. Dabo, D., Raimbault, L., Badreddine, R., Chaurand, P., Rose, J., De Windt, L.: Characterisation of glassy and heterogeneous cementing phases of municipal solid waste of incineration (MSWI) bottom ash. Austr. Inst. Min. Metall. Publ. Ser. (2008)
12. Saffarzadeh, A., Shimaoka, T., Wei, Y., Gardner, K.H., Musselman, C.N.: Impacts of natural weathering on the transformation/neoformation processes in landfilled MSWI bottom ash: a geoenvironmental perspective. Waste Manag. (2011). https://doi.org/10.1016/j.wasman.2011.07.017
13. Čarnogurská, M., Lázár, M., Puškár, M., Lengyelová, M., Václav, J., Širillová, U.: Measurement and evaluation of properties of MSW fly ash treated by plasma. Meas. J. Int. Meas. Confed. **62**, 155–161 (2015). https://doi.org/10.1016/j.measurement.2014.11.014
14. Luo, H., Cheng, Y., He, D., Yang, E.H.: Review of leaching behavior of municipal solid waste incineration (MSWI) ash. Sci. Total Environ. (2019). https://doi.org/10.1016/j.scitotenv.2019.03.004
15. Weibel, G.: Optimized Metal Recovery from Fly Ash from Municipal Solid Waste Incineration. Bern University (2017)
16. Weibel, G., Eggenberger, U., Kulik, D.A., Hummel, W., Schlumberger, S., Klink, W., et al.: Extraction of heavy metals from MSWI fly ash using hydrochloric acid and sodium chloride solution. Waste Manag. **76**, 457–471 (2018). https://doi.org/10.1016/j.wasman.2018.03.022
17. Yao, Z.T., Ji, X.S., Sarker, P.K., Tang, J.H., Ge, L.Q., Xia, M.S., et al.: A comprehensive review on the applications of coal fly ash. Earth-Sci. Rev. (2015). https://doi.org/10.1016/j.earscirev.2014.11.016
18. Gartner, E.: Industrially interesting approaches to "low-CO2" cements. Cem. Concr. Res. (2004). https://doi.org/10.1016/j.cemconres.2004.01.021
19. Maddalena, R., Roberts, J.J., Hamilton, A.: Can Portland cement be replaced by low-carbon alternative materials? A study on the thermal properties and carbon emissions of innovative cements. J. Clean. Prod. (2018). https://doi.org/10.1016/j.jclepro.2018.02.138
20. Kikuchi, R.: Recycling of municipal solid waste for cement production: pilot-scale test for transforming incineration ash of solid waste into cement clinker. Resour. Conserv. Recycl. (2001). https://doi.org/10.1016/S0921-3449(00)00077-X
21. Garcia-Lodeiro, I., Carcelen-Taboada, V., Fernández-Jiménez, A., Palomo, A.: Manufacture of hybrid cements with fly ash and bottom ash from a municipal solid waste incinerator. Constr. Build. Mater. **105**, 218–226 (2016). https://doi.org/10.1016/j.conbuildmat.2015.12.079
22. Sharifikolouei, E., Canonico, F., Salvo, M., Baino, F., Ferraris, M.: Vitrified and nonvitrified municipal solid wastes as ordinary Portland cement (OPC) and sand substitution in mortars. Int. J. Appl. Ceram. Technol. (2020). https://doi.org/10.1111/ijac.13447
23. Ferraris, M., Salvo, M., Ventrella, A., Buzzi, L., Veglia, M.: Use of vitrified MSWI bottom ashes for concrete production. Waste Manag. **29**, 1041–1047 (2009). https://doi.org/10.1016/j.wasman.2008.07.014
24. Xu, G., Shi, X.: Characteristics and applications of fly ash as a sustainable construction material: a state-of-the-art review. Resour. Conserv. Recycl. **136**, 95–109 (2018). https://doi.org/10.1016/j.resconrec.2018.04.010
25. Tang, P., Xuan, D., Poon, C.S., Tsang, D.C.W.: Valorization of concrete slurry waste (CSW) and fine incineration bottom ash (IBA) into cold bonded lightweight aggregates (CBLAs): feasibility and influence of binder types. J. Hazard Mater. (2019). https://doi.org/10.1016/j.jhazmat.2019.01.112
26. Lynn, C.J., Ghataora, G.S., Dhir, O.B.E.R.K.: Municipal incinerated bottom ash (MIBA) characteristics and potential for use in road pavements. Int. J. Pavem. Res. Technol. (2017). https://doi.org/10.1016/j.ijprt.2016.12.003
27. Ponsot, I., Bernardo, E., Bontempi, E., Depero, L., Detsch, R., Chinnam, R.K., et al.: Recycling of pre-stabilized municipal waste incinerator fly ash and soda-lime glass into sintered glass-ceramics. J. Clean. Prod. **89**, 224–230 (2015). https://doi.org/10.1016/j.jclepro.2014.10.091

28. Xiao, Y., Oorsprong, M., Yang, Y., Voncken, J.H.L.: Vitrification of bottom ash from a municipal solid waste incinerator. Waste Manag. **28**, 1020–1026 (2008). https://doi.org/10.1016/j.wasman.2007.02.034
29. Waste Gasification Technology (Direct Melting System): Waste gasification with Direct Melting System We Make The World A Cleaner Place (n.d.)
30. Shibaike, H., Hoshizawa, Y., Tanaka, H., Nishi, T., Takamiya, K., Kato, Y., et al.: Development of high-performance direct melting process for municipal solid waste. Nippon Steel Tech. Rep. 22–29 (2005)
31. Huang, Q., Cai, X., Alhadj-Mallah, M.M., Du, C., Chi, Y., Yan, J.: Thermal plasma vitrification of MSWI fly ash mixed with different biomass ashes. IEEE Trans. Plasma Sci. **42**, 3549–3554 (2014). https://doi.org/10.1109/TPS.2014.2358626
32. Faik, A., Guillot, S., Lambert, J., Véron, E., Ory, S., Bessada, C., et al.: Thermal storage material from inertized wastes: evolution of structural and radiative properties with temperature. Sol. Energy **86**, 139–146 (2012). https://doi.org/10.1016/J.SOLENER.2011.09.014
33. Sharifikolouei, E., Baino, F., Salvo, M., Tommasi, T., Pirone, R., Fino, D., et al.: Vitrification of municipal solid waste incineration fly ash: an approach to find the successful batch compositions. Ceram Int. (2020). https://doi.org/10.1016/j.ceramint.2020.11.118
34. Baino, F., Ferraris, M.: Production and characterization of ceramic foams derived from vitrified bottom ashes. Mater. Lett. (2019). https://doi.org/10.1016/j.matlet.2018.10.122
35. Rincon Romero, A., Salvo, M., Bernardo, E.: Up-cycling of vitrified bottom ash from MSWI into glass-ceramic foams by means of 'inorganic gel casting' and sinter-crystallization. Constr. Build. Mater. (2018). https://doi.org/10.1016/j.conbuildmat.2018.10.135
36. Bernardo, E., Albertini, F.: Glass foams from dismantled cathode ray tubes. Ceram. Int. (2006). https://doi.org/10.1016/j.ceramint.2005.04.019
37. Tulyaganov, D.U., Fernandes, H.R., Agathopoulos, S., Ferreira, J.M.F.: Preparation and characterization of high compressive strength foams from sheet glass. J. Porous Mater. (2006). https://doi.org/10.1007/s10934-006-7014-9
38. Salgot, M., Folch, M.: Wastewater treatment and water reuse. Curr. Opin. Environ. Sci. Heal. (2018). https://doi.org/10.1016/j.coesh.2018.03.005
39. Levine, A.D., Asano, T.: Recovering sustainable water from wastewater. Environ. Sci. Technol. (2004). https://doi.org/10.1021/es040504n
40. Guerra-Rodríguez, S., Oulego, P., Rodríguez, E., Singh, D.N., Rodríguez-Chueca, J.: Towards the implementation of circular economy in the wastewater sector: challenges and opportunities. Water (Switzerland) (2020). https://doi.org/10.3390/w12051431
41. Rashidi, H., Ghaffarianhoseini, A., Ghaffarianhoseini, A., Nik Sulaiman, N.M., Tookey, J., Hashim, N.A.: Application of wastewater treatment in sustainable design of green built environments: a review. Renew. Sustain. Energy Rev. (2015). https://doi.org/10.1016/j.rser.2015.04.104
42. Mo, W., Zhang, Q.: Energy-nutrients-water nexus: Integrated resource recovery in municipal wastewater treatment plants. J. Environ. Manage. (2013). https://doi.org/10.1016/j.jenvman.2013.05.007
43. Theregowda, R.B., González-Mejía, A.M., Ma, X., Garland, J.: Nutrient recovery from municipal wastewater for sustainable food production systems: an alternative to traditional fertilizers. Environ. Eng. Sci. (2019). https://doi.org/10.1089/ees.2019.0053
44. Neczaj, E., Grosser, A.: Circular economy in wastewater treatment plant–challenges and barriers. In: Proceedings (2018). https://doi.org/10.3390/proceedings2110614
45. Zhang, Q., Hu, J., Lee, D.J., Chang, Y., Lee, Y.J.: Sludge treatment: current research trends. Bioresour. Technol. (2017). https://doi.org/10.1016/j.biortech.2017.07.070
46. Marks, J.S., Zadoroznyj, M.: Managing sustainable urban water reuse: structural context and cultures of trust. Soc. Nat. Resour. (2005). https://doi.org/10.1080/08941920590947995
47. Sobhanardakani, S., Tayebi, L., Farmany, A.: Toxic metal (Pb, Hg and As) contamination of muscle, gill and liver tissues of Otolithes rubber, *Pampus argenteus*, *Parastromateus niger*, *Scomberomorus commerson* and *Onchorynchus mykiss*. World Appl. Sci. J. (2011)

48. Wuana, R.A., Okieimen, F.E.: Heavy metals in contaminated soils: a review of sources, chemistry, risks, and best available strategies for remediation. Heavy Met. Contam. Water Soil Anal. Assess., Remediat. Strateg. (2014). https://doi.org/10.1201/b16566
49. Bjuhr, J.: Trace Metals in Soils Irrigated with Waste Water in a Periurban Area Downstream Hanoi City, Vietnam. Semin Pap (2007)
50. Kobya, M., Gebologlu, U., Ulu, F., Oncel, S., Demirbas, E.: Removal of arsenic from drinking water by the electrocoagulation using Fe and Al electrodes. Electrochim. Acta (2011). https://doi.org/10.1016/j.electacta.2011.03.086
51. Abdullah, N., Yusof, N., Lau, W.J., Jaafar, J., Ismail, A.F.: Recent trends of heavy metal removal from water/wastewater by membrane technologies. J. Ind. Eng. Chem. (2019). https://doi.org/10.1016/j.jiec.2019.03.029
52. Mafu, L.D., Msagati, T.A.M., Mamba, B.B.: The enrichment and removal of arsenic (III) from water samples using HFSLM. Phys. Chem. Earth (2012). https://doi.org/10.1016/j.pce.2012.08.018
53. Smedley, P.L., Kinniburgh, D.G.: A review of the source, behaviour and distribution of arsenic in natural waters. Appl Geochem. (2002). https://doi.org/10.1016/S0883-2927(02)00018-5
54. Raouf, M.S.A., Raheim, A.R.M.A.: Removal of heavy metals from industrial waste water by biomass-based materials: a review. J. Pollut. Effect Control (2016). https://doi.org/10.4172/2375-4397.1000180
55. Obaid, S.S., Gaikwad, D.K., Sayyed, M.I., Al-Rashdi, K., Pawar, P.P.: Heavy metal ions removal from waste water bythe natural zeolites. Mater. Today Proc. (2018). https://doi.org/10.1016/j.matpr.2018.06.122
56. Zanin, E., Scapinello, J., de Oliveira, M., Rambo, C.L., Franscescon, F., Freitas, L., et al.: Adsorption of heavy metals from wastewater graphic industry using clinoptilolite zeolite as adsorbent. Process Saf. Environ. Prot. (2017). https://doi.org/10.1016/j.psep.2016.11.008
57. Hong, M., Yu, L., Wang, Y., Zhang, J., Chen, Z., Dong, L., et al.: Heavy metal adsorption with zeolites: the role of hierarchical pore architecture. Chem. Eng. J. **359**, 363–372 (2019). https://doi.org/10.1016/J.CEJ.2018.11.087
58. Cao, F., Lian, C., Yu, J., Yang, H., Lin, S.: Study on the adsorption performance and competitive mechanism for heavy metal contaminants removal using novel multi-pore activated carbons derived from recyclable long-root *Eichhornia crassipes*. Bioresour. Technol. (2019). https://doi.org/10.1016/j.biortech.2019.01.007
59. Wahby, A., Abdelouahab-Reddam, Z., El Mail, R., Stitou, M., Silvestre-Albero, J., Sepúlveda-Escribano, A., et al.: Mercury removal from aqueous solution by adsorption on activated carbons prepared from olive stones. Adsorption (2011). https://doi.org/10.1007/s10450-011-9334-6
60. Karnib, M., Kabbani, A., Holail, H., Olama, Z.: Heavy metals removal using activated carbon, silica and silica activated carbon composite. Energy Procedia (2014). https://doi.org/10.1016/j.egypro.2014.06.014
61. Ghassabzadeh, H., Mohadespour, A., Torab-Mostaedi, M., Zaheri, P., Maragheh, M.G., Taheri, H.: Adsorption of Ag, Cu and Hg from aqueous solutions using expanded perlite. J. Hazard Mater. (2010). https://doi.org/10.1016/j.jhazmat.2010.01.010
62. Xu, M., Yin, P., Liu, X., Tang, Q., Qu, R., Xu, Q.: Utilization of rice husks modified by organomultiphosphonic acids as low-cost biosorbents for enhanced adsorption of heavy metal ions. Bioresour. Technol. (2013). https://doi.org/10.1016/j.biortech.2013.09.075
63. Shalaby, N.H., Ewais, E.M.M., Elsaadany, R.M., Ahmed, A.: Rice husk templated water treatment sludge as low cost dye and metal adsorbent. Egypt J. Pet. **26**, 661–668 (2017). https://doi.org/10.1016/J.EJPE.2016.10.006
64. Ajmal, M., Ali Khan Rao, R., Anwar, S., Ahmad, J., Ahmad, R.: Adsorption studies on rice husk: removal and recovery of Cd(II) from wastewater. Bioresour. Technol. **86**, 147–149 (2003). https://doi.org/10.1016/S0960-8524(02)00159-1
65. Chiu, A.C.F., Akesseh, R., Moumouni, I.M., Xiao, Y.: Laboratory assessment of rice husk ash (RHA) in the solidification/stabilization of heavy metal contaminated slurry. J. Hazard Mater. (2019). https://doi.org/10.1016/j.jhazmat.2019.02.051

66. Williams, P.T., Nugranad, N.: Comparison of products from the pyrolysis and catalytic pyrolysis of rice husks. Energy (2000). https://doi.org/10.1016/S0360-5442(00)00009-8
67. Soltani, N., Bahrami, A., Pech-Canul, M.I., González, L.A.: Review on the physicochemical treatments of rice husk for production of advanced materials. Chem. Eng. J. (2015). https://doi.org/10.1016/j.cej.2014.11.056
68. Senthil Kumar, P., Ramakrishnan, K., Dinesh Kirupha, S., Sivanesan, S.: Thermodynamic and kinetic studies of cadmium adsorption from aqueous solution onto rice husk. Brazil. J. Chem. Eng. **27**, 347–355 (2010). https://doi.org/10.1590/s0104-66322010000200013
69. Romano, J.S., Rodrigues, F.A.: Cements obtained from rice hull: encapsulation of heavy metals. J. Hazard Mater. (2008). https://doi.org/10.1016/j.jhazmat.2007.11.051
70. Sharifikolouei, E., Baino, F., Galletti, C., Fino, D., Ferraris, M.: Adsorption of Pb and Cd in rice husk and their immobilization in porous glass-ceramic structures. Int. J. Appl. Ceram. Technol. **17**, 105–112 (2020). https://doi.org/10.1111/ijac.13356
71. Mao, L., Wu, Y., Zhang, W., Huang, Q.: The reuse of waste glass for enhancement of heavy metals immobilization during the introduction of galvanized sludge in brick manufacturing. J. Environ. Manage. (2019). https://doi.org/10.1016/j.jenvman.2018.10.120
72. Song, H., Wei, L., Ji, Y., Cao, L., Cheng, F.: Heavy metal fixing and heat resistance abilities of coal fly ash-waste glass based geopolymers by hydrothermal hot pressing. Adv. Powder Technol. (2018). https://doi.org/10.1016/j.apt.2018.03.013

The manufacturer's authorised representative in the EU is Springer Nature Customer Service Centre GmbH, Europaplatz 3, 69115 Heidelberg, Germany. If you have any concerns regarding our products, please contact ProductSafety@springernature.com

Printed and bound by CPI Group (UK) Ltd, Croydon, CR0 4YY
28/09/2026
02224086-0011